Computer Communications and Networks

Series Editor
A.J. Sammes
Centre for Forensic Computing
Cranfield University, Shrivenham Campus
Swindon, UK

The **Computer Communications and Networks** series is a range of textbooks, monographs and handbooks. It sets out to provide students, researchers, and non-specialists alike with a sure grounding in current knowledge, together with comprehensible access to the latest developments in computer communications and networking.

Emphasis is placed on clear and explanatory styles that support a tutorial approach, so that even the most complex of topics is presented in a lucid and intelligible manner.

More information about this series at http://www.springer.com/series/4198

Dietmar P.F. Möller

Guide to Computing Fundamentals in Cyber-Physical Systems

Concepts, Design Methods, and Applications

Dietmar P.F. Möller
IASOR
Clausthal University of Technology
Clausthal-Zellerfeld
Germany

ISSN 1617-7975   ISSN 2197-8433   (electronic)
Computer Communications and Networks
ISBN 978-3-319-79747-2   ISBN 978-3-319-25178-3   (eBook)
DOI 10.1007/978-3-319-25178-3

Printed on acid-free paper

This Springer imprint is published by Springer Nature
The registered company is Springer International Publishing AG Switzerland

Foreword

Changing market dynamics are reviving up the manufacturing and automotive industry. Consumers, today, are not looking only for a car, but in fact seek a bundled experience package. Manufacturers are turning to automation as a means to improve quality, productivity, safety, speed, and competitiveness while reducing costs. The evolving technology investment environment has a deep influence and impact on how value chain collaboration, internal operations, and customer experience for both products and services will emerge in the immediate future. The automotive industry has always been at the forefront in defining new paradigms for the manufacturing industry. This book outlines internal research and experience gained over multiple engagements of digitization of manufacturing industry.

Digital Manufacturing/Industry 4.0 is the transformation of manufacturing under conditions of adequately adapted automated manufacturing systems. Thus, digital technological trends are in the focus that is aimed at a novel method of manufacturing automation. The chapters in the book involve the use of digital models and methods of manufacturing planning and control and linking them to real manufacturing subsystems, manufacturing components (hardware), and tools (software).

The central feature of the book is networking the networked virtual computer world (cyber) with the manufacturing components world (physical) through cyber-physical systems. Therefore, cyber-physical systems in this sense can be introduced as a strong digital platform – well structured and well integrated – and only as complex as absolutely necessary with regard to the designated use in manufacturing. Thus, cyber-physical systems-based organized manufacturing systems will be able to largely control, depending on external requirements independently and autonomously, optimize, and configure what outruns on an outstanding automation level.

The core research areas are advanced methodologies to study systems as physical components and the Internet as the cyber part of cyber-physical systems which are essential. This book *Guide to Computing Fundamentals in Cyber-Physical Systems: Concepts, Design, Methods, and Applications* is a showcase of creative ideas of ongoing research work and fundamentals focusing on systems and software engineering of cyber-physical systems. The book shows how to analyze the intrinsic complexity of cyber-physical systems accurately and under varying operation conditions and scenarios to predict its behavior for engineering and planning purposes to provide adequate academic answers for today's emerging technology management questions in Digital Manufacturing/Industry 4.0. The chapters are well written showing academic rigor and professionalism of the author. Therefore, the book can be stated as an important reading for new researchers entering this field of cyber-physical systems research. It offers new perspectives and documents important progress in Digital Manufacturing/Industry 4.0 analysis and development.

I strongly recommend Prof. Dietmar P.F. Moeller's scholarly writing to students, academicians, and industrialists who are keen to learn advance methodologies in manufacturing. I can say without reservation that this book, and, more specifically, the method it espouses, will change the fundamental ideas for better innovation and digital disruption of manufacturing and automotive industry. Which will be a new wave of digital disruption-led consumerization as a result of the power and impact of converging technologies like big data, high-performance computing, cloud, mobility, and social media, enabling automotive enterprises to explore new business models and differentiation opportunities, both on the production side and the customer engagement side of the business? I failed to mention how much more I enjoy reading and reviewing the book. I think that the author can be confident that there will be many grateful readers who will have gained a broader perspective of the disciplines of digital manufacturing and cyber-physical system as a result of their efforts.

Department of Management Studies   K.B. Akhilesh
Indian Institute of Science
Bangalore, India

Preface

The goal of this book is to provide a comprehensive, in-depth, and state-of-the-art summary of cyber-physical systems and their applications. It describes the cyber-physical systems approach, clearly showing where the multitude of cyber-physical systems activities fit within the overall effort and providing an ideal framework for understanding the complexity of cyber-physical systems. For this reason, some choices have been made in selecting the material for this book. A top-down approach was taken that introduces the fundamentals of systems and embedded computing systems and focuses on the requirements of cyber-physical systems and the Internet of Things, the most important subject areas. Furthermore, ubiquitous computing is introduced, describing how current technologies, such as smart things and services with some kind of attachment, embedment, blend of tiny computers, sensors, tags, networks, smart devices, and others, relate to and support a vision for a greater range of tiny computer devices, used within the greater scope of cyber-physical systems. This provides a framework within which the reader can assimilate the associated requirements. Without such a reference, the practitioner is left to ponder the plethora of terms, standards, and practices that have been developed independently and that often lack cohesion, particularly in nomenclature and emphasis. Therefore, this book is intended to both cover all aspects of cyber-physical systems and to provide a framework for the consideration of the many issues associated with cyber-physical systems in Digital Manufacturing/Industry 4.0. These subjects are discussed with regard to individualized production, networked manufacturing, and concurrent open and closed product lines as part of cyber-physical systems applications and their respective methods in systems and software engineering.

First, an overview on the study of systems is given introducing four basic steps: (1) modeling, (2) setting up mathematical equations to describe systems using the standard forms of input–output and state variable descriptions, (3) analyzing systems, and (4) designing systems. In addition, the mathematical background of the expansion of systems, in regard to embedded computing systems, is introduced. Embedded computing systems are dedicated, computer-based systems for specific applications or products, and their importance as a platform for cyber-physical systems is discussed.

Second, cyber-physical systems, a new generation of engineered systems, are described in detail. They are the most important component within the Digital Manufacturing/Industry 4.0 paradigm shift, together with the Internet of Things, a global system of interconnected computer networks that use the standard Internet Protocol Suite (TCP/IP) to serve billions of users worldwide. Based on that foundation, ubiquitous computing (also referred to as pervasive computing) is introduced. How current technologies (smart things or objects), with some kind of attachment, embedment, blending of computers, sensors, tags, networks, and others (smart devices (mobile, wearable, wireless), smart environments (embedded computing systems, sensor-actor networks), and smart interaction (tight integration of and coordination between devices and environments, anything with everything), relate to and support a computing vision of greater availability and range of computer devices is described. With regard to the intrinsic complexity of the aforementioned approaches, systems and software engineering are the interdisciplinary approaches required to design complex technical systems based on certain thought patterns and basic principles of targeted design in terms of cyber-physical systems as intelligent and networked components in Digital Manufacturing Systems/Industry 4.0, a smart factory approach.

However, a textbook cannot describe all of the innovative aspects of cyber-physical systems and Digital Manufacturing/Industry 4.0 in detail. For this reason, the reader is referred to specific supplemental material, such as textbooks, reference guides, user manuals, etc., as well as Internet-based information which addresses several of the topics selected for the book.

Third, some actual case studies from different kinds of industrial and academic research and practice are presented to illustrate the actual state of the art and the ongoing research aspects in the context of Digital Manufacturing/Industry 4.0.

This book can serve as textbook or a reference book for college courses on cyber-physical systems and can be offered in computer science, electrical and computer engineering, information technology and informations systems, applied mathematics, and operations research as well as business informatics and management departments. The contents of the book are also very useful to researchers who are interested in the design of cyber-physical systems. Company engineers in the private sector can use the principles described in the book for their product designs.

The material in the book can be difficult to comprehend if the reader is new to such an approach. This is also due to the fact that cyber-physical systems and Digital Manufacturing/Industry 4.0 is a multidisciplinary domain, founded in computer science, engineering, mathematics, operations research, and more. The material may not be read and comprehended quickly or easily. Therefore, specific case studies have been included with related topics to help the reader master the material. It is assumed that the reader has some knowledge of basic calculus-based probability and statistics and some experience with systems and software engineering.

The book can be used as the primary text in a course in various ways. It contains more material than can be covered in detail in a quarter-long (30-h) or semester-long (45-h) course. Instructors may elect to choose their own topics and add their own case studies. The book can also be used for self-study as a reference for engineers, scientists, and computer scientists for on-the-job training, for study in graduate schools, and as a reference for cyber-physical systems and Digital Manufacturing/Industry 4.0 practitioners and researchers.

For instructors who have adopted the book for use in a course, a variety of teaching support materials are available for download from www.springer.com/book/9783319251769. These include a comprehensive set of PowerPoint slides to be used for lectures and all video-recorded classes.

The book is divided into eight chapters which can be read independently or consecutively.

Chapter 1, "Introduction to Systems," covers the study of systems based on the four basic steps: (1) modeling, (2) setting up mathematical equations to describe systems using the standard forms of input–output and state variable descriptions, (3) analyzing systems, and (4) designing systems. It also introduces the concept of component analysis of linear systems based on the theory of controllability, observability, and identifiability, as well as analytical solutions of linear systems by analyzing their behavior and/or composite structure to examine the system response to an input demand. Finally, the approach determining the steady-state error of systems, an analysis method which defines the difference between input and output of a system in the limit as time goes to infinity, is described.

Chapter 2, "Introduction to Embedded Computing Systems," contains a brief overview of embedded computing systems and their hardware architectures and an approach for determining the design metrics of embedded computing systems, a method which defines the preciseness of a design with regard to the requirement specifications. Furthermore, the concept of embedded control with regard to the respective mathematical notation of the different control laws and the principal methodological approach to hardware-software codesign is introduced in detail.

Chapter 3, "Introduction to Cyber-Physical Systems," summarizes the knowledge from Chaps. 1 and 2 to introduce cyber-physical systems and ensure that readers from several engineering and scientific disciplines have the same understanding of the term cyber-physical systems. These systems use computations and communication deeply embedded in and interacting with physical processes by adding cyber capabilities to physical systems. Therefore, Chap. 3 concentrates on recommendations with regard to cyber-physical systems design with a focus on the cyber-physical systems requirements used to emphasize disciplined approaches to their design. Cyber-physical systems cover an extremely wide range of application areas, which allows systems to be designed more economically by sharing abstract knowledge and design tools. This allows the design of more dependable cyber-physical systems by applying best practices to the entire range of cyber-physical applications. The technological and economic drivers create an environment that enables and requires a range of new capabilities. The specific topics of smart cities and the Internet of Everything are described in more detail. Smart cities are based on digital strategies which introduce how to build more and efficient infrastructure and services by making use of the Internet of Everything.

Chapter 4, "Introduction to the Internet of Things," begins with a brief introduction of the global system of interconnected computer networks that use the standard Internet Protocol Suite (TCP/IP) to serve billions of users worldwide and identifies the enabling technologies for its use. Furthermore, radio frequency identification (RFID), a wireless automatic identification technology, is described in detail as well as the concept of wireless sensor network technology, which has important applications, such as remote environmental monitoring and target tracking. This technology has been enabled by the availability of sensors that are smaller, cheaper, and intelligent. The importance of power line communication technology, enabling data to be sent over existing power cables, is introduced with regard to the smart home application domain.

Chapter 5, "Ubiquitous Computing" (also referred to as pervasive computing), describes how current technologies (smart things or objects), with some kind of attachment, embedment, blending of computers, sensors, tags, networks, and others (smart devices (mobile, wearable, wireless), smart environments (embedded computing systems, sensor-actor networks), and smart interactions (tight integration of and coordination between devices and environments, anything with everything), relate to and support a computing vision for a greater availability and range of computer devices. Therefore, it covers the important topics of tagging, sensing, and controlling in ubiquitous computing and possible applications, such as autonomous systems, for which their behavior and composite structure is analyzed in regard to a fault-tolerant behavior.

Chapter 6, "Systems and Software Engineering," introduces, from a general perspective, the intrinsic complexity of the aforementioned approaches to systems and software engineering as an interdisciplinary field of engineering that primarily focuses on how to successfully design, implement, evaluate, and manage complex engineered systems over their life cycles. It discusses the design challenges in cyber-physical systems and their impact on systems engineering with reference to requirements definition and management using Cradle®. Cradle® is a requirements management and systems engineering tool that integrates the entire project life cycle into one, massively scalable, integrated, multiuser software product. Furthermore this chapter introduces the principal concept of software engineering with special focus on the V-model and the Agile software development methodology. It also introduces different requirements in software design in cyber-physical systems.

Chapter 7, "Digital Manufacturing/Industry 4.0," begins with a brief introduction to manufacturing and the enabling technologies and their opportunities with regard to the sequence of industrial revolutions. It also introduces digital manufacturing in reference to smart and Agile manufacturing and smart factories, one of the major concepts of Digital Manufacturing/Industry 4.0. Based on that knowledge, Chap. 7 introduces the principal concept of individualized production, an important application in the area of smart factories, and refers to networked manufacturing-integrated industry and the idea of smart supply chains that enable product data to be sent over the Internet for service purposes and more. Furthermore, the paradigm of open and closed manufacturing lines is discussed along with the important topic of cyber security in Digital Manufacturing/Industry 4.0. More-over, insight into Digital Manufacturing/Industry 4.0 projects in the industrial and academic research areas is given for six use cases.

Chapter 8, "Social Impact on Working Lives of the Future," gives a brief introduction to the social impact on work lives in the future by introducing the changes in skills that will occur due to the modern globalized, digital work environment as compared to the historical development of manufacturing. There-fore, it refers to the economic, social, and organizational challenges of the future of work with regard to the requirements of the digitized and automated industry. It also introduces the changing demands in the world of work in regard to the effects of Digital Manufacturing/Industry 4.0. The reader is introduced to the principal concept of greater product individualization and shifting factors of global influence with regard to the digital transformation.

Besides the methodological and technical content, all of the chapters in the book contain chapter-specific comprehensive questions to help students determine if they have gained the required knowledge, identify possible knowledge gaps, and con-quer them. Moreover, all chapters include references and suggestions for further reading.

I would like to express my special thanks to Patricia Worster, University of Nebraska–Lincoln, for her excellent assistance in proofreading, and to Simon Rees, Springer Publ., for his help with the organizational procedures between the pub-lishing house and the author. Furthermore, I thank Dr. Alexander Herzog, Simula-tion Science Center Clausthal-Göttingen, for drawing the illustrations for this book from the sketches I drafted. Moreover, I sincerely thank all of the authors who have published cyber-physical systems material and directly and/or indirectly contributed to this book through citations.

Finally, I would like to deeply thank my wife, Angelika; my daughter Christina; and my grandchildren, Hannah and Karl, for their encouragement, patience, and understanding during the writing of this book.

Clausthal-Zellerfeld, Germany   Dietmar P.F. Möller

Contents

About the Author

Dietmar P.F. Möller is a Professor in the Institute of Applied Stochastics and Operations Research at Clausthal University of Technology (TUC), Germany; a Member of the Simulation Science Center Clausthal-Göttingen, Germany; and an Adjunct Professor in the Department of Electrical and Computer Engineering at the University of Nebraska-Lincoln (UNL), USA. His other publications include the Springer title *Introduction to Transportation Analysis, Modeling and Simulation*.

Introduction to Systems

This chapter begins with a brief overview of the study of systems, in Sect. 1.1, which is based on four steps: modeling, setting up mathematical equations, analysis, and design. Section 1.2 introduces the second step of the study of systems, setting up mathematical equations to describe the system, referring to the standard forms of input-output and state-variable descriptions of systems. Section 1.3 introduces the principal concept of component analysis of linear systems based on the theory of controllability, observability, and identifiability of systems with regard to their mathematical forms of notation, showing what can be achieved under their assumptions. Section 1.4 introduces the analytical solutions of linear systems by analyzing their behavior and/or composite structure to examine the system response to an input demand. Section 1.5 refers to the approach determining the steady-state error of systems, an analysis method which defines the difference between input and output of a system in the limit as time goes to infinity. Section 1.6 provides a case study on the implications of the concept of system stability analysis. Section 1.7 contains comprehensive questions from the introduction to systems topics, and followed by references and suggestions for further reading.

1.1 Study of Systems

The study and design of physical and cyber-physical systems consists of the following foundations:

- Modeling
- Setting up mathematical equations
- Analysis
- Design

It represents a new approach to engineering and science that investigates how relationships between parts and/or components give rise to the collective behavior

© Springer International Publishing Switzerland 2016
D.P.F. Möller, *Guide to Computing Fundamentals in Cyber-Physical Systems*,
Computer Communications and Networks, DOI 10.1007/978-3-319-25178-3_1

of a system and how the system interacts and forms relationships with its environment.

The first step, modeling, involves identifying a model that resembles the system in its salient features but is easier to study. A system, in general, is a way of working, organizing, or doing one or many tasks according to a fixed plan (Kamal 2008). Moreover, a system can be introduced as an object in the real world whereby its precise characteristics are often unknown. By applying test signals to the system's inputs, it is often possible to determine the characteristics of the system from the data measured as the system's output. Therefore, there are two types of deterministic test signals used, the unit step and the ramp function, which are related through the expressions (Moeller 2003)

$$u_{us}(t) = \begin{cases} 0 & \text{for } t < t_0 \\ 1 & \text{for } t_0 < t < t_1 \\ 1 & \text{for } t > t_1 \end{cases}$$

for the unit step and

$$u_R(t) = \begin{cases} 0 & \text{for } t < t_0 \\ a\dfrac{t - t_0}{t_1 - t_0} & \text{for } t_0 < t < t_1 \\ a & \text{for } t > t_1 \end{cases}$$

for the ramp function.

The unit step and the ramp function, as well as many other functions, prove to be particularly valuable in problems in which successive switching characteristics occur. As described in the equations above, it is possible to specify the order of the switching events. This will be the same when the given functions are delayed in time.

More generally, modeling refers to the process of developing a model of a system, which is an abstraction of the real-world object under investigation. Most mathematical models are used to study systems for many different purposes, to explain system behavior or data, to provide a compact representation of data, and more. Therefore, modeling is a powerful method for dealing with complex problems in the physical and cyber-physical systems world. The various methods used for modeling systems usually combine the following information:

- Goals and purpose
- Boundaries
- Necessary level of detail
- A priori knowledge
- Nonmeasurable data and/or state-space variables
- Data sets through experimentation
- Relevant components
- System inputs and outputs

It is important to be aware that the scope of use of systems models in today's scientific and engineering disciplines can be different. Control systems engineering is concerned with understanding and controlling segments of systems in order to provide useful economic products for society. This chiefly requires system synthesis and optimization. Engineers are primarily interested in mathematical models of systems operating under normal conditions. The goal of engineers in using models is to optimally control systems or at least keep them from drifting outside the margins of safe operating conditions. In contrast, in the life sciences, medical scientists are not solely interested in mathematical models of vital systems, such as the circulatory system, operating under normal conditions. Life science scientists prefer mathematical models that adequately describe the system behavior outside of the normal operating range, which can be interpreted, in medical terms, as an indicator of disease, such as hypertension in the case of the circulatory system. For this reason, hypertension represents a system that operates outside of normal conditions.

With respect to the spectrum of available models, the appropriate level of conceptual and mathematical representation depends on the goals and purposes for which the model was intended, the availability of a priori knowledge, data gathered through experimentation and measurements of the system, estimates of system parameters for initial and other constraints, and system states. Hence, the mathematical representation of a system is based on the constraints of the required level, which are:

- *Behavior level* at which the system can be described in a form where measurements are recorded in a chronological manner based on a set of trajectories which characterize the system's behavior. Therefore, the behavior level is important because experimentation with systems addresses this level, due to the input-output relationship, which can be expressed as follows:

$$\underline{y}(t) = F\left(\underline{u}, t\right),$$

with $u(t)$ as the input set, $y(t)$ as the output set, and F as the transfer function, described at the state-structure level. This results by iteration over time, in a set of trajectories known as system behavior. The internal state sets represent the state-transition function that provides the rules for computing future states, depending on the current ones, as follows:

$$\underline{y}(t) = G\left(\underline{u}(t), \underline{x}(t), t\right).$$

Thus, the state of a system represents the smallest collection of numbers, specified at time $t = t_0$ in order to uniquely predict the behavior of the system for any time $t \geq t_0$ for any input belonging to the given input set, provided that every element of the input set is known for $t \geq t_0$. Such numbers are called state variables.

- *Composite structure level* at which the system can be described by connecting elementary black boxes that can be introduced as a network description. The elementary black boxes are components, and each one must be introduced by a system description at the state-structure level. Moreover, each component must have input and output variables as well as a specification that determines the interconnection of the components, interfacing the input and output variables.

Difficulties in developing mathematical systems models may occur because systems are extremely complex and may have inadequate sets of data that do not allow precise descriptions of the system (Mitchel 2009). Thus, building a mathematical model for a real-world system first requires the selection of a model structure and, thereafter, some form of parameter estimation to determine model parameter values from the non-available ones. For this reason, it is sometimes better to develop simplified models that neglect some of the characteristics of the system (Luhmann 2012). An overcomplicated mathematical model will cause difficulties. From a more general point of view, two major factors are important when developing mathematical models of real-world systems:

1. A model is always a simplification of reality but should never be so simple that its answers are not true.
2. A model has to be simple enough to be easy to work with and study.

Hence, a suitable systems model is a compromise between mathematical difficulty, with regard to complicated equations, and the accuracy of the final result. The corresponding relationships are shown in Fig. 1.1.

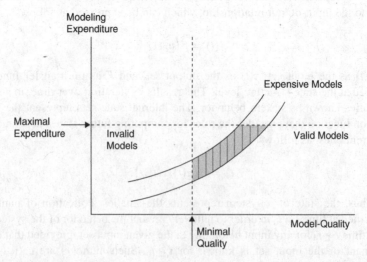

Fig. 1.1 Dependence of the modeling expenditure (costs) on the degree of accuracy (model quality) (Möller 2014)

From Fig. 1.1, it can be seen that there is no reason to develop expensive models because the quality gained is less than the increase in cost. This is important because a mathematical model is a very compact way to describe complex real-world systems. A complex model not only describes the relations between the system inputs and outputs, it also provides detailed insight into the system's structure and internal relationships. This is because the main relations between the variables of the real-world system to be modeled are mapped into appropriate mathematical expressions. For instance, the relationship between a system's input and output variables can be described, depending on the complexity of the system, by a set of ordinary differential equations.

In general, there are two different approaches to obtaining a mathematical model of a real-world system:

- *Deductive method of theoretical or axiomatic modeling*, a bottom-up approach starting at a high level of well-established a priori knowledge of system objects representing the mathematical model. In real-world situations, problems occur in assessing the range of applicability of these models. The deductive-modeling methodology is supplemented by an empirical model validation proof step, as shown in Fig. 1.2. Afterward, the model can be validated by comparing the simulation results with the known data from the system whether they match an error criterion or not.
- *Empirical method of experimental modeling* is based on measures available on the inputs and outputs of a system. Based on these measurements, the empirical model allows model building of the real-world system, as shown in Fig. 1.3. The characteristic signal-flow sequence of the experimental modeling process is used to determine the model structure for the mathematical description, based on a priori knowledge, which has to fit with the used error criterion, chosen in the same way as the performance criterion for the deductive-modeling method.

Let e be an error margin, which depends on the difference between measures of the real-world system output y_{RWS} and data from the simulation of the mathematical model output y_{MM}, as follows:

$$e := e\left(\underline{y}_{RWS}(t), \underline{y}_{MM}(t)\right).$$

The error criterion can be determined by minimizing a performance criterion:

$$J = \int_0^t e^2 \cdot dt \rightarrow \text{Min.}$$

The model fits the chosen performance criterion when the results obtained from simulation compared with the results from the system data are within the margin of error criterion. If the model developed does not fit the chosen performance criterion,

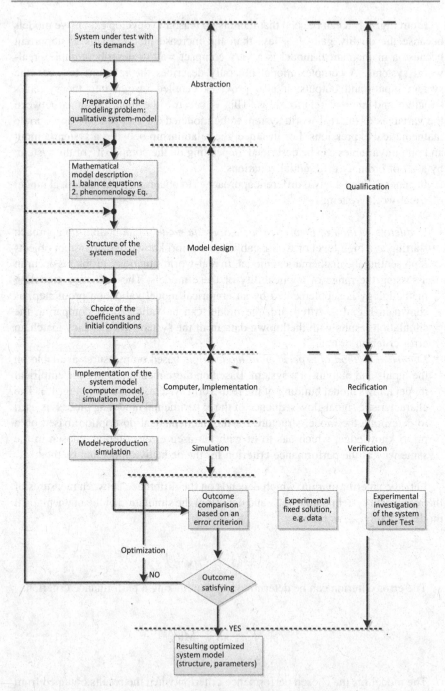

Fig. 1.2 Block diagram of empirical modeling expanded by deductive-modeling methodology (Möller 2014)

Fig. 1.3 Block diagram of the empirical modeling methodology (Möller 2014)

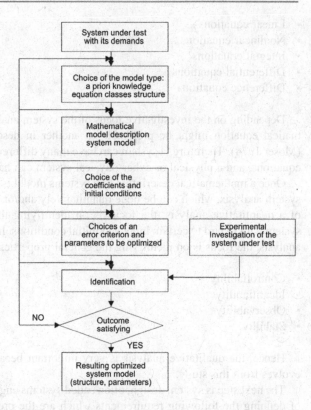

a modification is necessary at different levels, as shown for the deductive-modeling scheme in Fig. 1.2. The result of the modification, which can be understood as a specific form of model validation, is a model that fits better than the previously developed model. It is essential to mention that a systems model not only describes relationships between its inputs and outputs, such as for black-box models, it also gives insight into the system structure and into intrinsic and internal system relationships at the respective level of representation of non-black-box models. This is due to the fact that the relationships between the variables of the system are mapped into appropriate mathematical expressions. For instance, the relationship between input and output variables of a system can be described—depending on its complexity—by sets of ordinary differential equations, or by sets of partial differential equations, which represent the mathematical notation of the system.

After a systems model is identified for a physical or cyber-physical system, the next step is to identify basic relationships, i.e., the mathematical equations to describe the system. With regard to the domain-specific physical laws, such as Kirchhoff's voltage, and current laws in electrical engineering or Newton's laws in mechanical engineering, the respective mathematical equations can be set up. The equations that describe systems may be represented in many forms, such as:

- Linear equations
- Nonlinear equations
- Integral equations
- Differential equations
- Difference equations

Depending on the investigative focus of the system analysis, one form of mathematical equation might be preferable to another in describing the same system (Moser 1974). Therefore, a system can have many different forms of mathematical equations, and a physical or cyber-physical system can have different models.

Once a mathematical description of a systems model is obtained, the next step is system analysis, which can be done quantitatively and/or qualitatively. In the case of a quantitative analysis, the focus is on identifying the exact response of the system with regard to certain input and initial conditions. In the case of a qualitative analysis, the focus is on identifying the general properties of the system, such as:

- Controllability
- Identifiability
- Observability
- Stability

Hence, the qualitative analysis is very important because system design often evolves from this study.

The next step is system design, often called systems engineering. It is the process of defining the following requirements which are the prerequisites of the system design.

- *Architecture*: Structure which contains the system-specific components, the behavior to fulfill the function or purpose of the system, and more
- *Components*: Parts that are directly or indirectly related to each other in the specific system
- *Data*: Abstract concept from which information and, thereafter, knowledge can be derived for the system design
- *Interfaces*: Shared boundary across which separate components of the system exchange information
- *Modules*: Self-contained components of a system which have an interface to other components of the system

Systems design can also be introduced as the application of systems theory to product development.

If the response of a designed system is found to be unsatisfactory, the system must be improved or optimized. In some cases, the response of systems can be improved or optimized by adjusting certain parameters; in other cases, the redesign of system components or modules has to be introduced to improve or optimize the system behavior satisfactorily (Meadows 2009).

With regard to the first step in the study of a system, it should be noted that the system design has been carried out on the model developed of the physical or cyber-physical system. However, if the model is properly chosen, the performance of the physical or cyber-physical system should be improved correspondingly by introducing the required adjustments or redesigns.

1.2   Standard Forms of System Description

The second step in the study of a system is to set up the mathematical equations which describe the system behavior. Because of different analytical methods or objectives, different mathematical equations may be used to describe the same system. These equations belong to two standard forms of mathematical system representations: (1) the transfer functions that describe the external or input-output properties of a system and (2) the sets of differential equations that describe the internal and input-output behavior of a system, called the internal or state-variable description of the system.

1.2.1   Input-Output Description

The concept of an input-output description of a system involves a mathematical relationship between the input and output of the system. Hence, the input-output relationship can be considered as a multivariable system (MVS), described by the input-output description of a linear, causal, relaxed system which has the form

$$Y(t) = \int_{-\infty}^{t} G(t, \tau), U(\tau) d\tau$$

where U_i and Y_i are the input and output, and G_i is the impulse response matrix of the system, which can be rewritten as follows:

$$G(s) = \frac{Y_i}{U_i}; \quad i = 1, \ldots, n.$$

The system shown in Fig. 1.4 has n inputs and m outputs. The inputs are denoted by $u_1, u_2, \ldots, u_n$ or by $n \times 1$ column vector $u = [u_1, u_2, \ldots, u_n]'$. The outputs are denoted by $y_1, y_2, \ldots, y_m$ or by $m \times 1$ column vector $y = [y_1, y_2, \ldots, y_m]'$. The time interval in which the inputs and outputs are defined is from $-\infty, \infty$. To denote a

Fig. 1.4 System with
n inputs and m outputs

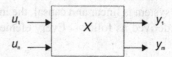

vector function defined over $(-\infty, \infty)$, u or $u(.)$ is used. To denote the value of u at time t, $u(t)$ is used. If the function u is defined only over (t_o, t), we can write $u(t_o, t)$.

If the output of a system at time t depends only on the input applied at time t_1, the system is called an instantaneous or zero-memory system. A network that consists only of resistors is such a system. Most systems of interest, however, have a memory behavior, i.e., the output at a time t_1 depends not only on the input applied at t_1 but also on the input applied before and/or after t_1. If an input $u_{(t0,t)}$ is applied to the system, the output $y_{(t1, \infty)}$ is generally not determinable unless we know that the input was applied before t_1. An input-output description that lacks a unique relationship is of no use in determining the input-output description. Before the input is applied, the system must be assumed to be relaxed or at rest and that the output is excited solely and uniquely by the input applied thereafter. If an input $u_{(-\infty, \infty)}$ is applied at $t = -\infty$, the corresponding output will be excited solely and uniquely by u. Hence, under the relaxedness assumption, it is legitimate to write

$$y = Gu$$

where G is some operator or function that uniquely specifies the output y in terms of the input u of the system. This equation is applicable only to a relaxed system.

In particular, an input-output-denoted relaxed systems model has the linearity characteristic given in Definition 1.1.

Definition 1.1 A relaxed system is said to be linear if, and only if

$$G\left(\alpha_1 u^1 + \alpha_2 u^2\right) = \alpha_1 Gu^1 + \alpha_2 Gu^2$$

for any inputs u^1 and u^2 and for real numbers α_1 and α_2. Otherwise, the relaxed system is said to be nonlinear ∎.

In particular, an input-output-denoted relaxed systems model has the causal characteristic given in Definition 1.2.

Definition 1.2 A relaxed system is said to be causal or nonanticipatory if the output of the system at time t does not depend on the input applied after time t_1; it depends only on the input applied before and at time t. This means that the past affects the future but not conversely. Hence, if a relaxed system is causal, its input and output relation can be written as

$$y(t) = Gu_{(-\infty,\infty)}$$

for all t in $(-\infty, \infty)$ ∎.

If a system is not causal, it is said to be noncausal or anticipatory. If a relaxed system is linear and causal, the impulse response matrix $G(t, \tau)$ of the system can be derived as follows. Every element of $G(t, \tau)$ is, by definition, the output due to a δ-function input applied at time τ. If a relaxed system is causal, the output is zero before any input is applied. Hence, for a linear, causal, relaxed system, we have

$$G(t,\tau) = 0$$

for all τ and all $t < \tau$.

Consequently, the input-output description of linear, causal, relaxed systems becomes

$$Y(t) = \int_{-\infty}^{t} G(t,\tau), u(\tau)d\tau.$$

1.2.2   State-Variable Description

The concept of the state of a system refers to a minimum set of variables, known as state variables that fully describe the system and its response to any given set of inputs. In particular, a state-determined systems model has the characteristics given in Definition 1.3.

Definition 1.3 A mathematical description of a system in terms of a minimum set of variables $x_i(t)$, $i = 1, \ldots, r$, together with knowledge of those variables at an initial time t_0 and the system inputs for time $t \geq t_0$, is sufficient to predict the future system state and outputs for all time $t > t_0$ ■.

This definition asserts that the dynamic behavior of a state-determined system is completely characterized by the response of the set of r variables $x_i(t)$, where the number r is defined as the order of the system.

If the system is state determined, knowledge of its state variables $(x_1(t_0), x_2(t_0), \ldots, x_r(t_0))$ at some initial time t_0 and the inputs $u_1(t), \ldots, u_n(t)$ for $n = r$ for $t \geq t_0$ are sufficient to determine all future behavior of the system.

The state variables are an internal description of the system which completely characterize the system state at any time t and from which any output variables $y_m(t)$ can be computed (Rockwell 2010).

A standard form for state equations used throughout system analysis can be expressed as a set of n-coupled, first-order ordinary differential equations, representing the state equations, in which the time derivative of each state variable is expressed in terms of the state variables $x_1(t), \ldots, x_r(t)$ and the system inputs $u_1(t), \ldots, u_n(t)$. In general, the form of the n state equation is

$$\dot{x}_1 = f_1(x, u, t)$$
$$\dot{x}_2 = f_2(x, u, t)$$

$$\dot{x}_i = f_i(x, u, t)$$

where $\dot{x}_1 = \frac{dx_i}{dt}$ and each of the functions $f_i(x,u,t)$, $i = 1, \ldots, r$, is a general nonlinear, time-varying function of the state variables, system inputs, and time.

It is common to express the state equations in vector form, in which the set of r state variables is written as a state vector:

$$x(t) = [x_1(t), x_2(t), \ldots, x_r(t)]^T$$

and the set of m inputs is written as an input vector:

$$u(t) = [u_1(t), u_2(t), \ldots, u_n(t)]^T.$$

Each state variable is a time-varying component of the column vector $x(t)$. This form of the state equations explicitly represents the basic elements contained in the definition of a state-determined system. For a set of initial conditions (the values of the x_i, $i = 1, 2, \ldots, r$, at some time t_0) and the inputs for $t \geq t_0$, the state equations explicitly specify the derivatives of all state variables. The value of each state variable at some time Δt later can then be found by direct integration (Rockwell 2010).

The system state at any instant may be interpreted as a point in an r-dimensional state space, and the dynamic state response $x(t)$ can be interpreted as a path or trajectory traced out in the state space.

The state of equations that describes the unique relations between the input, output, and state is called a dynamical equation which holds the form for the state equation

$$\dot{x}(t) = f(x(t), u(t), t)$$

and for the output equation

$$y(t) = g(x(t), u(t), t)$$

or more explicitly

$$\dot{x}_1(t) = f_1(x_1(t), x_2(t), \ldots, x_r(t), u_1(t), u_2(t), \ldots, u_n(t), t)$$
$$\dot{x}_2(t) = f_2(x_1(t), x_2(t), \ldots, x_r(t), u_1(t), u_2(t), \ldots, u_n(t), t)$$

$$\dot{x}_r(t) = f_r(x_1(t), x_2(t), \ldots, x_r(t), u_1(t), u_2(t), \ldots, u_n(t), t)$$
$$y_1(t) = f_1(x_1(t), x_2(t), \ldots, x_r(t), u_1(t), u_2(t), \ldots, u_n(t), t)$$
$$y_2(t) = f_2(x_1(t), x_2(t), \ldots, x_r(t), u_1(t), u_2(t), \ldots, u_n(t), t)$$

$$y_m(t) = f_m(x_1(t), x_2(t), \ldots, x_r(t), u_1(t), u_2(t), \ldots, u_n(t), t)$$

where $x = [x_1 x_2 \ldots x_r]'$ is the state, $y = [y_1 y_2 \ldots y_m]'$ is the output, and $u = [u_1 u_2 \ldots u_n]'$ is the input. The input u, the output y, and the state x are real vector functions of

t defined over $[-\infty, \infty]$. The unexplicitly written state and output equations are
linear functions of x and u. It is easy to show that if f and g are linear functions of
x and u, then they are of the form

$$f(x(t), u(t), t) = A(t)x(t) + B(t)u(t)$$

$$g(x(t), u(t), t) = C(t)x(t) + D(t)u(t).$$

where A, B, C, and D are, respectively, $n \times n$, $n \times p$, $q \times n$, and $q \times p$ matrices
Henceforth, the form of an n-dimensional linear dynamical equation is

$$\dot{x} = A(t)x + B(t)u \quad \text{(state   equation)}$$

$$y = C(t)x + D(t)u \quad \text{(output   equation)}.$$

For this equation to have a unique solution, every entry of $A(.)$ must be a
continuous function of t defined over $[-\infty, \infty]$. For convenience, the entries of
B, C, and D are also assumed to be continuous functions in t defined over $[-\infty, \infty]$.
Since the values A, B, C, and D change with time, the dynamical equation is then
more precisely called a linear time-varying dynamical equation. If the values $A, B,$
C, and D are independent of time t, then the dynamical equation is more precisely
called a linear time-invariant dynamical equation.

Hence, the form of an n-dimensional, linear time-invariant dynamical equation is

$$\dot{x} = Ax + Bu$$

$$y = Cx + Du$$

where A, B, C, and D are, respectively, $n \times n$, $n \times p$, $q \times n$, and $q \times p$ real constant
matrices. In the time-invariant case, the characteristics of the equation do not
change with time; hence, there is no loss of generality in choosing the initial time
t_0 to be 0. The time interval of interest then becomes $(0, \infty)$.

Many systems in the physical world are oscillatory systems, with conversion of
energy from one form to another. In a pendulum, the potential energy of the bob in
the gravitational force field is converted to kinetic energy as the bob swings from its
highest position to the neutral position. Electrical RCI networks allow energy to be
changed between components, and oscillation may result. If no energy loss occurs,
the oscillation continues at constant amplitude; however, most real-world systems
lose energy, i.e., through damping, and the oscillations eventually cease. A simple
oscillatory system, such as the simple pendulum, can be described by nonlinear
differential equations of second order in the form

$$x'' = \frac{dx'}{dt} = k \cdot F(x) \cdot x' - x$$

where x'' is the acceleration of the displacement, x' is the rate of change of displacement over time, x is the displacement, and k is the damping term. $F(x)$ is an algebraic function of x, which controls the nature of the oscillation.

The second term of the differential equation $(-x)$, when not dominant, causes acceleration of the oscillating object toward the neutral point.

A special form for $F(x)$ was suggested by the Dutch physicist van der Pol when he investigated how to maintain the oscillations in a circuit that depends on continuous oscillations. The equation for maintaining the energy of the oscillating system becomes positive when $|x|$ is <1.0 and negative when $|x|$ is >1.0. Thus, the second-order differential equation for the van der Pol oscillator can be written as

$$x'' + k \cdot \left(x^2 - 1\right) \cdot x' + x = 0$$

Alternatively, one can describe the van der Pol oscillator by utilizing a set of two first-order differential equations as follows, the solution of which can be obtained by simulation:

$$x = x_1$$
$$x' = x_1' = x_2$$
$$x'' = x_2'$$

which results in

$$x_2 = -k \cdot \left(x_1^2 - 1\right) \cdot x_2 - x_1,$$

or

$$\frac{d}{dt} = \begin{bmatrix} x_1 \\ x_2 \end{bmatrix} = \begin{bmatrix} 0 & 1 \\ -1 & k \end{bmatrix} \begin{bmatrix} x_1 \\ x_2 \end{bmatrix} - k \begin{bmatrix} x_2 & 0 \\ 0 & 0 \end{bmatrix} \begin{bmatrix} x_1^2 \\ x_2^2 \end{bmatrix}.$$

The state-vector $x(t)$ is defined as a minimal set of state variables which uniquely determines the future state of a dynamic system if its present values are given. Thus, if $x(t_0)$, the state at $t = t_0$, is known, then the state vector at any future time $x(t)$, for $t > t_0$, is uniquely determined by differential equations such that

$$x' = A(t) \cdot x + B(t) \cdot u; \ t > 0.$$

This equation set may be rewritten more compactly in matrix form:

$$x' = Ax + Bu,$$

which is in state-vector form, with x as the state vector, u as the source or input vector, and A and B as the respective system and input matrix.

We may write an output vector, which gives the output variables as linear combinations of the state variables and the inputs. The output vector has the general form of

$$y = Cx + Du,$$

where C is the output matrix, and D is the transition matrix.

The state-variable description for the time-varying case was discussed above. Thus, the corresponding time-invariant vector-matrix notation yields

$$\dot{x} = Ax + Bu$$

$$y = Cx + Du$$

By using the Laplace transforms for the above equations and by setting all initial conditions to zero, one obtains

$$sX(s) = A \cdot X(s) + B \cdot U(s)$$
$$Y(s) = C \cdot X(s) + D \cdot U(s).$$

Solving the first equation above for $X(s)$ and substituting into the second equation above yields

$$Y(s) = \left[C(sI - A)^{-1}B + D\right]U(s) = G(s) \cdot U(s),$$

where $G(s)$ is the system transfer matrix of dimension n by m. For a single-input, single-output system, a so-called SISO system, the system matrix in the above equation becomes a form of a system transfer function given by

$$G(s) = \frac{Y(s)}{U(s)} = c^T(sI - A)^{-1}b + dc$$

which is shown in Fig. 1.5.

Example 1.1 Let us assume that the transfer matrix of a linear system can be described by

$$x'_1 = x_2$$
$$x'_2 = x_3$$
$$x'_3 = -2x_1 - 4x_2 - 6x_3 + u(t)$$
$$y = x_1 + x_2 + u(t).$$

Fig. 1.5 Transfer characteristic of a SISO system

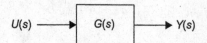

$U(s) \longrightarrow \boxed{G(s)} \longrightarrow Y(s)$

Solving for $(sI - A)^{-1}$ yields

$$(sI - A)^{-1} = \left[\begin{pmatrix} s & 0 & 0 \\ 0 & s & 0 \\ 0 & 0 & s \end{pmatrix} - \begin{pmatrix} 0 & 1 & 0 \\ 0 & 0 & 1 \\ -2 & -4 & -6 \end{pmatrix} \right]^{-T} = \begin{pmatrix} s & -1 & 0 \\ 0 & s & \\ 2 & 4 & s+6 \end{pmatrix}^{-1}.$$

Rewriting this equation yields the form

$$G(s) = \frac{Y(s)}{U(s)} = \begin{pmatrix} 1 \\ 1 \\ 0 \end{pmatrix}^T (sI - A)^{-1} \begin{pmatrix} 0 \\ 0 \\ 1 \end{pmatrix} + 1.$$

A system is defined by sets of first-order differential equations. To prove the existence and uniqueness of solutions of first-order differential equations, we introduce a systematic procedure for this proof.

Definition 1.4 The set $Ax = b$ with m equations and n unknowns has solutions if, and only if, rank[A] = rank [Ab]. Let r = rank [A]. If condition rank [A] = rank [Ab] is satisfied and if $r = n$, then the existence of solutions is unique. ∎

Definition 1.5 The set $Ax = b$ with m equations and n unknowns has solutions if, and only if, rank[A] = rank [Ab]. Let r = rank [A]. If condition rank [A] = rank [Ab] is satisfied and if $r < n$, an infinite number of solutions exists and r unknown variables can be expressed as linear combinations of the other $n - r$ unknown variables, whose values are arbitrary. ∎

The proof for existence and uniqueness of solutions requires that one form the augmented matrix [Ab]. The first n columns of the augmented matrix are the columns of A. The last column of the augmented matrix is the column vector b.

Determine whether the following set has a unique solution:

$$6x + 3y + 2z = 18$$
$$-6x + 3y + 4z = 12$$
$$6x + 3y + 4z = 24$$

The matrices A and b and v are

$$A = \begin{bmatrix} 6 & 3 & 2 \\ -6 & 3 & 4 \\ 6 & 3 & 4 \end{bmatrix}$$

$$b = \begin{bmatrix} 18 \\ 12 \\ 24 \end{bmatrix}$$

$$[Ab] = \begin{bmatrix} 6 & 3 & 2 & 18 \\ -6 & 3 & 4 & 12 \\ 6 & 3 & 4 & 24 \end{bmatrix}$$

$$v = \begin{bmatrix} x \\ y \\ z \end{bmatrix}.$$

Obviously, from Definitions 1.4 and 1.5, the rank of A and $[Ab]$ has to be proved. Rank $[A] = 3$ and rank $[Ab] = 3$. Because A and $[Ab]$ have the same rank, a solution exists. This rank equals the number of unknowns, the solution is unique, and $x = 1$, $y = 2$, and $z = 3$.

1.3   Controllability, Observability, and Identifiability

Analyzing and designing physical and cyber-physical systems requires a priori knowledge as to whether the system being analyzed or designed can be assumed to be controllable, observable, and/or identifiable. Controllability, observability, and identifiability are important properties of systems, described in state-variable notation. With regard to this analysis, a linear system can be said to be state controllable when the system input u can be used to transfer the system from any initial state to any arbitrary state in finite time. Moreover, a linear system can be said to be observable if the initial state $x(t_0)$ can be determined uniquely for a given output $y(t)$ for $t_0 \leq t \leq t_1$ for any $t_1 > t_0$. If a mathematical model of a system can be written in the state notation, the method of controllability, observability, and identifiability analysis can be used for model predictions.

From a more general point of view, the description of controllability can be given for the time-varying case as follows.

Definition 1.6   A linear dynamic system

$$x' = A \cdot x + B \cdot u$$
$$y = C \cdot x + D \cdot u$$

is said to be:

- Controllable at time $t_0 \in T$, if for a finite $t_1 > t_0$, $t_1 \in T$ exists
- Completely controllable, if for each $t_0 \in T$ a finite time $t_1 > t_0$, $t_1 \in T$ exists
- Differential or particularly controllable, if for each $t_0 \in T$ and each finite $t_1 > t_0$, $t_1 \in T$, the matrix

$$w(t_0, t_1) = \int\limits_{t_0}^{t_1} \Phi(t_0, \tau)B(\tau)B^T(\tau)\Phi^T(t_0, \tau)d\tau$$

is regular. ∎

The different notations for controllability represent the diverse characteristics of systems for which the corresponding mathematical systems model has the same result, due to controllability, as the real-world system itself.

An important task in system controllability analysis refers to determining a system characteristic which allows the prediction of how the system may behave with regard to different scenarios.

Definition 1.7 A linear dynamic system

$$x' = A \cdot x + B \cdot u$$
$$y = C \cdot x + D \cdot u$$

is said to be completely state controllable if a control signal u exists, defined over the finite interval $t_0 < t < t_F$, which transfers the system from any initial state $x(t_0) = x_Q$ to any desired final state $x(t_F) = x_E$ in the defined time interval. ∎

Definition 1.7 is said to be true, if and only if the (n, np) controllability matrix

$$Q_C := \begin{bmatrix} B & AB & A^2B & \ldots & A^{n-1}B \end{bmatrix}$$

has full row rank n, which means that the vector elements B, AB, $\ldots$, $A^{n-1}B$ of Q_C are linearly independent, meaning the controllability matrix Q_C has a nonzero determinant.

Example 1.2 A dynamic system can be described by the state-equation model:

$$x'(t) = A \cdot x(t) + B \cdot u(t)$$
$$= \begin{bmatrix} -3 & 1 \\ -2 & 1.5 \end{bmatrix} \begin{bmatrix} x_1 \\ x_2 \end{bmatrix} + \begin{bmatrix} 0 \\ 1 \end{bmatrix} [u],$$

with

$$A = \begin{bmatrix} -3 & 1 \\ -2 & 1.5 \end{bmatrix}$$

$$B = \begin{bmatrix} 0 \\ 1 \end{bmatrix}.$$

The dynamic system given in Example 1.2 is completely state controllable if B and AB are linearly independent and the rank of the controllability matrix Q_C: [B, AB] = 2, with

$$B = \begin{bmatrix} 0 \\ 1 \end{bmatrix}$$

and

$$AB = \begin{bmatrix} 1.0 \\ 1.5 \end{bmatrix},$$

hence

$$Q_C := [B, AB] = 2,$$

which means the dynamic system is completely state controllable.

Example 1.3 Suppose a dynamic system can be described in the Laplace domain notation:

$$X_1 = \frac{1}{s+1} \cdot U$$

and

$$X_2 = \frac{1}{s+2} \cdot (X_1 + U),$$

which can be rewritten in the state-equation notation as

$$x_1' = -x_1 + u$$

$$x_2' = x_1 - 2 \cdot x_2 + u.$$

Assuming

$$b = \begin{bmatrix} 1 \\ 1 \end{bmatrix}$$

and

$$Ab = \begin{bmatrix} -1 & 0 \\ 1 & -2 \end{bmatrix} \begin{bmatrix} 1 \\ 1 \end{bmatrix} = \begin{bmatrix} -1 \\ -1 \end{bmatrix},$$

the dynamic system given above is not state controllable, while $b + Ab = 0$, which means that the vectors b and Ab are linearly dependent.

Definition 1.8 A linearly dynamic system

$$x' = A \cdot x + B \cdot u$$
$$y = C \cdot x + D \cdot u$$

is said to be:

- Observable at time $t_0 \in T$, if for a finite $t_1 > t_0$, $t_1 \in T$ exists
- Completely observable, if for each $t_0 \in T$ and each finite $t_1 > t_0$, $t_1 \in T$ exists
- Differential or particularly observable, if for each $t_0 \in T$ and each finite $t_1 > t_0$, $t_1 \in T$, the matrix

$$m(t_0, t_1) = \int_{t_0}^{t_1} \Phi^T(t_1, t_0) C^T(t) C(t) \Phi(t_1, t_0) dt$$

is a regular one. ∎

The different ideas of observability are due to the properties of the dynamic system and not due to the properties of the mathematical model. But the mathematical model has the same results, due to observability, as the dynamic system.

Let us next consider that the system is completely observable; hence, predictions with it are possible, and we may write:

Definition 1.9 A linear dynamic system

$$x' = A \cdot x + B \cdot u$$
$$y = C \cdot x + D \cdot u$$

is said to be completely observable within the finite interval $t_0 < t < t_F$, if any initial state $x(t_0) = x_0$ can be determined from the output y, observed over the same interval. ∎

Definition 1.9 is said to be true if, and only if, the (n, nr) observability matrix

$$Q_O := \left[C^T C T A^T \ldots C^T (A^T)^{n-1} \right]$$

has full rank n, with C^T as the transpose of C.

Example 1.4 The state-variable description of a dynamic system is given for the time-varying case as follows:

$$x'(t) = A \cdot x(t) + B \cdot u(t)$$
$$y(t) = C \cdot x(t).$$

The corresponding parameters are

$$\frac{d}{dt}x = \begin{bmatrix} 0 & 1 & 0 \\ 0 & 0 & 1 \\ -6 & -11 & 6 \end{bmatrix} \begin{bmatrix} x_1 \\ x_2 \\ x_3 \end{bmatrix} + \begin{bmatrix} 0 \\ 0 \\ 1 \end{bmatrix} [u]$$

$$y(t) = \begin{bmatrix} 20 & 9 & 1 \end{bmatrix} \begin{bmatrix} x_1 \\ x_2 \\ x_3 \end{bmatrix},$$

with

$$A = \begin{bmatrix} 0 & 1 & 0 \\ 0 & 0 & 1 \\ -6 & -11 & -6 \end{bmatrix}$$

and

$$C = \begin{bmatrix} 20 & 9 & 1 \end{bmatrix},$$

or

$$C^T = \begin{bmatrix} 20 \\ 9 \\ 1 \end{bmatrix},$$

which yields

$$C^T A^T = \begin{bmatrix} 20 \\ 9 \\ 1 \end{bmatrix} \begin{bmatrix} 0 & 0 & -6 \\ 1 & 0 & -11 \\ 0 & 1 & -6 \end{bmatrix} = \begin{bmatrix} -6 \\ 9 \\ 3 \end{bmatrix}$$

and

$$C^T \left(A^T\right)^2 = \begin{bmatrix} 20 \\ 9 \\ 1 \end{bmatrix} \begin{bmatrix} 0 & -6 & 36 \\ 0 & -11 & 60 \\ 1 & -6 & 25 \end{bmatrix} = \begin{bmatrix} -18 \\ -39 \\ -9 \end{bmatrix}.$$

The vectors C^T, $C^T A^T$, and $C^T \left(A^T\right)^2$ are linearly independent, and the rank of the observability matrix is

$$Q_O := [C^T, C^T A^T, C^T (A^T)^2] = 3.$$

thus, the dynamic system, given in Example 1.4, is completely observable.

Example 1.5 Suppose the output equation is

$$y(t) = [4 \quad 5 \quad 1] \, [x_1, x_2, x_3]^T$$

instead of

$$y(t) = [20 \quad 9 \quad 1] \, [x_1, x_2, x_3]^T$$

the vectors C^T, $C^T A^T$, and $C^T (A^T)^2$ are linearly dependent, and the rank of the observability matrix has the value:

$$Q_O := [C^T, C^T A^T, C^T (A^T)^2] = 2.$$

Thus, the dynamic system is not observable.

Definition 1.10 A dynamic system is said to be identifiable in its parameters, within the time interval $t_0 < t < t_F$, if the parameter vector Θ may be determined from the output y, observed over the same time interval $t_0 < t < t_F$. ∎

Definition 1.11 A dynamic system is said, for the true model parameter vector Θ_T, to be:

1. Parameter identifiable if there exists an input sequence $\{u\}$ such that Θ and Θ_T are distinguished for all $\Theta \neq \Theta_T$
2. System identifiable if there exists an input sequence $\{u\}$ such that Θ and Θ_T are distinguishable for all $\Theta \neq \Theta_T$ but are a finite set
3. Unidentifiable in all other cases ∎

The state-variable concept of dynamic systems completely characterizes the system's past, since the past input is not required to determine the future output of the dynamic system. This seemingly elementary notation of state equations is of importance in the systems state-variable approach. In fact, this mathematical notation, describing dynamic systems, is fundamentally based upon the state-variable concept.

Depending on the particular form of equations used to describe the system, the state equations can be in one of several mathematical forms. It is possible to classify state equations based on their mathematical structure.

Time is usually an independent variable in a state-variable model:

• Sometimes the time variable is considered to be a discrete variable; in this case, the state-variable model typically will be described by recursive equations.
• In other cases, the independent variable time is considered to be a real value.
• In the case of additional independent variables in the state-variable model, they are said to be distributed; such state-variable models are described by partial differential equations.

- If time is the only independent variable, then the state-variable model is said to be lumped. The actual physical size will not really serve as a measure of lumpiness; moreover, it has to be considered that occasions are frequently available to distinguish between lumped and distributed elements.
- The state-equation model can include random effects; in such cases, the state-equation model is said to be stochastic.
- If none of these features are included, the state-equation model is said to be deterministic.

1.4   Analytical Solutions of Linear Systems Models

Analytical solutions of linear systems models are required to examine their behavior with regard to the system response to an input demand. Let the state-variable equations in vector-matrix notation yield a diagonal matrix A in the form

$$x' = A \cdot x + b \cdot u \tag{1.1}$$

and

$$y = c^T \cdot x, \tag{1.2}$$

which are quite simple to solve for $x_i(t)$, where $i = 1, 2, \ldots, n$. Consider a linear system with $u = 0$, described by

$$x' = A \cdot x. \tag{1.3}$$

The time solution of (1.3) has the form

$$x(t) = \Phi(t, t_0)x(t_0)$$

where $\Phi(t, t_0)$ is referred to as the state-transition matrix. The initial condition $x(t_0)$ is transferred to the state x at time by the matrix $\Phi(t, t_0)$. It is very obvious that $\Phi(t, t_0) = I$, the identity matrix, since the state $x(t)$ is equal to $x(t_0)$ at $t = t_0$. The transition characteristic of the state-transition matrix can be written as

$$\Phi(t_2, t_0) = \Phi(t_2, t_1)\Phi(t_1, t_0), \tag{1.4}$$

which indicates that if an initial state-vector $x(t_0)$ is transferred to $x(t_1)$ by $\Phi(t_1, t_0)$ and if $x(t_1)$ is then transferred to $x(t_2)$ by $\Phi(t_2, t_1)$, then $x(t_0)$ can be transferred in a direct way to $x(t_2)$ by $\Phi(t_2, t_0)$, which is the product of the two state-transition matrices. If matrix A in (1.3) is constant with time, then the state-transition matrix $\Phi(t, t_0)$ is a function only of the distance t and t_0, that is

$$\Phi(t, t_1) = \Phi(t - t_0). \tag{1.5}$$

Consider that the matrix-exponential solution of the linear dynamic system, as shown in (1.3), yields

$$x(t) = e^{At} \cdot x(0).$$ (1.6)

This obviously means that the time response of the dynamic system is equal to the exponential function if the matrix differentiation rule is used to form

$$x' = \frac{d}{dt}[x(t)] = \frac{d}{dt}\left[e^{At} \cdot x(0)\right] = A \cdot e^{At} \cdot x(0).$$ (1.7)

Substituting (1.6) into (1.7) yields $x' = A \cdot x$. Comparing (1.6) and (1.3) yields

$$\Phi(t) = e^{At}.$$ (1.8)

We may now express the term e^{At} in a Taylor series about $t = 0$ to give

$$\Phi(t) = e^{At} = \sum_{k=0}^{\infty} \frac{1}{k!} A^k t^k = I + A \cdot t + \frac{1}{2!} A^2 \cdot t^2 + \ldots\ldots,$$ (1.9)

which is the equation for solving the state-transition equation by series expansion.

Moreover, we may premultiply each term of the state variables (1.3) by the exponential expression, which gives

$$x'(t) \cdot e-^{At} = A \cdot x(t) \cdot e-^{At} + B \cdot u(t) \cdot e-^{At},$$ (1.10)

that is, dropping the explicit notation for time dependence and rearranging the equation to

$$x' \cdot e-^{At} - A \cdot x \cdot e-^{At} = B \cdot u \cdot e-^{At}$$ (1.11)

or

$$\frac{d}{dt}\left(e-^{At}x\right) = B \cdot u \cdot e-^{At}.$$ (1.12)

Multiplying by dt and integrating over the time interval t_0 to t, as well as changing the variables, we may write

$$\int_{t_0}^{t} d/d\tau\left(e-^{A^\tau}x\right)d\tau = \int_{t_0}^{t} B \cdot u(\tau) \cdot e-^{A^\tau}d\tau,$$ (1.13)

or

$$x(t) \cdot e-^{At} - x(t_0) \cdot e-^{At}{}_0 = \int_{t_0}^{t} B \cdot u(\tau) \cdot e-^{A^\tau}d\tau.$$ (1.14)

Premultiplying all terms in this equation gives

$$x(t) = x(t_0) \cdot e^{-A(t-t_0)} + \int_{t_0}^{t} B \cdot u(\tau) \cdot e^{-A(t-t_0)} d\tau, \qquad (1.15)$$

that is, rewritten in terms of the state-transition matrix

$$x(t) = x(t_0) \cdot \Phi(t - t_0) = \int_{t_0}^{t} B \cdot u(\tau) \cdot \Phi(t - t_0) \cdot d\tau, \qquad (1.16)$$

which is the matrix form of the convolution integral. The convolution integral in (1.16) involves the impulse response function and the superposition integral in terms of the state-transition matrix.

1.4.1 Solution of State Equations Using the Laplace Transform

The state-transition matrix is used to yield the complete solution of the linear state-variable equations. Let the time-dependent behavior of a linear state-differential equation system be written in the notation:

$$x'(t) = A \cdot x(t) + B \cdot u(t) \qquad (1.17)$$

$$y(t) = C \cdot x(t) + D \cdot u(t), \qquad (1.18)$$

which means the dependence of the output vector $y(t)$ from the input vector $u(t)$, with system matrix A and input matrix B, assumed to be constant matrices. To determine the solution of this state-differential equation, the Laplace transform L[f (t)] = $F(s)$ can be used as follows:

$$s \cdot X(s) = A \cdot X(s) + X(0) + B \cdot U(s), \qquad (1.19)$$

where $X(0)$ denotes the initial state vector, presumably a known quantity, and s denotes the Laplace operator, which is the first derivative. Solving (1.19) gives

$$X(s) = (sI - A)^{-1} X(0) + (sI - A)^{-1} B \cdot U(s), \qquad (1.20)$$

with $X(s)$ as the Laplace transform of $x(t)$, and I as the (n, n) unit matrix, which is defined by $AI = IA = A$, that is

$$\begin{bmatrix} a_{11} & a_{12} \\ a_{21} & a_{22} \end{bmatrix} \begin{bmatrix} 1 & 0 \\ 0 & 1 \end{bmatrix} = \begin{bmatrix} 1 & 0 \\ 0 & 1 \end{bmatrix} \begin{bmatrix} a_{11} & a_{12} \\ a_{21} & a_{22} \end{bmatrix} = \begin{bmatrix} a_{11} & a_{12} \\ a_{21} & a_{22} \end{bmatrix}. \qquad (1.21)$$

Using A^{-1} as the inverse matrix of A, that is $A\,A^{-1}=A^{-1}A=I$, and $(sI-A)$ as a matrix called the characteristic matrix, gives

$$(sI-A)X(s) = X(0) + B\cdot U(s), \tag{1.22}$$

with $L(s) = (sI-A)$ and $L^{-1}(s) = (sI-A)^{-1}$, where $L(s)$ is the adjoint of the characteristic matrix and $\Delta(s)$ the determinant of the matrix, called the characteristic polynomial of matrix A, we can write

$$l^{-1}(s) = (sI - A)^{-1} = \frac{1}{\Delta(s)} \cdot\cdot l(s). \tag{1.23}$$

The roots of the polynomial above are called eigenvalues of the dynamic system. For a linear dynamic system with constant coefficients, they will be simple relationships.

Obviously, the time-domain solution can be obtained by convolving the inverse Laplace transform of (1.23) for (1.20), that is

$$X(s) = \frac{1}{\Delta(s)} \cdot l(s) \cdot X(0) + \frac{1}{\Delta(s)} \cdot l(s) \cdot U(s). \tag{1.24}$$

As for ordinary differential equations, we may expand $X(s)$ into a partial fraction expansion:

$$X(s) = X_1(s) \cdot \frac{1}{(s - s_1)} + X_2(s) \cdot \frac{1}{(s - s_1)} + \ldots \tag{1.25}$$

with the specialized form of the solution

$$X(s)_i = l(s) \cdot X(0) + l(s) \cdot U(s) = X_i + U_i,$$

and the corresponding results

$$X_1 = \lim_{s \to s_1} \left\{ L(s) \cdot \frac{1}{[(s - s_2)(s - s_3)\ldots]} \right\} \cdot X(0) \tag{1.26}$$

$$\Rightarrow X_1 = \left\{ L(s) \cdot \frac{1}{[(s_1 - s_2)(s_1 - s_3)\ldots]} \right\} \cdot X(0) \tag{1.27}$$

$$X_2 = \lim_{s \to s_2} \left\{ L(s_1) \cdot \frac{1}{[(s - s_1)(s - s_3)\ldots]} \right\} \cdot X(0) \tag{1.28}$$

$$\Rightarrow X_2 = \left\{ L(s_2) \cdot \frac{1}{[(s_2 - s_1)(s_2 - s_3)\ldots]} \right\} \cdot X(0) \tag{1.29}$$

and then form X_i as

$$X_i = \lim_{S \to s_i} \left[\frac{(s - s_i)}{\Delta(s)}\right] \cdot L(s) \cdot X(0)$$

$$i = 1, 2, \ldots, n,$$

(1.30)

for each root of $L(s)U(s)\frac{1}{\Delta(s)}$. The resultant state-vector $x(t)$ can be written as

$$x(t) = X_1 \cdot e^{s_1 t} + X_2 \cdot e^{e_2 t} + , \ldots, + X_n \cdot e^{s_n t} + U_1 \cdot e^{s_1 t} + , \ldots, + U_m \cdot e^{s_m t}.$$

Therefore, we can write the complete solution $X(t)$ by using the Laplace transform:

$$X(t) = L^{-1}\left\{(sI - A)^{-1}\right\}X(0) + L^{-1}\left\{(sI - A)^{-1}B \cdot U(s)\right\}.$$

(1.31)

Using the notation of

$$X(t) = L^{-1}\left\{(sI - A)^{-1}\right\}X(0) = e^{At}$$

(1.32)

and the respective correspondence of the Laplace transforms

$$\Phi(t) = L^{-1}\left\{(sI - A)^{-1}\right\}X(0) = e^{At}$$

(1.33)

we obtain the state-transition matrix for the linear system. The transition matrix determines the transient behavior of the dynamic system over all time that is between time t_0 and time t_1:

$$\Phi(t) = e^{At}.$$

(1.34)

Assuming that the initial time is denoted $t = t_0$ instead of $t = 0$, we may write

$$x(t) = x(t_0) \cdot e^{A(t - t_0)}$$

(1.35)

as well as

$$x(t) = x(0) \cdot \Phi(t - t_0).$$

(1.36)

1.4.2 Eigenvalues of the Linear Vector-Equation Systems

Consider a linear system with $u = 0$, described in (1.3), that has the form

$$x' = A \cdot x,$$

(1.37)

where A holds an $n \times n$ matrix. Using Laplace transforms and solving for $X(s)$, we receive

$$sX(s) = A \cdot X(s) \tag{1.38}$$

with

$$0 = (A - sI)X(s) \tag{1.39}$$

whereby s is a scalar. The s scalars are called eigenvalues of A, and the vectors $X(s)$ are called eigenvectors of A. The complete set of all eigenvalues is called the spectrum of A. Equation 1.39 yields for nontrivial cases of $X(s)$:

$$\det(A - sI) = 0, \tag{1.40}$$

the characteristic equation, with $\det(A - sI)$ as the characteristic polynomial of A. It is an nth-degree polynomial in s. The characteristic equation is given by

$$det(A - sI) = \begin{bmatrix} a_{11} - s & a_{12} & a_{13} \\ a_{21} & a_{22} - s & a_{23} \\ a_{31} & a_{32} & a_{33} - s \end{bmatrix} \tag{1.41}$$

$$= (-1)^n \left(s^n + a_1 s^{n-1} + \ldots + a_{n-2} s^2 + a_{n-1} s + a_n \right) = 0. \tag{1.42}$$

The n roots of the characteristic in Eq. 1.41 are the so-called eigenvalues of A. They are also called characteristic roots. Note that a real $n \times n$ matrix A does not necessarily possess real eigenvalues. But since $\det(A - sI) = 0$ is a polynomial with real coefficients, any complex eigenvalues must occur in conjugate pairs; namely, if $\alpha + j\beta$ is an eigenvalue, then $\alpha - j\beta$ is an eigenvalue of A.

Example 1.6 Assume a system can be described by system matrix A in the form

$$A = \begin{bmatrix} 1 & 2 & -1 \\ 1 & 3 & 2 \\ 2 & -1 & 0 \end{bmatrix}. \tag{1.43}$$

With regard to the Laplace transform, we can write

$$L(sI - A) = \begin{bmatrix} s-1 & -2 & 1 \\ -1 & s-3 & -2 \\ -2 & 1 & s \end{bmatrix}. \tag{1.44}$$

Hence, we find the eigenvalues of A as follows:

$$\Delta(s) = (s-1)(s-3)s - 8 - 1 + 2(s-3) - 2s + 2(s-1)$$
$$= s^3 - 4s^2 + 5s - 17 \tag{1.45}$$

which can be rewritten as

$$L^{-1}(s) = \frac{1}{\Delta(s)} \cdot L(s). \tag{1.46}$$

Example 1.7 Consider a linear system described by the differential-equation system:

$$x_1' = -k_1 \cdot x_1 + u \tag{1.47}$$

and

$$x_2' = k_1 \cdot x_1 - k_2 \cdot x_2. \tag{1.48}$$

If the system matrix A has the form:

$$A = \begin{bmatrix} -k_1 & 0 \\ k_1 & -k_2 \end{bmatrix}, \tag{1.49}$$

then the eigenvalues of A can be calculated as follows:

$$\det(A - \lambda I) = 0 \tag{1.50}$$

$$\det \begin{bmatrix} -k_1 - \lambda & 0 \\ k_1 & -k_2 - \lambda \end{bmatrix} = (-k_1 - \lambda)(-k_2 - \lambda) = 0 \tag{1.51}$$

whereby the eigenvalues can be found as $\lambda_1 = -k_1$ and $\lambda_2 = -k_2$.

1.5 Steady-State Errors of Systems

A steady-state error is the difference between the output response and the input signal when all of the transients have decayed. To determine the steady-state error, it is necessary to know the characteristics of the input signal, but often the input is not known ahead of time. This, for example, will be case if the input varies in a random form with respect to time. For example, speed and direction of a drone target can become unpredictable for the drone control system if the target system is maneuvering to evade the attacking drone. Therefore, it can be stated that no single mathematical expression can suit all of the different ways by which input to the drone control system can change. Hence, in the case of a drone controller design, the designer can try to overcome the stochastic influence by using test signals which change in a particular manner, depending on for what the control system is designed. Irrespective of the units of the input signals, typically used test signals, such as unit step response, ramp change (see Sect. 1.1 for details), and the parabolic

Fig. 1.6 Unity feedback
control system (for details see
text)

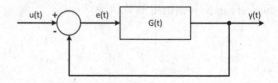

change, can be used to study the steady-state error characteristics of a system.
Therefore, the information gained from studying a system's response to these types
of input signals can be used to predict the system performance when more disper-
sive inputs are encountered.

Let us consider a unity feedback system with error $e(t)$ which is the difference
between the input $u(t)$ and the output $y(t)$ signals, as shown in Fig. 1.6.

Let the input be $u(t)$ and the output be $y(t)$; then, assuming a steady-state
condition exists, $y(t)$ can eventually adapt to behave in a steady-state form. If at
this stage, $y(t)$ is not equal to u(t), that is

$$y(t) \neq u(t),$$

then the resulting error $e(t)$ is termed the steady-state error, where

$$e(t) = u(t) - y(t).$$

System steady-state errors can be grouped into three categories (McDonalds and
Lowe 1975):

  (i) *Zero error*: output follows input without error.
 (ii) *Finite and constant error*: output follows input with some fixed magnitude of
     error.
(iii) *Infinite error*: output diverges from input with ever-increasing magnitude of
     error. This really means that the system cannot follow the input at all.

The steady-state error is the value of $e(t)$ when t is large enough for all system
transients to decay. The steady-state error is defined by the expression:

$$\lim_{t \to \infty} [e(t)] = \lim_{t \to 0} [se(s)] \qquad (1.52)$$

where $e(s)$ is the Laplace transform of $e(t)$. If an expression for $e(s)$ exists, $e(t)$ can
easily be determined without evaluating the system response.

Figure 1.7 shows a block diagram of the general form of a typical control system,
where $e(s)$ is given in the form

$$e(s) = u(s) - H(s)y(s) = u(s) - H(s)G(s)e(s).$$

Therefore,

$$e(s)[1 + H(s)G(s)] = u(s)$$

Fig. 1.7 Block diagram of a
typical control system

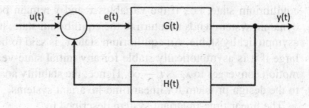

and

$$e(s) = \frac{u(s)}{1 + H(s)G(s)}. \tag{1.53}$$

Hence, from Eq. 1.52, the steady-state error for any control system can be calculated directly by evaluating

$$e(s) = \lim_{s \to 0} \left[\frac{su(s)}{1 + H(s)G(s)} \right] \tag{1.54}$$

Example 1.8 The open-loop control system transfer function has the form

$$H(s)G(s) = \frac{90}{s + 10}.$$

What is the steady-state error of the system when the input signal is a step function of unity magnitude?

From Eq. 1.54, we receive

$$e(s) = \lim_{s \to 0} \left[s \frac{u(s)}{1 + \frac{90}{s+10}} \right] = \lim_{s \to 0} \left[s \frac{u(s)(s + 10)}{s + 10 + 90} \right]$$

$$= \lim_{s \to 0} \left[s \frac{(s + 10)}{s(s + 100)} \right] = \lim_{s \to 0} \left[\frac{s + 10}{s + 100} \right] = 0.1$$

Therefore, the steady-state error is 0.1 or 10 %.

1.6   Case Study in Systems Stability Analysis

A system exhibits a tendency toward instability when it oscillates about a desired output level. Hence, a system without that behavior is said to be stable. Therefore, the transient response of a system is of primary interest and part of investigation in systems analysis. Thus, a system is said to be stable if the system remains near the equilibrium state, i.e., if the variables x and y remain bounded as $t \to \infty$. If the dynamic system tends to return to the equilibrium state, it is said that the system is asymptotically stable. An equilibrium state x_e is said to be asymptotically stable at large if it is asymptotically stable for any initial state-vector $x(0)$, such that every motion converges to x_e as $t \to \infty$. Hence, the stability analysis is inherently related to the design problem for linear time-invariant systems.

The linear time-invariant system described by

$$\frac{dx}{dt} = f\,(x, y),$$

$$\frac{dy}{dt} = g\,(x, y),$$

where f and g are continuous functions of x and y and have continuous partial derivatives, will have a unique solution $x = \Phi(t)$, $y = \Psi(t)$ for $t \geq 0$. The concept of stability can be illustrated as a curve in the x-y plane, called the phase plane, with t as the parameter. The solutions are indicated as curves, referred to as trajectories.

Consider a mass m moving on a horizontal level and attached by a spring to a fixed point on a wall. Neglecting the resistive element and external force, the resulting equation will be

$$\frac{d^2x}{dt^2} = -w^2 \cdot x.$$

Defining

$$y = \frac{dx}{dt},$$

we have a system of differential equations:

$$\frac{dx}{dt} = y,$$

$$\frac{dy}{dt} = -w^2 \cdot x,$$

which gives

$$x = \alpha \cdot \cos\,(\omega \cdot t + \beta)$$

$$y = -\alpha \cdot \omega \cdot \sin\,(\omega \cdot t + \beta)$$

These equations define the trajectories:

$$x^2 + \left(\frac{y}{\omega}\right)^2 = \alpha^2$$

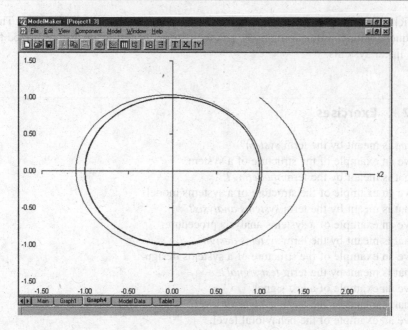

Fig. 1.8 Simple trajectories in the x-y plane

in the x-y plane, as shown in Fig. 1.8, as a simulation result using the ModelMaker simulation software with the x-axis as the displacement and the y-axis as the speed, starting at the initial conditions $x(0) = 1$ and $y(0) = 1$.

Obviously, the trajectories shown in Fig. 1.8 can be given directly in the form

$$\frac{dy}{dx} = \frac{\frac{dy}{dt}}{\frac{dx}{dt}} = -\frac{\omega^2 \cdot x}{y}.$$

In order to investigate the behavior of a dynamic system near equilibrium, we may first assume that the equilibrium points are $x_0 = 0$ and $y_0 = 0$. Hence, we may define the equilibrium points as:

  (i) *Stable* if x and y remain bounded as $t \to \infty$
 (ii) *Asymptotically stable* if $x, y \to 0$ as $t \to \infty$
(iii) *Unstable* in any other case

As mentioned earlier, stability analysis of a dynamic system is very important. For this reason, we present different criteria to determine the stability of dynamic systems. A procedure for determining the stability of a linear time-invariant system can be achieved by examining its characteristic polynomial, called Routh-Hurwitz criterion. If the time-domain formulation is transformed into the frequency domain by such transformations as the simple exponential function, the stability of the system in the frequency domain can be studied by means of the Nyquist criteria, which is the imaginary or frequency axis for the particular system function. The frequency-domain transformations, and their subsequent use, are also restricted to the linear systems.

1.7   Exercises

What is meant by the term *system*?
Give an example of the structure of a system.
What is meant by the term *modeling*?
Give an example of the structure of a systems model.
What is meant by the term *systems analysis*?
Give an example of a systems analysis procedure.
What is meant by the term *systems design*?
Give an example of the structure of a systems design.
What is meant by the term *test signal*?
Give an example of a test signal.
What is meant by the term *behavioral* level?
Give an example of the behavioral level.
What is meant by the term *composite structure level*?
Give an example of the composite structure level.
What is meant by the term *input-output description*?
Give an example of the input-output description of a system.
What is meant by the term *state-variable description*?
Give an example of the state-variable description of a system.
What is meant by the term *Laplace transform*?
Give an example of the Laplace transform in systems analysis.
What is meant by the term *eigenvalue*?
Give an example of the eigenvalue of a system.
What is meant by the term *controllability*?
Give an example of the controllability of a system.
What is meant by the term *observability*?
Give an example of the observability of a system.
What is meant by the term *identifiability*?
Give an example of the identifiability of a system.
What is meant by the term *stability*?
Give an example of the stability of a system.

References

(Kamal 2008) Kamal R.: Embedded Systems: Architecture, Programming, and Design, McGraw Hill Publ. 2008.
(Luhmann 2012) Luhmann, N.: Introduction to Systems Theory, Polity Press, 2013

(McDonalds Lowe 1971) McDonalds, A. C., Lowe, H.: Feedback and Control Systems, Prentice Hall, 1971

(McDonalds Lowe 1975) McDonalds, A. C., Lowe, H.: Feedback and Control Systems, Prentice Hall, 1975

(Meadows 2009) Meadows, D. H.: Thinking in Systems. Earthscan Publ. 2009.

(Mitchel 2009) Mitchel, M.: Complexity- A Guided Tour, Oxford University Press, 2009.

(Moeller 2003) Moeller, D. P. F.: Mathematical and Computational Modeling and Simulation, Springer Verlag, Berlin Heidelberg, 2003

(Möller 2014) Möller D. P. F.: Introduction to Transportation Analysis, Modeling and Simulation, Springer Publ. Series Simulation Foundations, Methods, and Applications, 2014.

(Moser 1974) Moser, J. Ed.: Dynamical Systems, Theorey and Applications, Springer Publ. 1974.

(Rockwell 2010) Rockwell, D.: State-Space Representation of LTI Systems, http://web.mit.edu/2.14/www/Handouts/StateSpace.pdf

Introduction to Embedded Computing Systems

2

This chapter begins with a brief overview of embedded computing systems in Sect. 2.1, taking into account the introduction to systems in Chap. 1. Thereafter, Sect. 2.2 introduces the hardware architecture of embedded computing systems. Section 2.3 is an introduction to the methodology for determining the design metrics of embedded computing systems, a method which defines the preciseness of a design with regard to the requirements specifications. Section 2.4 introduces the concept of embedded control with regard to the respective mathematical notation formulations of the different control laws. Section 2.5 introduces the principal concept of hardware-software codesign. Since the expected growth rate of design productivity in the traditional way is far below that of system complexity, hardware-software codesign has been developed as a new design methodology during the past decade. Section 2.6 presents a case study of the concept of system stability analysis. Section 2.7 contains comprehensive questions from the system theory domain, followed by references and suggestions for further reading.

2.1 Embedded Computing Systems

In Chap. 1, an introduction to systems was given to ensure that readers from all engineering and scientific disciplines have the same understanding of the term "system" and the mathematical background necessary for the study of systems. Embedded computing systems (ECSs) are dedicated systems which have computer hardware with embedded software as one of their most important components. Hence, ECSs are dedicated computer-based systems for an application or a product, which explains how they are different from the more general systems introduced in Chap. 1.

As implementation technology continues to improve, the design of ECS becomes more challenging due to increasing system complexity, as well as relentless time-to-market pressure. Moreover, ECSs may be independent, part of a larger system, or a part of a heterogeneous system. They perform dedicated functions in a

© Springer International Publishing Switzerland 2016

D.P.F. Möller, *Guide to Computing Fundamentals in Cyber-Physical Systems*,

Computer Communications and Networks, DOI 10.1007/978-3-319-25178-3_2

huge variety of applications, although these are not usually visible to the user. Some examples are:

- Automotive assistance systems
- Aircraft electronics
- Home appliances/systems
- Medical equipment
- Military systems
- Navigation systems
- Telecommunication systems
- Telematics systems
- Consumer electronics, such as DVD players
- High-definition digital television

A variety of networking options exist for ECS, as compared to autonomous embedded subsystems that have been implemented on special microcontrollers and optimized for specific applications. In general, ECSs have three main components:

- *Hardware*, which consists of the microprocessor or microcontroller, timers, interrupt controller, program and data memory, serial ports, parallel ports, input devices, interfaces, output devices, and power supply.
- *Application software* that concurrently performs a series of tasks or multiple tasks.
- *Real-time operating systems* that supervise the application software and provide a mechanism for the processor to run a scheduled process and do the context switch between various processes (tasks). The real-time operating system defines the way in which the ECS works. It organizes access to a resource consisting of a series of tasks in sequence and schedules their execution by following a plan to control the latencies and to meet the deadlines. A small ECS may not need a real-time operating system (Kamal 2008).

Embedded computing systems can be classified into three types:

- *Small-scale embedded computing systems*, which are designed with a single 8- or 16-bit microcontroller based on complex instruction set computer (CISC) architecture, such as 68HC05, 68HC08, PIC16FX, and 8051
- *Medium-scale embedded computing systems*, which are are designed with a single or a few 16- or 32-bit microcontrollers which are based on a CISC architecture, such as 8051, 80251, 80x86, 68HC11xx, 68HC12xx, and 80196; digital signal processor (DSP); or reduced instruction set computer (RISC) architecture
- *Large-scale embedded computing systems*, which are designed based on scalable processors or configurable processors, which are based on CISC with a RISC core .or RISC architectures, such as 80960CA, ARM7, and MPC604, and

programmable logic arrays, which for the most part involve enormous hardware and software complexity

In addition to microprocessors and microcontrollers ECSs may also consist of application-specific integrated circuits (ASICs) and/or field-programmable gate arrays (FPGAs) as well as other programmable computing units such as DSPs. Since ECSs interact continuously with an environment that is analog in nature, there must typically be components that perform analog-to-digital (A/D) and digital-to-analog (D/A) conversions.

A significant part of the ECS design problem consists of selecting the software and hardware architecture for the system, as well as deciding which parts should be implemented in software running on the programmable components and which should be implemented in more specialized hardware. Therefore, the design of ECSs should be based on the use of more formal models, i.e., abstract system representation from requirements, to describe the system behavior at a high level of abstraction before a decision on its hardware and software composition is made. But embedded computing systems design is not a straightforward process of either hardware or software design. Rather, design theories and practices for hardware and software are tailored toward the individual properties of these two domains, often using abstractions that are diametrically opposed.

In hardware systems design, a system is composed from interconnected, inherently parallel building blocks, logic gates, and functional or architectural components, such as processors. Although the abstraction level changes, the building blocks are always deterministic or probabilistic, and their composition is determined by how data flows among them. A building block's formal semantics consists of a transfer function, typically specified by equations (see Chap. 1). Thus, the basic operation for constructing hardware models is the composition of transfer functions. This type of equation-based model is an analytical model.

Software systems design uses sequential building blocks, such as objects and threads, whose structure often changes dynamically. Within the design, one can create, delete, or migrate blocks, which can represent instructions, subroutines, or software components. An abstract machine, known as a virtual machine or automaton, defines a block's formal semantics operationally. Abstract machines can be nondeterministic, and the designer defines the blocks' composition by specifying how control flows among them. Thus, the basic operation for constructing software models is the product of sequential machines. This type of machine-based model is a computational model that can include programs, state machines, and other notations for describing the embedded computing systems dynamics.

In contrast, the traditional software design derives a program from which a compiler can generate code; and the traditional hardware design derives a hardware description from which a computer-aided design (CAD) tool can synthesize a circuit. In both domains, the design process usually mixes bottom-up activities, such as reuse and adaptation of component libraries, and top-down activities, such as successive model refinement, to meet the requirements. The final implementation of the system should be made, as much as possible, using automatic synthesis from this high level of abstraction to ensure an implementation that is correct in construction.

The evolution in embedded systems design shows how design practices have moved from a close coupling of design and implementation levels to relative independence between the two (Henzinger and Sifakis 2007).

The first generation of methodologies traced their origins to one of two sources: language-based methods belonging to the software tradition and synthesis-based methods resulting from the hardware tradition. The language-based method is centered on a particular programming language with a particular target runtime system (often fixed-priority scheduling with preemption). Early examples include Ada and, more recently, RT-Java. Synthesis-based methods have evolved from circuit design methodologies. They start from a system description in a tractable, often structural, fragment of a hardware description language, such as VHSIC Hardware Description Language (VHDL) (Ashenden 2008; Perry 2002) and Verilog (Vahid and Lysecki 2007), and automatically derive an implementation that obeys a given set of constraints.

The second generation of methodologies introduced a semantic separation of the design level from the implementation level to obtain maximum independence from a specific execution platform during early design phases. There are several forms. The synchronous programming languages embody abstract hardware semantics (synchronicity) within software. Implementation technologies are available for different platforms, including bare machines and time-triggered architectures. SystemC combines synchronous hardware semantics with asynchronous execution mechanisms from software (C++). Implementations require partitioning into components that will be realized in hardware on the one side and in software on the other. Semantics of common dataflow languages, such as MATLAB's Simulink (Klee and Allen 2011), are defined through a simulation engine. Hence, implementations focus on generating efficient code. Languages for describing distributed systems, such as the Specification and Description Language (SDL), generally adopt asynchronous semantics.

The third generation of methodologies is based on modeling languages, such as the Unified Modeling Language (UML) (Booch et al. 2005; Rumbaugh et al. 2004) and the Architecture Analysis and Design Language (AADL) (Feiler and Gluch 2012), and goes beyond implementation independence. They attempt to be generic not only in the choice of an implementation platform but even in the choice of the execution and interaction semantics for abstract system descriptions. This leads to independence from a particular programming language as well as to an emphasis on the system architecture as a means of organizing computation, communication, and resource constraints. Much recent attention has focused on frameworks for expressing different models of computation and their interoperation (Balarin et al. 2003; Balasubramanian et al. 2006; Eker et al. 2005; Sifakis 2005). These frameworks support the construction of systems from components and high-level primitives for their coordination. They aim to offer not just a disjointed union of models within a common metalanguage but also to preserve properties during

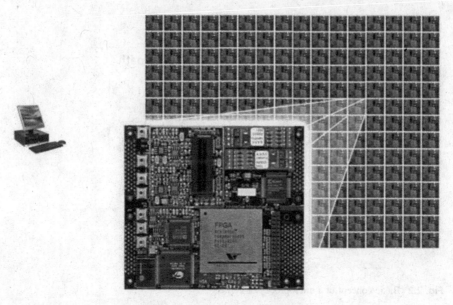

Fig. 2.1 PC vs. FPGA-based ECS

model composition and to support meaningful analyses and transformations across heterogeneous model boundaries.

The market for embedded computing systems is growing on average which means that the worldwide ECS market is expected to increase with growth estimated at 14 % rate of p.a. A statement from the automotive industry by Patrick Hook Associates says: "Embedded software systems today and in the future are responsible for 90 % of all automotive innovations." A comparison between desktop computers and laptops with ECS shows that in only a year, millions of desktop computers and laptops are manufactured but billions of ECSs are produced.

The trend in embedded computing systems is that they become more complex; have more resources (processing power, memory, bandwidth); can be programmed in higher programming languages, such as C/C ++, Java, etc.; can communicate with other systems; and more often than not are based on component industry standards. Therefore, based on different computing components, ECSs are integrated into many important and innovative diversified application fields in science and engineering (Fig. 2.1).

2.2   Hardware Architectures of Embedded Computing Systems

Embedded computing systems are usually based on standard and application-specific components, which are based on dedicated hardware constructs in addition to software, making microprocessors and microcontrollers of particular importance.

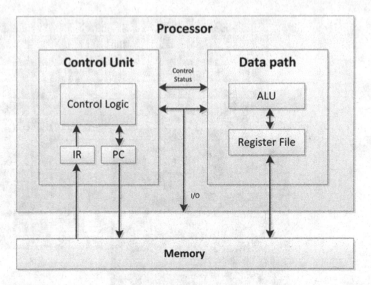

Fig. 2.2 Basic concept of a microprocessor kernel

Microprocessors and/or microcontrollers are used to implement the desired system functionality. For example, the following function can be implemented

$$total = 0$$
$$for\, i = 1\, to\, N\, loop$$
$$total+ = M[i]$$
$$end\, loop$$

on a microprocessor (μP), also called a general-purpose processor (GPP), microcontroller (μC), digital signal processor (DSP), single-purpose processor (SPP), application-specific processor (ASP), or programmable logic device (PLD).

The basic architectural concept of the microprocessor (μP/GPP) is shown in Fig. 2.2. It includes:

- Programmable units that can be used in many applications
- Typical features of the μP/GPP are:
  - Program memory
  - Generalized dataflow path with a great arithmetic logic register file and unit (ALU)
- Advantages for the user
  - Short time to market
  - Low nonrecurring engineering costs (NRE)
  - Great flexibility

When compared, the microprocessor differs from the microcontroller in essential ways. Besides the standard processor core, the microprocessor has more

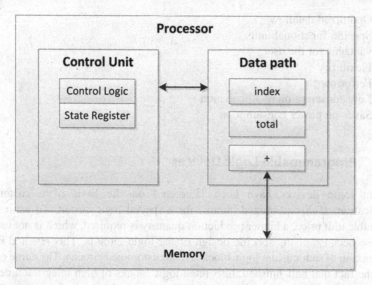

Fig. 2.3 Basic concept of a single-purpose processor kernel

independently operating units for specific tasks, which are integrated on its chip. This unit can be, for example, an analog-to-digital converter converting input analog signals and a digital-to-analog converter converting the processed digital information into analog output information. The several types of microcontrollers have different types of special components integrated to adapt to the application domain of the embedded computing system.

The basic architecture concept of a single-purpose processor (SPP) is shown in Fig. 2.3. It includes a digital circuit capable of performing a single dedicated program.

- Technical characteristics:
  - Contains only components required to execute a single, dedicated program
  - Contains no program memory
- Advantages for the user
  - Fast
  - Low power consumption
  - Small silicon area

The basic architecture concept of an application-specific processor (ASP) is similar to the one shown in Fig. 2.2. It includes a programmable processor optimized for a class of particulate applications with the same characteristics.

- Compromise between μP and SPP
- Technical characteristics of the ASP
  - Program memory

– Optimized dataflow
– Specific functional units
• Advantages for the user
– Flexibility
– Performance
– Low consumption of silicon area
– Saves on power consumption

2.2.1 Programmable Logic Devices

Custom logic devices have been fabricated on the basis of a customer's specifications (full customized circuit) for a special logic circuit. To achieve an acceptable unit price, a higher production quantity is required, which is not usually the case because of the specific design of the logic circuits. This resulted in the introduction of semicustom logic blocks (semicustomized circuit). The name comes from the fact that half-finished, integrated logic blocks of high integration density were produced in large quantities with final programming by the customer. These programmable logic devices (PLD) achieved very high growth rates because they are very flexible and are a less expensive solution due to the large quantities produced. In addition, the rapid technological progress has resulted in noticeable advanced technical specifications such as:

• Gate density
• Speed
• Cost
• Architectural flexibility
• Technology
• Housing dimensions
• Development tools

Thus, the gate density increased from an initial 100 to 500 simple-sized circuits to more than 100,000 usable gates of complex-sized circuits—and has now reached gate densities of more than one million in complex field programmable gate arrays (FPGAs)—where the initial signal delays of 45 ns could be reduced to less than 10 ps. The initial set, consisting of a dual in-line package (DIP) with 20 terminals, was gradually replaced by a pin grid array with 175 or more connections; and plastic quad flat packs were replaced with 154 or 160 ports and more. Programming these semicustom logic devices can be accomplished in two different ways:

• Mask programming (manufacturer)
• Field programming (user)

The mask-programming process is part of the entire logic circuit, usually developed by the user and supplied to the manufacturer for implementation.

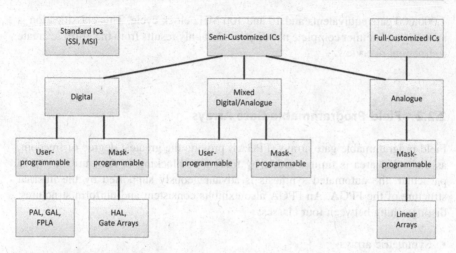

Fig. 2.4 Manufacturing technology of logic devices

Leveraging an existing partial design or complete disclosure of the customer's circuit design to the manufacturer makes this a sensitive issue if competitors of the circuit designer have their products manufactured at the same site. Therefore, designers started to move away from this design technology by introducing field-programmable logic hardware components. Figure 2.4 gives an overview of various standard logic components and PLD. In particular, as indicated in this illustration, digital semicustom integrated circuits are field programmable, as their connection technology is relatively simple. This technological advantage enables an early market launch, so that in the final product, the proportion of field-programmable logic devices compared to mask-programmable logic devices has increased significantly and is trending upward.

The acronyms/initialisms used in Fig. 2.4 are defined as follows:

FPGA   Field-programmable gate array
FPLA   Field-programmable logic array
GAL   Generic array logic
HAL   Hardware array logic
LCA   Logic cell array
PAL   Programmable array logic
pASIC   Programmable application-specific integrated circuit
PLD   Programmable logic device

The SPLD are referred to as monolithic PLD which have, for example, in between 100 and 500 gate equivalents and 60 and 200 MHz clock cycle. The CLPD are referred to as block segmented PLD and have by way of example in between 500 and 20,000 gate equivalents and 25 and 200 MHz clock cycle. FPGAs, also known as channel array PLD, have, for instance, in between 1,000 and

1,000,000 gate equivalents and 10 and 100 MHz clock cycle. This classification is arbitrary and neither complete nor disjointed; it only results from the effort to create a structural overview.

2.2.2 Field-Programmable Gate Arrays

Field-programmable gate arrays (FPGAs) provide the greatest degree of freedom, as their chip area is limited either by folding or blocking in their utilization. In particular, the automated synthesis is advantageously supported by the internal structure of the FPGA. An FPGA also exhibits consistent and uniform structures, distinguishing between four classes:

- Symmetric arrays
- Series array
- Sea-of-gates
- Hierarchical PLD

Thus, FPGAs include more than 106 programmable elements on one chip and must satisfy the following properties for programmable elements:

- Minimum chip area
- Low- and high-resistance OFF
- Low parasitic capacitances, wherein the programming by static random-access memory (SRAM), erasable programmable read-only memory (EPROM), electrically erasable programmable read-only memory (EEPROM), or antifuse technology is realized

The architecture of the Xilinx FPGA, shown in Fig. 2.5, consists of:

- Configurable logic blocks (CLB)
- Input/output blocks (I/O block)

The CLB are composed of the basic elements, input block, output block, and logic operation, which are the smallest programmable units in FPGA. Internally, there are CLB from combinatorial sequential logic elements, so that inputs i_a–i_e and outputs Q_X and Q_Y can be combined with each other twice in any way. The signal appearing at the outputs can be created from the logic operations in the CLB, combinatorial or clocked, or represent the latched signal di. Control signals for the CLB are ce (clock enable), k (clock), and rd (asynchronous reset directly). These signals control the CLB with appropriate internal concession. The number of CLB in a FPGA depends on the device. The area occupied by a CLB can be estimated as a function of the technology used, as follows:

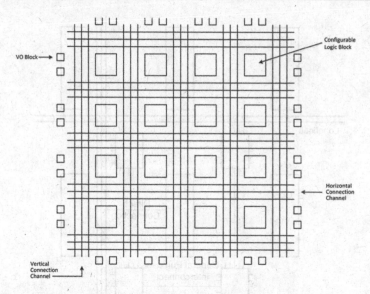

Fig. 2.5 General architecture of the Xilinx FPGA concept

$$A_{\mathrm{CLB}} = A_{\mathrm{RLLB}} + \left(M^* A_{\mathrm{SRAM}}^* 2^K \right)$$

with A_{CLB} as the occupied area of the logic block, A_{RLLB} as the surface of a logic block without flip-flop, M as the bit plane of a flip-flop, A_{SRAM} as a bit area subject to the SRAM technology, and K as the surface for the logic function.

A two-dimensional interconnection network exists between CLB which consists of horizontal and vertical connecting elements. The connections are programmed with each other. The connections themselves are subdivided into:

- *Short lines* for short connections, that is, direct connections (direct interconnect) and general interconnections (general-purpose interconnect)
- *Long lines* for long lines of communication that are chip-wide connections

The reason for this classification can be seen against the background of an optimal solution in terms of flexibility and signal propagation on the chip. Thus, Fig. 2.6 shows the structure of the CLB, including a switch matrix (switching matrix) and the links between the programmable connections (direct connection), the general connections (general-purpose interconnect), and the long communication lines (long lines).

Direct links are the shortest and, therefore, the most effective connections; however, certain restrictions are present as these connections can only produce certain outputs at a CLB from certain inputs to an adjacent CLB. Direct interconnects are used for connections from a CLB to its adjacent CLB, and they are switchable independent of the resources of the general-purpose links. For compounds comprising more than one CLB, the general connections (general-

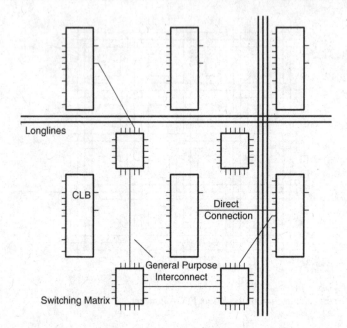

Fig. 2.6 Connectivity structure of the FPGA XC3000 (Xilinx)

purpose interconnects) have horizontal and vertical pipes with four lines per row and five lines per column. Here, the switch matrices are inserted (switching matrix), which enables the connection of CLB which can be configured in different ways. The compounds of general connections show delays related to the data path because each of these line segments has to pass the data path switch (routing switch) of the switching matrix. The general connections proceed from one interconnect switch to the next and can be switched in accordance with their programming. The concept of programmable connections offers great flexibility but has the disadvantage of appropriate length and signals of indefinite duration. The available on-chip long lines, which run alongside the general connections over the entire length of the chip, are used for low-latency connections with more distance from each CLB; but their number is limited. The variability of CLB that result from the internal interconnects allows very high flexibility in the design process. However, this flexibility requires that the user provide graphical input with special input devices for these blocks, the only way an efficient use of CLB is guaranteed. The I/O blocks represent the connection with the outside. Each I/O block can be programmed for data input, data output, or for bidirectional data exchange. The data in the I/O block can also be buffered. In Fig. 2.7, an example of the I/O block (IOB) structure for the XC3000 FPGA from Xilinx is shown.

The Xilinx FPGA is configured in a so-called start-up phase. In this case, the configuration's descriptive program is written from an external ROM in a chip internal matrix of static RAM cells. This matrix is a layer below the user program

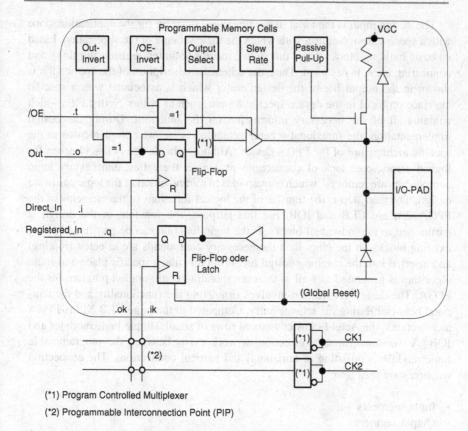

(*1) Program Controlled Multiplexer
(*2) Programmable Interconnection Point (PIP)

Fig. 2.7 I/O block structure of the FPGA XC3000 (Xilinx)

minimizing CLB. During configuration, one can choose between two options: serial or parallel.

The way of describing the application-specific logic function depends on the FPGA device used. It cannot be readily direct transferred to other device modules. For describing and programming the application-specific function of a Xilinx FPGA, design tools are available, such as Xilinx Automatic CAE Tool, which can easily be used. These tools include a design manager, responsible for calling the necessary design steps for the various programs, a design editor at the CLB level, and programmable interconnect paths to perform a finer more optimized design. Hence, the design process for a specific application with a Xilinx FPGA is divided into three steps:

- Design input
- Design implementation
- Design verification

The design input is the input of the functional diagram for the application, done with a specific design editor, with which the logic scheme of the application, based on basic building blocks (logic functions), macros (counter, register, flip-flop), and connecting lines, is designed. Thus, the schematic description of the application is shown in the output file of the design editor which is associated with a specific interface protocol in the device-specific format, such as Xilinx Netlist File, which contains all of the necessary information on the blueprint. During the design implementation, the functional scheme prompted is stepwise implemented to the specific architecture of the FPGA device. Although the design is first checked for logical errors, e.g., lack of connections at a gate, thereafter, unnecessary logic components are removed, which corresponds to minimization of the logic hardware design. The next step is the transfer of the logical functions of the elements of the FPGA that are CLB and IOB. For this purpose, the function of the design is partitioned to the individual blocks. If the logic function can be partitioned to the existing blocks on the chip, then the necessary compounds are selected (routing) and inserted into the resulting design file. From this file, a specific place-and-route algorithm is generated as well as the corresponding configuration program for the FPGA. The design verification involves simulating the functionality and the temporal behavior during the scheme entry. Compared to the large CLB Xilinx FPGA architectures, the Actel FPGA consists of rows of small, simple logic modules and IOB. A two-dimensional connection network exists between the programmable logic modules, consisting of horizontal and vertical connections. The connection resources are split into:

- Input segments
- Output segments
- Clock lines
- Connecting segments

A connecting segment is comprised of four input segments connected via logic modules. The output segment connects the output of a logic module with connecting channels above and below the block. The clock lines are special low-delay connections for connecting several logic modules. The connection segments consist of metallic conductors of different lengths, which can be assembled using antifuse technology to create longer lines. In contrast to the previously discussed Actel FPGA concept, the Crosspoint FPGAs belong to the class of serial block structure FPGA architectures similar to the class of hierarchical Altera PLD. Field-programmable gate arrays satisfying the hierarchical PLD architecture are also available from Advanced Micro Devices (AMD). The architecture is based on the hierarchical grouping of programmable logic array blocks (LAB) in EPROM technology and consists of two types of cells:

- Programmable LAB
- IOB

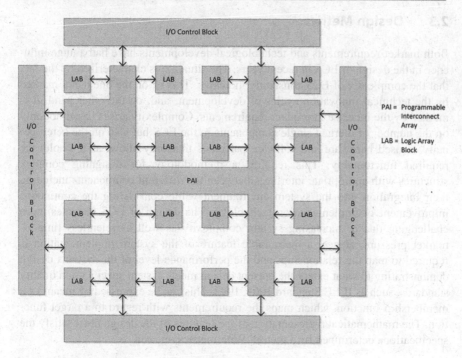

Fig. 2.8 Block structure of the Altera FPGA architecture

It can be seen from Fig. 2.8 that the LAB programmable logic blocks are connected both internally and externally by the programmable interconnect array (PIA). The compounds are, in turn, implemented by horizontal and vertical interconnect lines.

The LAB programmable logic block includes a macrocell array and a product term array. Each macrocell of the array consists of three AND gates that operate on an OR gate which is connected to an XOR gate that generates the output signal of the macrocell. In addition, the macrocell includes a flip-flop. The product term array contains wired AND connections available to the macrocell for logical gates.

The fourth FPGA concept is based on the structure of the so-called sea-of-gates, which includes a high number of blocks in an array of blocks; and each of these blocks can be connected only to its direct four neighbor blocks. The connections in this FPGA structure occur via multiplexers which are generally implemented in the technology of programmable static random-access memory (SRAM). Longer compounds can be implemented by looping the compound through the multiplexer into the blocks.

The four classes of FPGA modules are useful for the manifold applications with regard to the appropriate programmable logic hardware components.

2.3   Design Metrics

Both market requirements and technological developments have had a huge influence in the design of ECS in recent years. The situation is characterized by the fact that the complexity of ECS constantly increases. This is, on the one hand, marked by the technical innovation cycles of development; and, on the other hand, it is marked by the increase in product requirements. Complexity arises based not only on the number of merged single components in the ECS but also on the heterogeneity of used hardware and software partitions that only allow, as a whole, the required functionality. This requires a methodology for designing complex structures with appropriate interfaces between the different components including their integration into the system environment while considering the continuous improvement of implementation technology. The design of ECS becomes more challenging due to increasing system complexity as well as relentless time-to-market pressure. Hence, a measurable feature of the system implementation is required to map the relationships and the performance level of the systems design demonstrating to what extent the present design meets system specification quality standards, such as IEEE Standard 1061, 1992. This can be expressed by means of a membership function, which maps the requirements with regard to a target function. The mathematical representation of the features of the design must satisfy the specifications determined by a metric. With metrics one can:

- Compare drafts of embedded computing systems designs with regard to the fulfillment of their specifications, i.e., a formal comparison and assessment option
- Deal with the increased system complexity and/or requirements in embedded computing systems
- Compare the development and test costs with regard to constraints of the hardware-software partitioning of the embedded computing system to identify the optimal match in the functionality breakdown among the hardware and software components
- Identify the risk of time-to-market constraints assumed in the development and production of embedded computing systems

Against this background, the manufacturer always has to comply with shorter product development in order to launch product innovations into the market more quickly. This requires product development and manufacturing focused on time to market, as the economic success of a product depends on its timely availability. This results in:

- Time pressure in the development and production of embedded computing systems and their optimal service and/or maintenance
- Reusing hardware and software components
- Continuity and penetrability in the design of embedded computing systems

Hence, a general systematic approach to creating quality models is essential and can be based on common metrics such as:

- *Flexibility*: ability to change the functionality of the embedded computing system without incurring heavy NRE cost
- *Maintainability*: ability to modify the system after its initial release
- *NRE cost*: one-time monetary cost of designing the embedded computing system
- *Performance*: execution time or throughput of the embedded computing system
- *Power*: amount of power consumed by the embedded computing system
- *Size*: physical space required by the embedded computing system
- *Time to market*: time required to develop a system to the point that it can be released and sold to customers
- *Time to prototype*: time needed to build a working version of the embedded computing system
- *Unit cost*: monetary cost of manufacturing each copy of the embedded computing system, excluding NRE cost

Hence, a design metric to fit with the foregoing constraints must take the hardware architecture used for the respective design into consideration. When using a general-purpose processor (GPP), it is appropriate to use a GPP-oriented metric. The goal is to identify functionalities that significantly rely on operations that involve conditional dependent control flows, complex data structures, and complex I/O management, as shown in Sciuto et al. 2003. In the case of an ASIC design, an ASIC-like metric is appropriate. The goal is to identify regular functionalities that significantly rely on operations that involve bit manipulation and which finally result in a respective metric.

As shown in Sciuto et al. 2003, the affinity function can be expressed by a normalization function applied to a linear combination of the metrics, with weights that depend on the executor class considered. Intuitively, the affinity toward a GPP executor depends primarily on the

- I/O ratio
- Conditional ratio
- Structure ratio
- Number of declared variables of GPP-compatible type

Hence, it is possible to evaluate the affinity for each hardware-compatible type.

Although many improvements for software development are proposed, the embedded computing systems designer faces a hard task in applying these improvements to software development, due to the strong dependence between software and hardware in embedded computing systems. A trade-off between software qualities, measured by traditional metrics and optimization for a specific platform, is needed. This requires an evaluation of the relationship between quality metrics for software products and physical metrics for embedded systems in order to guide a designer in selecting the best design alternative during design space exploration at the model level (Oliveira et al. 2008).

A metric which describes the general procedure for creating a quality model includes:

- Describe the environment of the company and the project as well as the task, and define the reviews
- Define the goals
- Define assessment objectives and associated metrics
- Define a workflow for data collection
- Collect, analyze, and interpret data
- Summarize and apply the experience to develop best practices

In order to describe the quality of the model, the elements of the evaluation have to be defined. These include, for example:

- Adaptability to changes.
- Efficiency of the hardware components used.
- Efficiency and quality of the written source code.
- Flexibility: functionality of the embedded computing system can greatly increase the need to change without NRE; software is typically very flexible.
- Functional safety
- Size: software is frequently expressed in bytes; hardware is often expressed as gates or transistors
- Correctness: the functionality and test functions of the embedded computing system have been implemented correctly
- Costs
- Power
- Terms of storage
- NRE: one-time development costs
- Performance
- Portability
- Responsiveness: real-time capability, that is, the reaction time to change
- Interface compatibility
- Security: given safe operation of the embedded computing system
- Silicon area
- Scalability
- Time to market: time required to produce a marketable version of the embedded computing system including development time, production time, testing, and evaluation
- Time to prototype: time to produce a working version of the embedded computing system
- Unit cost (UC): unit costs without NRE
- Availability
- Maintainability: ability to modify the embedded system after the first release
- Reliability

Therefore, the goal node metric enables the assessment on the basis of the specified range of elements. These can question, for example, the development process of ECS with a goal of functional safety with respect to the efficiency of the source code, from the perspective of the customer and in the context of the NRE which can be described as follows:

Assessment element	Metric	Description
1	1.1	Adaptation to known process
	1.2	New process
2	2	High requirement
3	3.1	Structogram
	3.2	XML
4	4	High
5	5.1	Low
	5.2	Moderate

For the above example, the degree to which membership functions are given can be expressed in the form of an n-tuple notation:

$$min E(\Psi) = E(\Psi M1.1, \Psi M1.2, \Psi M2, \Psi M3.1, \Psi M3.2, \Psi M4, \Psi M5.1, \Psi M5.2)$$

where M1.x = process functional, M2 = requirements functional, M3.x = functional code, M4 = customer functional, and M5.x = project requirements functional, whereby

$$\Psi(M1.x, M2, M3.x, M4, M5.x) = 0 \, for \, M1.x = M2 = M3.x = M4 = M5.x$$
$$\Psi(M1.x, M2, M3.x, M4, M5.x) = 1 \, for \, M1.x \neq M2 \neq M3.x \neq M4 \neq M5.x$$

with

$$M1.x, M2, M3.x, M4, M5.x \in \Pi$$

where Ψ is a metric of Π. Thus, the degree of fulfillment of the membership function can be exemplified as follows:

$$min E(\Psi) = E(1, 0, 1, 0, 1, 1, 1, 0)$$

that is, the development process of the ECS has been optimally adapted to the known simple process; a new process must not be created; the high requirement for functional safety is optimally fulfilled; requirements on the efficiency and the quality of the written source code are optimally fulfilled; the design of the ECS optimally fit from the perspective of the customer; and the NRE has been optimally fulfilled due to the low requirements on the project. In addition to the specified values 0 and 1 for the degree of fulfillment of the membership function, intermediate values are possible, such as 0.1, 0.2, 0.3, ..., 0.9, and 1.0.

2.4   Embedded Control Systems

The concept of control systems represents a very common class of embedded systems. The concept describes a process in which the control system seeks to make a system's output, whose dynamic depends on the chosen system's plant model, track a desired input without feedback from the system's output, which is called open-loop control. Thus, the design of an embedded control system improves systems performance with regard to:

- System accuracy
- Speed of system response, allowable overshoot, and maximum duration of settling time
- Stability

A closed-loop control system provides feedback from the system's output to the reference input for further processing. Effects of disturbances can be detected and compensated for by appropriate control actions, which are characterized both in terms of strength and in terms of time sequences. This compensation is based on feedback with a negative sign. Therefore, the following relationships can be derived:

- Control process is triggered by disturbances acting on the plant of the embedded closed-loop control system.
- Disturbance occurs and is identified in an embedded closed-loop control system which continuously observes and compares the system's output with the reference input.
- A deviation of the actual system output from the reference input in a closed-loop embedded control system releases an adaptation to the set value. Therefore, the impact of disturbances is clearly eased. To achieve this effect, a reversal sign has to be introduced in the closed-loop action.
- The closed action of an embedded control operation is self-contained. Hence, the temporal sequence of reactions of the control loop is determined by the used parts described by their transfer functions. They follow the principle of causality, that is, the control circuit generates the cause toward the associated response (effect). For this purpose, passing through the control loop, the input excitation of a block element's transfer function is transferred to the output according to the transfer function of the controlled system. The information shall always be directed, i.e., in one direction. Hence, directed arrows (called action lines) are identified in the representation of a control system.

2.4.1   Control

The control refers to the directed influence of a process whose properties correspond to the observed block transfer elements. Activities in the control system

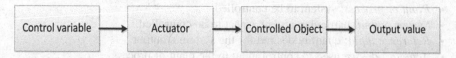

Fig. 2.9 Block diagram structure of a control system

which influence one or more variables as input variables and other variables as output variables are based on the system's intrinsic laws. Moreover, in control systems, the system's output not only depends on the unilateral impact of the arrangement of the reference input (set value) but also depends on the disturbances occurring. The reference input acts as a control input for the output block transfer element according to physical laws and links and/or timing so that the desired behavior is established. Although, the system's output has no influence on the reference input (missing feedback), the system's output may differ due to external disturbances from the desired target value. Hence, an embedded control system can be introduced as an open-loop-block-based transfer function consisting of a number of transfer block components connected in series. The control principle in its conceptual annotation is shown in Fig. 2.9.

In real control systems, disturbances frequently occur at any time and in any amplitude through which their influence on the system's output may be significantly displaced from the reference input. Against this background, it is useful to capture the system's output by a separate transfer block. In case of deviations of the system's output from the reference input, the influence of the disturbance on the plant can be compensated for through the principle of feedback control. Thus, with a simple open-loop control system, one cannot act against foreseeable disturbance. Hence, a system is required which in the simplest case has transfer components for observing the system's output and comparing it with the reference input to calculate the error between them, forcing the system's output to follow the reference input. This principle is the closed-loop control system.

2.4.2 Feedback Control

The concept of feedback describes a control system in which the system's output, whose dynamic depends on the chosen system's plant model, is forced to follow a reference input while remaining relatively insensitive to the effects of disturbances. In the case of a difference between both signals, the summing point of the feedback loop generates an error signal which is transferred to the controller input. The controller acts on the error with regard to a control strategy and manipulates the plant model to make it track the reference input. Moreover, this closed-loop feedback forces the system's output to follow the reference input with regard to present disturbance inputs. Thus, closed-loop control contains more transfer elements than open-loop control. The transfer elements of a closed-loop control are:

- *Plant or process*: system to be controlled.
- *System output*: particular system aspect to be controlled.
- *Reference input*: quantity desired for the system's output.
- *Actuator*: device used to control input to the plant or process.
- *Controller*: device used to generate input to the actuator or plant to force the system's output to follow the reference input. Therefore, the controller contains the control strategy to make the desired output track the desired reference input.
- *Disturbance*: additional undesirable input to the plant imposed by the environment that may cause the system's output to differ from the expected output with regard to the reference input.

To this point, these transfer elements are the same in an open-loop control system. The closed-loop control system has the following additional transfer elements:

- *Sensor*: device to measure system output
- *Error detector*: determines the difference between the measured system output and reference input

Therefore, a closed-loop controller continuously detects and compares the potential difference between the reference input and the system's output by making use of the sensor and error detector. The resulting error value of the error detector is read by the controller's input which then computes a setting for the actuator to manipulate the actuator and, thereafter, the plant of the closed-loop embedded control system. The controller uses the feedback from the error detector to force the system's output based on the control law of the controller which has been implemented. The actuator modifies the input to the plant with regard to the requirements based on the error detector output and the controller transfer function.

In Fig. 2.10, the block diagram of a closed-loop control system is shown, with its conceptual annotation, referring to the transfer elements described above.

The terms occurring from the closed action sequence in the control loop cycle are summarized in the following table:

Symbol	Denomination
$u(t)$	Reference input or set value
$x_d(t)$	Error detection or control deviation
$y(t)$	Control output or correcting input
$r(t)$	Actuatingoutput
$z(t)$	Disturbance input
$x(t)$	System output or control variable
$x_R(t)$	Measured system output or measured control variable

In Fig. 2.10, the structure of control systems is presented in block diagram form, depicted as an interconnection of symbols representing certain basic mathematical operations in such a way that the overall diagram obeys the system's mathematical

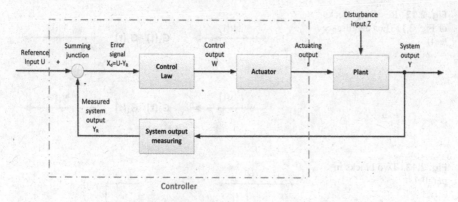

Fig. 2.10 Closed action of the control loop in block diagram form

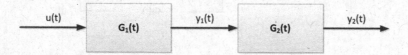

Fig. 2.11 Two blocks in series

model, as described in Chap. 1. The interconnecting lines between blocks represent the variables describing the system's behavior, such as input and state variable (see Sect. 1.2). For a fixed linear system with no initial energy, the output $y(t)$ is given by

$$y(t) = G(t) \cdot u(t)$$

where $G(t)$ is the transfer function and $u(t)$ is the input. Hence, a block diagram is merely a pictogram representation of a set of algebraic equations, allowing blocks to be combined by calculating the equivalent transfer function and, thereby, simplifying the diagram.

Two blocks are said to be in series when the output of one goes only to the input of the other, as shown in Fig. 2.11.

The transfer functions of the individual blocks are:

$$G_1(t) = \frac{y_1(t)}{u_1(t)}$$
$$G_2(t) = \frac{y_2(t)}{y_1(t)}.$$

Therefore, if $u_1(t) \cdot G_{1(t)}$ is substituted for $y_1(t)$, we find $y_2(t) = G_2(t) \cdot u_1(t) \cdot G_1(t)$. These equations, which are different forms of the same equation, demonstrate the important idea that simple, linear blocks connected together in a series can be multiplied together, a fact that allows the reduction of just such a two-block series to single blocks, as shown in Fig. 2.12.

Fig. 2.12 Resulting blocks
of Fig. 2.11 (For details, see
text)

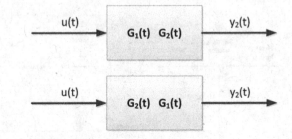

Fig. 2.13 Two blocks in
parallel

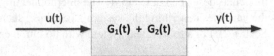

Fig. 2.14 Resulting block of
Fig. 2.13 (For details, see
text)

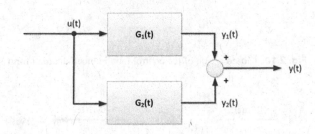

Two blocks are said to be in parallel when they have a common input and their
outputs are combined by a summing point. If, as indicated in Fig. 2.13, the
individual blocks have the transfer functions $G_1(t)$ and $G_2(t)$ and the signs at the
summing point are both positive, the overall transfer function $y(t)/u(t)$ will be the
sum $G_1(t) + G_2(t)$, as shown in Fig. 2.14.

To prove this statement, note that

$$y(t) = y_1(t) + y_2(t)$$

where $y_1(t) = G_1(t) \cdot u(t)$ and $y_2(t) = G_2(t) \cdot u(t)$. Substituting for $y_1(t)$ and $y_2(t)$, we
have

$$y(t) = [G_1(t) + G_2(t)]u(t).$$

The block diagram of a feedback system that has a forward path from the
summing point to the output and a feedback path from the system's output back
to the summing junction (closed loop) is shown in Fig. 2.15.

Fig. 2.15 Feedback loop

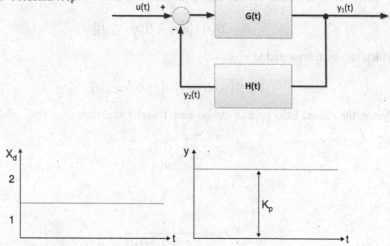

Fig. 2.16 Unit step of an ideal proportional controller

The block diagram shows the simplest form of a feedback control system. The transforms of the control system's input and output are $u(t)$ and $y_1(t)$, respectively. The transfer function

$$G_1(t) = \frac{y_1(t)}{u(t)}$$

is introduced as the forward loop gain or forward transfer function, and

$$H(t) = \frac{y_2(t)}{y_1(t)}$$

as the feedback loop gain or feedback transfer function.

Let the model of a feedback system be given in terms of its forward and feedback transfer functions G(t) and H(t). It is often necessary to determine the closed-loop gain or closed-loop transfer function

$$F(t) = \frac{y_1(t)}{u(t)}$$

This function can be derived from the block algebra equations for the closed-loop system, shown in Fig. 2.5, by solving them for the ratio $y_1(t)/u(t)$. This yield, in a set of equations that corresponds with the block diagram, is as follows:

$$V(t) = u(t) - y_2(t)$$
$$y_1(t) = G(t) \cdot V(t)$$
$$y_2(t) = H(t) \cdot y_1(t)$$

Let us combine these equations to eliminate $V(t)$ and $y_2(t)$ yields

$$y_1(t) = G(t) \cdot [u(t) - H(t) \cdot y_1(t)]$$

which can be rearranged to give

$$[1 + G(t) \cdot H(t)]y_1(t) = G(t) \cdot u(t)$$

Hence, the closed-loop gain or closed-loop transfer function

$$F(t) = \frac{y_1(t)}{u(t)}$$

is

$$F(t) = \frac{G(t)}{1 + G(t) \cdot H(t)}.$$

It is clear that the sign of the feedback signal at the summing point is negative. Assuming that the sign at the summing point is positive for the feedback signal, then the closed-loop gain or closed-loop transfer function will become negative. Assuming a commonly used simplification occurs when the feedback transfer function is unity, which means that $H(t) = 1$, this control system is called a unity feedback system, yielding:

$$F(t) = \frac{G(t)}{1 - G(t)}.$$

2.4.3 Feedback Components of Embedded Control Systems

In practice, specific feedback transfer functions are used when designing embedded control systems. These closed-loop transfer function characteristics can be described by the:

- Transient behavior or static characteristic curves
- Mathematical methods

The mathematical notation of the respective feedback law for the dynamic behavior of embedded closed-loop control system transfer functions depends on the chosen characteristic of the specific controller block. In practice, the following elements are of importance:

- Proportional control
- Integral control
- Derivative control

2.4.3.1 Proportional Control

The proportional control (P-feedback) is the most straightforward feedback, where the output of the controller varies directly as the input (or system error) $x_d = u - x_R$ which results in

$$y(t) = K_P \cdot x_d(t)$$

where K_P is the gain factor of the proportional control. Increasing K_P will increase the closed-loop gain of the control system and can, therefore, be used to increase the speed of the control system response and to reduce the magnitude of any error. The embedded control system with a proportional feedback is referred to as a system zero order or a system without memory element. The graph in Fig. 2.16 shows the response of the proportional control using the step response as input (see Sect. 1.1) with a fixed gain of K_P.

The proportional control alone, however, is often not good enough because increasing K_P not only makes the system more sensitive but also tends to destabilize it. Consequently, the amount by which K_P can be increased is limited; and this limit may not be high enough to achieve the desired response. In practice, when trying to adjust K_P, conflicting requirements may occur. On one hand, it is intended to reduce any steady-state error as much as possible; but to attempt this by increasing K_P is likely to cause the response to oscillate, resulting in a prolongation of the settling time. On the other hand, the response to any change of the input signal should be as fast as possible but with little overshoot or oscillation. Fast control system response can be achieved by increasing K_P, but the increase is likely to destabilize the control system.

To solve the conflicting requirements with regard to the control system gain, a P-controller is required that has a:

- K_P value that is high in order to reduce the control system error
- K_P value that is high to ensure a rapid response
- K_P value that is low enough to ensure that the dynamic response does not overshoot excessively and that any tendency to oscillate is damped fast enough

To fulfill these requirements, the P-controller has to be expanded by adding, to the proportional part, one or two other control terms, such as integral control or differential or integral and differential control.

2.4.3.2 Integral Control

The prime purpose of adding an integral control part to a controller is to remove any steady-state error, which can be achieved by an integral gain term that effectively has an infinite value at zero, representing the steady-state condition. The integral controller is usually used together with proportional and derivative control and in cases where speed of response and instability are not a problem.

An integral control dependence exists for which the output signal x and time integral of input x_d are proportional. Time integration of the control deviation e with

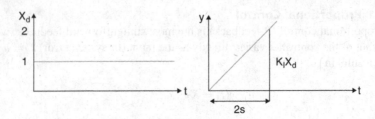

Fig. 2.17 Unit step of an ideal integral controller

the system's output or the actuated variable y acts with a reset time T_N. The reset time is called the integration factor or integration time constant. This means that for a reset time of $T_N = 2$ s at time $t = 0$, the output value y after 2 s has reached the value of the constant input x_d. In the case of an integral controller, the actuator variable r, apart from the initial value, is proportional to the time integral of the control deviation

$$y(t) = 1/T_N * \int_{t_0}^{t_1} x_d(t) \mathrm{d}t$$

If the input to the integral control element is zero, the output value does not change. By choosing a constant input value unequal to zero, the integral controller output changes with a constant increase. If the input value increases uniformly, then the integral controller output always changes faster. The integral controller has no steady-state error like the proportional controller. The integral controller is relatively slow in comparison to the proportional controller. By choosing a reset time T_N (proportional factor $K_I = 1/T_N$) that is too large, there is, however, an overshoot of the control variable; and the controller becomes extremely unstable Technically, the software-based version of an integral controller is implemented by summation over a time interval. The graph in Fig. 2.17 shows the response of the integral controller for a unit step response at time $T_N = 1$ with $K_I = 1/T_N$ and $x_d = 1$.

2.4.3.3 Derivative Control
Derivate control is used in the controller to speed up the transient response of embedded control systems. Derivative action is always accompanied by proportional control. Integral control is used only if necessary. Embedding derivative action in the controller has a stabilizing effect on the embedded control system by virtue of the addition of phase lead to the closed-loop control system by reducing the phase lag of the gain factor of the derivative control.

For a derivative control, the output u is proportional to the time derivative of the input signal x_d. Therefore, the actuating variable y is proportional to the rate of change of the control deviation x_d which yields

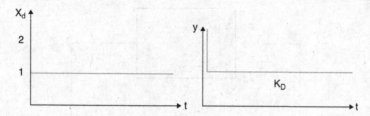

Fig. 2.18 Unit step of an ideal derivative controller

$$y(t) = T_V \frac{\mathrm{d}x_d t}{\mathrm{d}t}$$

In the case of sudden changes in the system's output (control variable), the actuated variable y increases immediately and, thereafter, goes back to its original value. Ideally, a derivative controller follows the Dirac pulse as a step response. A pure derivative controller cannot be realized in practice because the differentiation eliminates the set point. Therefore, the derivative controller is used in combination with the proportional controller, or integral controller, to achieve a quick response to sudden changes in the system's output (control variable) x.

Technically, the software-based version of a derivate controller is implemented by differentiation over a time interval. The constant T_V is called derivative action time. The graph in Fig. 2.18 shows the unit step response of a derivative controller, for a gain factor of $K_D = T_V = 1$.

2.4.3.4 Proportional, Integral, and Derivative Control

These controls, as mentioned earlier, are widely used for controlling the response of embedded control systems. The derivative action is used to increase the speed of response, while the integral part prevents steady-state errors from occurring in the flow rate or actuator position.

The integral behavior of the proportional-integral-derivative (PID) controller is usually used when the controller is trying to maintain the system's output at its nominal working range and where changes in the system's output only occur as a result of changes in the load.

In a case where the input to a PID controller is changed significantly, the integral part of the controller is usually turned off or suppressed until the system's output is close to its nominal working range. If the integral part is not suppressed, then the large change in the input to the PID controller causes large oscillations to be superimposed onto the response of the embedded control system. Hence, the oscillating response interacts with the two other control elements, the proportional and the derivative; and the result is a very cyclic response of the embedded control system with a very long settling time.

A general constraint for using integral control is that it should only be used if steady-state errors exist that cannot be tolerated in the embedded control system strategy. Even the contribution of the integral behavior used should be just enough

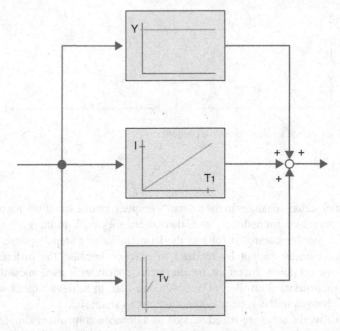

Fig. 2.19 Block diagram of the PID controller (For details, see text)

to remove the steady-state error without causing the steady response to oscillate. Where steady-state errors either do not exist or can be tolerated, then a proportional-derivative controller will be sufficient enough.

The PID controller contains all three control laws, the proportional, the integral, and the derivative. The input to the PID controller is the error signal x_d which is connected with the three parallel input ports of the controller, as shown in Fig. 2.19. The output signals of the proportional, the integral, and the derivative controller elements are merged into a summing point, as shown in Fig. 2.19. The output of the summing point is the weighted sum of the proportional, the integral, and the derivative controller outputs. The three outputs have the same positive sign, and the weighting factors of the summing inputs of the summing point are assumed to have a value of 1. In Fig. 2.19, the constant T_I represents the reset time of the integral element; and T_V represents the derivative action time of the differential element.

From Fig. 2.19, the following equation can be derived:

$$y(t) = K_P * x_d + K_I * \int_{t_0}^{t_1} x_d(\tau) d\tau + K_D * \frac{dx_d}{dt} + x_d(0)$$

where $x_d = (0)$ is the initial value, K_P is the gain factor of the proportional term, and $T_I = 1/T_N$ is the integral controller gain factor with T_N as the reset time; and T_V is the derivative controller gain factor. After excluding K_P and with regard to the boundary condition $x_d(0) = (0)$, it follows

$$y(t) = K_P \left(x_d + \frac{T_I}{K_P} * \int_{t_0}^{t_1} x_d(\tau) d\tau + \frac{K_D}{K_P} * \frac{dx_d}{dt} \right).$$

With

$$\frac{K_P}{T_I} = T_N$$

and

$$\frac{K_D}{K_P} = T_V$$

we receive:

$$y(t) = K_P \left(x_d + \frac{1}{T_N} * \int_{t_0}^{t_1} x_d(\tau) d\tau + T_V * \frac{dx_d}{df} \right).$$

Using the Laplace transform, the above equation can be written as follows with s as the Laplace operator

$$G(s) = K_P \left(1 + \frac{1}{s*T_N} + s*T_V \right).$$

For a number of calculations, it may be more appropriate to rewrite the above additive form into the following multiplicative form

$$G(s) = K_P * \frac{(1 + s + T_1) * (1 + s*T_2)}{s + T_N}.$$

Comparison of coefficients yields

$$T_1 = \frac{T_N}{2} \left(1 + \sqrt{1 - \frac{4T_V}{T_N}} \right)$$

$$T_2 = \frac{T_N}{2} \left(1 - \sqrt{1 - \frac{4T_V}{T_N}} \right)$$

where $T_N > 4*T_V$. From $T_N > 5*T_V$, the following relations can be found

$$T_1 = T_N$$
$$T_2 = T_V.$$

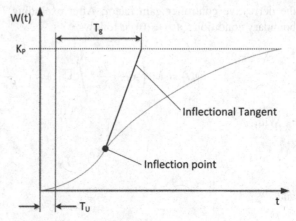

Fig. 2.20 Transient behavior of a step response (For details, see text)

It can be seen from the above equations that the PID controller has two zero elements and a pole at the origin of the s-plane. The gain factors, K_P, T_N, and T_V, of the PID controller can be calculated using the tangent at the inflection point of the step response with the abscissa as the lower auxiliary variable T_U and the intersection of the tangent with the 5τ value of the step response as the top auxiliary variable T_g, as shown in Fig. 2.20.

From Fig. 2.20, the corresponding values for T_U and T_g, pictured on the abscissa time t, can be read. Let the PID controller overshoot the quotient of the auxiliary variable O_{max}, for the maximum overshoot height at T_{95} describes the default values for the PID controller

$$K_P = \frac{T_{95}}{\ddot{U}_{max}} * \frac{T_g}{T_U}$$

Assuming that the PID controller is not allowed to overshoot results in the following equation with regard to the above-introduced auxiliary variables

$$K_P = \frac{T_{60}}{K_S} * \frac{T_g}{T_U}$$
$$T_N = T_g$$
$$2 \cdot T_V = T_U$$

The ideal PID controller was introduced as a parallel connection of an ideal proportional-integral-derivative controller, which is represented by the addition of individual transfer functions as follows

$$g(t) = K_P + \frac{K_P}{T_N} * t + K_P * T_v * \delta(t).$$

The transfer function g(t) of the ideal PID controller given above can be illustrated as shown in Fig. 2.21.

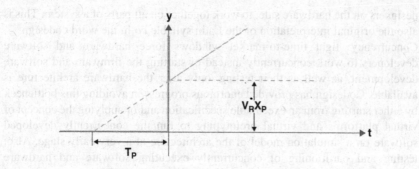

Fig. 2.21 Transfer function of the ideal PID controller

When designing a controller, simulation programs are very often used to optimize the controller design. This can be done based on the industry standard software package, MATLAB Simulink (Chaturvedi 2010).

2.5 Hardware-Software Codesign

Hardware-software codesign originated in the early 1990s as a new method of designing complex digital systems. At that time, the concurrent design of hardware and software was already a daily business; and designers worked carefully in deciding how to design the interface between hardware and software. This involved definition and implementation of the hardware architecture that had not been consciously treated as a task of codesign. Nevertheless, it motivates and stimulates the research goals that today's codesign methodologies already try to accomplish (Teich 2012):

- Satisfying the need for system-level design (SLD) automation
- Allowing development of correct digital systems comprising hundreds of millions of transistors running programs with millions of lines of code

Due to technological advances, hardware-software codesign has become a key technology for successful digital systems design today and is used more and more in development, e.g., embedded computing systems or in more general information systems with specific target architectures.

The major purpose and intention of hardware-software codesign can be explained by different interpretations of the prefix *co* in codesign (Teich 2012):

- *Co*ordination: codesign techniques are used to coordinate the design steps of interdisciplinary design groups with regard to firmware operating system application developers on the software side, and hardware developers and chip designers on the hardware side, to work together on all parts of a system. This is also the original interpretation of the Latin syllable *co* in the word codesign.

- *Co*ncurrency: tight time-to-market windows force hardware and software developers to work concurrently instead of starting the firmware and software development, as well as their testing, only after the hardware architecture is available. Codesign has provided enormous progress in avoiding this bottleneck by either starting from an executable specification and/or applying the concept of virtual platforms and virtual prototyping to run the concurrently developed software on a simulation model of the architecture at a very early stage. Also, testing and partitioning of concurrently executing software and hardware components require special cosimulation techniques to reflect concurrency and synchronization of subsystems.
- *Co*rrectness: correctness challenges of complex hardware and software techniques require not only verification of the correctness of each individual subsystem but also coverification of their correct interactions after their integration.
- *Co*mplexity: codesign techniques are mainly driven by the complexity of today's digital systems designs and serve as a means to close the well-known design gap and produce correctly working, highly optimized (e.g., with respect to cost, power, or performance) system implementations.

Moreover, the methodology of hardware-software codesign can be explained by additional interpretations of the syllable *co*:

- *Co*synthesis: marks the fields where hardware or software could be used based on the possible minimization of communication between application areas
- *Co*simulation: permits early review of the system's logic functionality and behavior based on partitioning review
- *Co*test: set by the user since neither the hardware test methods nor the software metrics contain application-specific testing methods

The potential results from the co-methods are:

- Abstract system level in the design phase
- Very complex system and high-performance standards
- Short time to market in design and production
- Systems with standard microprocessormicrocontroller components, PCs, one-chip solutions, DSP, and more
- Systems with application-specific hardware, such as ASIC, ASP; DSP, FPGA, and more
- Systems with specific software
- Many comprehensive applications

This requires that the available techniques support the complexity management in hardware and software codesign by:

- Hardware-software partitioning
  - Decisions are postponed that place constraints on the design when possible.
- Abstractions and decomposition techniques
- Incremental development
  - Growing software requires top-down design
- Description languages
- Simulation
- Standards
- Design methodology management framework

The current hardware-software codesign process includes:

- Basic features of the design process
  - System immediately partitioned to hardware and software components
  - Hardware and software developed separately
  - Hardware as first approach is often adopted
- Implications of these features are:
  - Hardware and software trade-offs are restricted.
    - ○ Impact of hardware and software on each other cannot be assessed easily.
  - Late system integration.
- Consequences with regard to these features are:
  - Poor-quality designs
  - Costly modifications
  - Schedule slippages

Therefore, the codesign process of embedded computing systems can be described as follows:

- Systems design process that combines hardware and software perspectives beginning at the earliest stages to exploit design flexibility and efficient allocation of functions
- Integrated design of systems implemented using both hardware and software components

Therefore, the key concepts of the codesign process can be introduced as:

- Concurrent: hardware and software can be developed at the same time on parallel paths.
- Integrated: interaction between hardware and software development to generate a design meeting performance criteriaand functional specifications.

With regard to the aforementioned, the requirements for the codesign process can be classified as:

- Unified representation
  - Supports uniform design and analysis techniques for hardware and software
  - Permits evaluation in an integrated design environment
- Iterative partitioning techniques
  - Allow evaluation of different designs (hardware and software partitions)
  - Aid in determining best implementation for a systems design
  - Partitioning applied to modules to best meet design criteria (functionality and performance goals)
- Continuous and/or incremental evaluation
  - Supports evaluation at several stages of the design process
  - Can be provided by an integrated modeling substrate

From this it follows for the codesign method at the system level that during specification an executable specification of the overall system has to be created. For the initial phase, the following action items are essential:

- Describe system functionality.
- Document all steps of the design process.
- Automatically verify the properties of critical system features.
- Analyze and explore implementation alternatives.
- Synthesize subsystems.
- Change/use already existing designs.

Apart from the necessity of specification, formal analysis, and cosimulation tools for performance and cost analysis, it was soon discovered and agreed upon that the major synthesis problem in the codesign of digital systems involves three major tasks (Teich 2012):

- *Allocation*: defined as selecting a set of system resources including processors/ controllers and hardware blocks and their interconnection network, thereby composing the system architecture in terms of resources. These resources could exist as library templates. Alternatively, the design flow should be able to synthesize them.
- *Binding*: defined as mapping the functionality onto processing resources, variables, and data structures onto memories and communications to routes between corresponding resources.
- *Scheduling*: defined as ensuring that functions are executed on proper resources including function execution, memory access, and communication. This can involve either the definition of a partial order of execution or the specification of schedulers for each processor/controller and communication and memory resources involved as well as task priorities, etc.

Therefore, codesign accomplishes the necessary design refinements automatically, saving development time and allowing a fast verification of the abovementioned designs at the system level (Lee and Seisha (2015). In the double-roof model introduced by (Teich 2012), shown in Fig. 2.22, according to

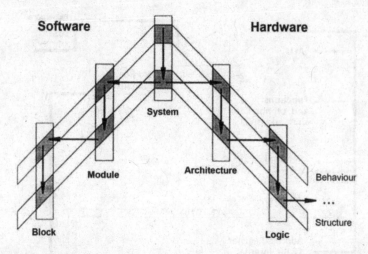

Fig. 2.22 Double-roof model of codesign introduced by Teich 2012, showing the system level at the *top* connecting the software (*left*) and hardware development chains (*right*) through successive synthesis refinements (*vertical arrows*). Each synthesis step maps a functional specification onto a structural implementation on the next lower level of abstraction

Haubelt and Teich (2010) and Teich (2000), the typical abstraction levels of digital design automation are depicted. From Fig. 2.22, it can be seen that:

- Models at the module level describe functions and interactions of complex modules.
- Models at the block level describe program instructions which execute on the hardware architecture elementary operations provided.

Besides the classification of models referring to the level of abstraction, differentiation in the view of the abstraction level results in:

- *Behavior*: describes the functionality independent of the concrete implementation
- *Structure*: describes the communicating components, their breakdown, and communication and represents the actual implementation

2.6 Case Study: FPGA-Based CPU Core

The case study focuses on an FPGA-based central processing unit (CPU) design. The design goal is an FPGA-based core of a CPU core whose data should have a width of 4 bits and an address bus of 12 bits wide. The CPU core has two user-accessible registers and an accumulator and an index register, both 4 bits wide. For the internal organization, more registers are required, such as the program counter (PC) and the address register, both 12 bits wide, and the command register which is

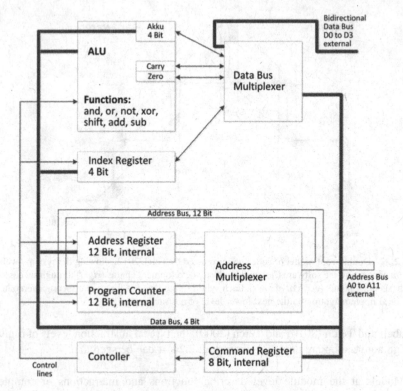

Fig. 2.23 Schematic representation of the FPGA-based CPU

8 bits wide (see Fig. 2.23). The CPU core cycles are controlled by a hardwired control unit. Since the selected FPGA does not have a tristate logic bus, it must be implemented by multiplexers. Through connector pins of the FPGA, the following signals are assumed to be externally accessible:

- Power supply (+5 V) and ground (GND)
- Address bus output signal for the memory output
- Bidirectional data bus for reading and writing to memory
- Reset wire
- Write memory wire to distinguish between reading and writing

For test purposes, the accumulator and the content of the index register and the internal counter are routed to the outside. The implementation of the CPU core in a single FPGA eliminates the otherwise expensive communication between the functional units of more than one device. Only the address bus and data bus are made available to all units. The controller can be implemented as a compact block

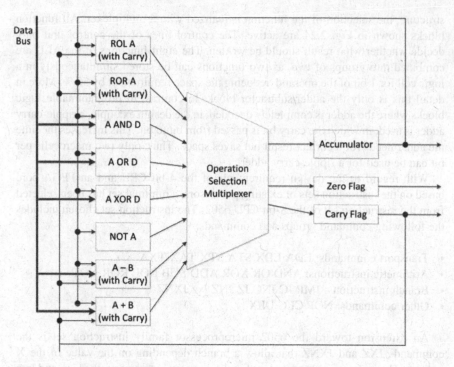

Fig. 2.24 Schematic representation of the ALU

because of the need to perform only the control lines for the registers and multiplexers. The arithmetic logic unit should have the following features:

- Rotation of the accumulator to the right through the carry flag
- Rotation of the accumulator to the left through the carry flag
- Logical exclusive OR of the accumulator in maintenance with the value present on the data bus
- Negation of battery contents
- Logical AND of the battery capacity with the value present on the data bus
- Logical OR of the battery capacity with the value present on the data bus
- Subtracting the value that is present on the data bus from the battery underloading taking into account the carry flag
- Addition of the value that is present on the data bus architecture taking into account the carry flag

The accumulator is always the first operand. The second operand is available on the data bus. To release the control unit, the accumulator is integrated into the ALU. The ALU ensures that upon completion of an arithmetic operation, the result is written to the accumulator. The control unit shows the beginning of an arithmetic operation on the control line. The design options in an FPGA structure are shown in the block diagram of Fig. 2.24. Since the used FPGA architecture has no bus structure, the selection of the function is realized with a multiplexer. All function blocks shown in Fig. 2.24 are active. The control lines of the control unit only decide whether/what result should be written. The eight functions of the ALU are combined into groups of two, as two functions can be stored simultaneously in a logic cell for 1 bit of the operand, reducing the space requirement by 50 %. More in detail this is only the adder/subtractor block. In contrast to programmable logic blocks where the adder is completely decoded, in the design example, a ripple-carry adder is used, in which the carry bit is passed from bit to bit. This increases the time the adder needs for a correct result but saves space. Thus, only two macrocells per bit can be used for a ripple-carry adder.

With regard to the design requirements of the 4-bit CPU and an FPGA core based on the instruction sets of existing processors, a minimal set has been selected from the instruction set of the 8-bit CPU 6502. The instruction set chosen includes the following command groups and commands:

- Transport commands: LDA LDX STA STX TAX TXA
- Arithmetic instructions: AND OR XOR ADD SUB ROL ROR NOT
- Branch instructions: JMP JC JNC JZ JNZ jxz JXNZ
- Other commands: NOP CLC DEX

An extension toward the 6502 microprocessor family instruction set is the commands JXZ and JXNZ that allow a branch depending on the value of the X register. The instruction word is stored in the memory in two parts, the high and low nibbles. In the three stages of instruction decode, the corresponding sequence of a 22-bit wide control word is generated. Table 2.1 contains the hexadecimal codes for each command (with all addressing modes), as well as information on the implementation and the duration (in cycles).

With the tabular program representation in Table 2.1, the conceptual, detailed design of the FPGA CPU core is complete. Thereafter, the prototypical realization by programming of the FPGA and the testing has to be done. The testing is necessary because the correct programming of the module in the next programming process may not always be fully verified. For this purpose, a simple test environment with the FPGA CPU core has to be built. The test environment consists of the FPGA CPU core, an EPROM, a static RAM chip, a GAL, and a quartz crystal. The reset of the input-output functions is implemented with the GAL. The test environment can be powered either by a quartz crystal oscillator or by a single-step circuit with a clock. The test hardware has a reset button to access the CPU core in a defined state. Through DIP switches, one of the test programs stored in the EPROM can be selected. In addition, the test environment has seven-segment displays to show the contents of the accumulator, index register, instruction register, address bus, and data bus containing the display, so that the processing of individual commands can be performed clock cycle by clock cycle.

Table 2.1 Instruction set of the 4-bit CPU

Instruction	Address mode	Storage mode		Execution	Duration
		Code	Operand		
LDA	inh	10h	–	$(O) \rightarrow A$	4
	imm	50h	i		5
	dir	90h	iii		7
	idx	D0h	iii		7
LDX	inh	11h	–	$(O) \rightarrow X$	4
	imm	51h	i		5
	dir	91h	iii		7
	idx	D1h	iii		7
STA	dir	92h	iii	$(A) \rightarrow O$	9
	idx	D2h	iii		9
TAX	inh	02h	–	(A)	6
TXA	inh	03h	–	$(X) \rightarrow A$	4
AND	inh	0CH	–	$(A) \cdot (O) \rightarrow A$	4
	imm	4Ch	i		5
	dir	8Ch	iii		7
	idx	CCh	iii		7
OR	inh	0Dh	–	$(A) + (O) \rightarrow A$	4
	imm	4Dh	i		5
	dir	8Dh	iii		7
	idx	Cdh	iii		7
XOR	inh	0Ah	–	$(A) \oplus (O) \rightarrow A$	4
	imm	4Ah	i		5
	dir	8Ah	iii		7
	idx	Cah	iii		7
ADD	inh	0Fh	–	$(A) + (O) \rightarrow A$	4
	imm	4Fh	i		5
	dir	8Fh	iii		7
	idx	CFh	iii		7
SUB	inh	0Eh	–	$(A) - (O) \rightarrow A$	4
	imm	4Eh	i		5
	dir	8Eh	iii		7
	idx	CEh	iii		7
ROL	inh	09h	–		4
	imm	49h	i		5
	dir	89h	iii		7
	idx	C9h	iii		7
ROR	inh	08h	–		4
	imm	48h	i		5
	dir	88h	iii		7
	idx	C8h	iii		7
NOT	inh	04h	–	$\neg (A) \rightarrow A$	4

(continued)

Table 2.1 (continued)

| Instruction | Address mode | Storage mode | | Execution | Duration |
		Code	Operand		
JMP	dir	98h	iii	(O) → PC	7
	idx	D8h	iii		7
JC	dir	9Ah	iii	? c = 1: (O) → PC	7
	idx	Dah	iii		7
JZ	dir	9Bh	iii	? z = 1: (O) → PC	7
	idx	Dbh	iii		7
JNC	dir	9Eh	iii	? c = 0: (O) → PC	7
	idx	DEh	iii		7
JNZ	dir	9Fh	iii	? z = 0: (O) → PC	7
	idx	Dfh	iii		7
JXZ	dir	99h	iii	? X = 0 h: (O) → PC	7
	idx	Ddh	iii		7
JXNZ	dir	9Dh	iii	? X ≠ 0 h: (O) → PC	7
	idx	Ddh	iii		7
NOP	inh	00h	–	–	4
CLC	inh	01h	–	0 → c	4
DEX	inh	07h	–	(X) -1 → X	4

2.7 Exercises

What is meant by the term *embedded computing system*?

Describe the architectural structure of an embedded computing system.

What is meant by the term *design metrics*?

Give an example of a design metric.

What is meant by the term *heterogeneity in ECS design*?

Give an example of heterogeneity in ECS design.

What is meant by the term *consistency in ECS design*?

Give an example of consistency in ECS design.

What is meant by the term *hard real-time requirements*?

Give an example of a hard real-time requirement.

What is meant by the term *soft real-time requirements*?

Give an example of a soft real-time requirement.

What is meant by the term *embedded control*?

Describe the structure of an embedded control system.

What is meant by the term *open-loop control system*?

Describe the impact of an open-loop control system.

What is meant by the term *closed-loop control system*?

Describe the impact of a closed-loop control system.

What is meant by the term *error detector*?

Describe the impact of an error detector.
What is meant by the term *proportional controller*?
Describe the structure of a proportional controller.
What is meant by the term *integral controller*?
Give an example of an integral controller.
What is meant by the term *derivative controller*?
Give an example of a derivative controller.
What is meant by the term *overshoot*?
Give an example of an overshoot.
What is meant by the term *PID controller*?
Describe the impact of the PID time response characteristic.
What is meant by the term *hardware/software codesign*?
Give an example of the workflow in hardware/software codesign.
What is meant by the term *codesign*?
Describe the codesign process.
What is meant by the term *target architecture*?
Give an example of target architecture.
What is meant by the term *current HW/SW design process*?
Give an example of the workflow of the current HW/SW design process.

References and Further Reading

(Adamski et al. 2005) Adamski, M., A., Karatkevich, A., Wegrzyn, M.: Design of Embedded
    Control Systems, Springer Publ. 2005
(Ashenden 2008) Ashenden P. J.: The Designers Guide to VHDL, Elsevier Publ. 2008
(Balarin et al. 2003) Balarin, F.: Metropolis: An Integrated Electronic \ Environment, Computer,
    pp. 45–52, 2003
(Balasubramanian et al. 2006) Balasibramanian, K.: Developing Applications Using Model-
    Driven Design Environments, Computer, pp. 33–40, 2006
(Booch et al. 2005) Booch, G., Rumbaugh, J., Jacobsen, I.: The Unified Modeling Language User
    Guide, Pearson Higher Education Publ. 2005
(Chaturvedi 2010) Chaturvedi, D. K.: Modeling and Simulation of Systems Using MATLAB and
    Simulink, CRC Press, 2010
(Colnaric and Verber 2008) Colnaric, M., Verber, D.: Distributed Embedded Control Systems,
    Springer Publ. 2008
(Eker et al.2003) Eker, J.: Taming Heterogeneity: The Ptolemy Approach, Proc. IEEE, Vol. 91, no.
    1, pp. 127–144, 2005
(Feiler and Gluch 2012) Feiler, P. H., Gluch, D. P.: Model-based Engineering with AADL: An.
    Introduction to the SAE Architecture Analysis and Design Language, Pearson Higher Educa-
    tion Publ., 2012
(Haubelt and Teich 2010) Haubelt, C., Teich, J.: Digital Hardware/Software-Systems: Specifica-
    tion and Verification (In German), Springer Publ. 2010
(Henzinger and Sifakis 2007) Henzinger T. A., Safakis, J.: Embedded Systems Design, IEEE
    Computer Society, pp. 32–40, 2007
(Kamal 2008) Kamal, R.: Embedded Systems: Architecture, Programming, and Design, McGraw
    Hill Publ. 2008
(Klee Allen 2011) Klee H., Allen, R.: Simulation of Dynamic Systems with Matlab and Simulink,
    CRC Press, 2011

(Lee and Seisha 2015) Lee, E. A.; Seshia, S. A.: Introduction to Embedded Systems: A Cyber-
    Physical Systems Approach, 2nd ed. Berkeley, CA, USA, 2015
(Moeller 2003) Moeller, D. P. F.: Mathematical and Computational Modeling and Simulation,
    Springer Verlag, Berlin Heidelberg, 2003
(Möller 2014) Möller D. P. F.: Introduction to Transportation Analysis, Modeling and Simulation,
    Springer Publ. Series Simulation Foundations, Methods, and Applications, 2014
(Oliveira, M. F. S., Redin, R. M.., Carro, L., da Cunha Lamb, L., Wagner, F. R.: Software Quality
    Metrics and their Impact on Embedded Software. In: IEEE MOMPES 5th Internat. Workshop,
    pp. 68–77, 2008,
(Perry 2002) Perry, D. L.: VHDL Programming by Example, McGraw Hill, 2002
(Plessel 2013) Plessel, C.: Hardware/Software Codesign; http://homepages.uni-paderborn.de/
    plessl/lectures/2010-Codesign/script/Skript-Codesign.pdf. Accessed August 18, 2013
(Rockwell 2010) Rockwell, D.: State-Space Representation of LTI Systems, http://web.mit.edu/2.
    14/www/Handouts/StateSpace.pdf
(Rumbaugh et al. 2004) Rumbaugh, J., Jacobsen, I., Booch, G.: The Unified Modeling Reference
    Language Manual, Pearson Higher Education Publ., 2004
(Sciuto et al. 2003) Sciuto, D., Salice, F., Pomante, L., Fornaciari, W. : Metrics for Design Space
    Exploration of Heterogeneous Multiprocessor Embedded Systems. http://pdf.aminer.org/000/
    106/293/metrics_for_design_space_exploration_of_heterogeneous_multiprocessor_embed
    ded_systems.pdf
(Sifakis 2005) Sifakis J.: A Framework for Component-Based Construction, Proc. Software Eng.
    and Formal Methods, IEEE, pp. 293–300, 2005
(Teich 2000) Teich, J.: Embedded system synthesis and optimization. In: Proc. Workshop Systems
    Design Automation, pp. 9–22, Rathen, Germany, 2000
(Teich 2012) Teich, J.: Hardware/Software Codesign: The Past, the Present, and Predicting the
    Future; Proceedings of the IEEE, Vol. 100, pp. 1411–1430, 2012
(Vahid and Lysecky 2007) Vahid, F., Lysecky, R.: Verilog for Digital Design, John Wiley Publ.
    2007
(Visioli 2006) Visioli, A.: Practical PID Control, Springer Publ. 2006
(Zurawski 2009) Zurak, R.: Embedded Systems Handbook, CRC Publ., 2009

Introduction to Cyber-Physical Systems

3

This chapter begins with an overview of cyber-physical systems in Sect. 3.1 taking into account the introduction to systems in Chap. 1 and embedded computing systems in Chap. 2. Thereafter, Sect. 3.2 concentrates to recommendations with regard to cyber-physical systems design. In Sect. 3.3 cyber-physical system requirements are described to emphasize disciplined approaches to their design. Section 3.4 introduces the opportunities created applying the cyber-physical technology in a wide range of domains, offering numerous opportunities in products and applications. Section 3.5 refers to smart cities and the Internet of Everything. The case study in Sect. 3.6 focuses on cyber-physical vehicle tracking. Section 3.7 contains comprehensive questions from the cyber-physical systems domain, followed by references and suggestions for further reading.

3.1   Cyber-Physical Systems

In Chap. 1 a specific introduction to systems was given to make sure that the readers from several engineering and scientific disciplines have the same understanding of the term system and the same mathematical background to the study of systems. Chapter 2 has introduced embedded computing systems which are systems with dedicated functions within larger and/or heterogeneous system architecture often in conjunction with real-time constraints. The integration of embedded computing systems—they represent physical systems where the computer is completely encapsulated by the device it controls and interacting with physical processes—with networked computing has led to the emergence of a new generation of engineered systems, the so-called cyber-physical systems (CPSs). These systems use computations and communication deeply embedded in and interacting with physical processes by adding new capabilities to physical systems. Therefore,

© Springer International Publishing Switzerland 2016
D.P.F. Möller, *Guide to Computing Fundamentals in Cyber-Physical Systems*,
Computer Communications and Networks, DOI 10.1007/978-3-319-25178-3_3

Chap. 3 summarizes the forgoing mentioned knowledge to introduce cyber-physical systems with regard to the existing manifold of definitions such as:

- Systems of collaborating computational elements controlling physical entities
- Networks of interacting elements with physical inputs and output instead of a stand-alone device
- Internet of Things, Data, and Services
- Interconnections between physical and virtual world models
- Ability for autonomous behavior such as self-control and/or self-optimization
- Internet-based business models, social networks, and communities
- System of systems
- New way of cooperation among distributed and intelligent smart networked devices as well as with humans

From these definitions, it can be stated that cyber-physical systems span a wide range from the miniscale, e.g., pacemakers, to the large scale, e.g., power grids, in their diversity of applications, because the ubiquity of computer devices is a huge source of economic leverage.

From the history in cyber-physical systems, it should be noted that in 2006, the National Science Foundation (NSF) in the USA has identified cyber-physical systems (CPSs) as one of the promising research themes of the future. In the following year, based on the recommendation of the President's Council of Advisors on Science and Technology (PCAST) (PCAST 2007), a research program was established by the NSF titled *Cyber-Physical Systems*, in which about 65 projects amounting to almost 60 million US$ have been funded. In a subsequent report of PCAST in 2010, further research needs for *cyber-physical systems* were expelled, and the related NSF program initially extended until 2013, with a budget of over 30 million US$.

In the year 2012, a study funded by the Federal Ministry of Education and Research in Germany (BMBF) was published by the German Academy of Science and Engineering (acatech) on the topic *information and communication technologies* (ICTs), which was tied to the megatrend *Internet of Things* which addresses the opportunities and challenges of the technology trends of *cyber-physical systems* (Geisberger and Broy 2012).

According to these studies, cyber-physical systems will play an essential role in industry and society leading to breakthroughs in all relevant areas by bridging the wide range of fields of action in which cyber-physical systems can be applied. The reason for that is that cyber-physical systems go beyond traditional systems (see Chaps. 1 and 2) employed in industry in their complexity, such as automation systems, which require a much closer networking of the appropriate systems and software engineering disciplines. Traditionally automation systems networking is carried out through specially developed network structures like PROFIBUS (Process Field Bus), DP (Decentralised Peripherals), used to operate sensors and actuators via a centralised controller in production automation applications, and PA (Process Automation), used to monitor measuring equipment via a process control system in process automation applications. In contrast, cyber-physical systems usually use open network technologies like the Internet. The massive importance of growth of the Internet is characterized by the Internet of Things (IoT) (see Chap. 4; Möller 2013) in which the real and the virtual worlds are converging. Thus, cyber-physical systems, which are based on converging real (physical) and virtual (cyber) systems, gain a growing importance in networking of embedded computing systems (see Chap. 2) and components of the information and communication technology (ICT) as well as with the Internet or more in general the *Internet of Things* as a dynamic global network infrastructure with self-configuring capabilities based on a standard and interoperable communication protocol, the Internet Protocol version 6 (IPv6). The IPv6 is the latest version that routes traffic across the Internet. It was developed by the Internet Engineering Task Force to replace the IPv4 to overcome with the long-anticipated problem of IPv4 address exhaustion. IPv6 allows to access 2^{128} addresses. The IPv6 specification can be found on the Web at http1 2015. The IPv6 addressing architecture can be found on the Web at http2 2015. The document includes the IPv6 addressing model, text representations of IPv6 addresses, and definition of IPv6 unicast addresses, anycast addresses, multicast addresses, and IPv6 node's required addresses.

Against this background cyber-physical systems go beyond the traditional systems employed in industry with regard to their complexity, requiring a close networking with the appropriate disciplines. This makes three basic technologies essential for cyber-physical systems:

- Embedded computing systems (ECSs)
- Networking, information, and communication technologies (NICT)
- Sensing and actuating technologies (SAT)

Therefore, the decreasing cost of computation, networking, and sensing provides the basic economic motivation embedding networking, information, and communication technologies in every industry and application. Moreover, the exponential growth in computing power has launched extremely sophisticated computers at consumer electronics prices into the market. The same trends have vastly improved sensing and actuation technologies. Thus, computers and communication have become the universal system integrator that keeps large systems together which enable the composition of the cyber-physical systems infrastructure. Hence, cyber-physical systems have an advanced and complex system architecture that connects computing, networking, and the physical and cyber or virtual environment within one paradigm. It provides services such as:

- Control
- Information feedback
- Real-time monitoring

as most essential actions to merge the interaction of the physical and the cyber worlds by integration and collaboration of computation, communication, and control (3C) (Ning 2013).

Against the background of technological innovations, it can be stated that cyber-physical systems and the Internet of Things become closer to each other with many similarities. Both are sensing, actuating, computing, transmitting information, and using interaction technologies to merge the cyber and the physical worlds. But nowadays also some differences can be recognized. The Internet of Things emphasizes the connection of things with networks, while cyber-physical systems emphasize the integration of computational and physical element information (Li et al. 2011). With regard to this remarks, ubiquitous computing (see Chap. 5) is of crucial interest, because it is not only supporting global ubiquity to interlink systems to produce an omnipresent service architecture but rather to support context-based ubiquity, which can be understood as situated access versus mass access which may reduce human interaction in automation. Thus, the term cyber-physical system refers to the integration of computation with the physical and the cyber world's whereby things can be sensed through ubiquitous sensors and preprocessed by ubiquitous computing. Therefore, the architecture of cyber-physical systems should be ubiquitous and/or an integration of paradigms such as:

- *Ambient intelligence*: referring to electronic environments sensitive and responsive to the presence of things/objects. In an ambient intelligent world, devices work together supporting things and/or people doing their tasks, everyday life activities, etc., in an easy way using information and intelligence hidden in the network connecting these devices through the Internet of Things requesting for ubiquitous computing which allows massive networking of computers, sensors, actuators, and radio modules where things/objects communicate with each other.
- *Disappearing computers*: referring to the miniaturization of devices and their integration into the environment. Thus, the devices disappear into our surroundings until only the user interface remains perceivable by users (Seitz et al. 2007).
- *Pervasive computing*: often synonymously called ubiquitous computing, describes an emerging field of research that brings in revolutionary paradigms for computing models in the twenty-first century.
- *Post-PC era*: market trend involving a decline in sales of PCs in favor of post-PC devices, which include mobile devices such as smartphones and tablet computers. These devices emphasize portability and connectivity, including cloud-based services, more focused apps to perform tasks, and the ability to synchronize information between multiple devices seamlessly.
- *Ubiquitous computing*: concept in software and computer engineering where computing is made to appear everywhere and anywhere. In contrast to desktop computing, ubiquitous computing can occur using any device, in any location, and in any format (see Chap. 5).

Hence, cyber-physical systems are usually fused with other related concepts such as sensing through sensors and sensor networks, actuating through actuators and actuator networks, algorithms to adopt the behavior of networked systems, ontologies to interlink the cyber-physical systems applications, as well as interoperability standards and human-machine interfaces (HMIs) with situational adequacy and ergonomic issues. Comparing embedded computing systems with cyber-physical systems, it can be stated that cyber-physical systems are the integration of computing and communication with physical and virtual processes with the objective to convey how to interact with the physical world to monitor and control the physical processes, usually with feedback loops where physical processes affect computations and vice versa. Thus, cyber-physical systems obtain many benefits: they can make systems safer and more efficient; they can reduce the cost of building and operating these systems; and they can allow individual machines to work together to form complex systems that provide new capabilities.

The design of such systems, therefore, requires understanding the dynamics of computers, software, networks, and physical and virtual or cyber processes (Pellizzoni 2015) allowing to create new machines with complex dynamics and high reliability. Moreover, it will allow being able to apply the principles of cyber-physical systems to new industries and applications in a reliable and economically efficient way.

Thus, the consolidating technological advances are that embedded computing systems created networked embedded computing systems which bear cyber-physical systems which finally converge to Internet of Everywhere, Data, and Services, as shown in Fig. 3.1 based on the report of Geisberger and Broy (2012).

The last two decades have launched a change in the direction of a digital revolution that has been transforming the industry. This change is not a matter of choice; it is driven by fundamental, long-term technological and economic trends which will continue. Thus, the challenges of cyber-physical systems as

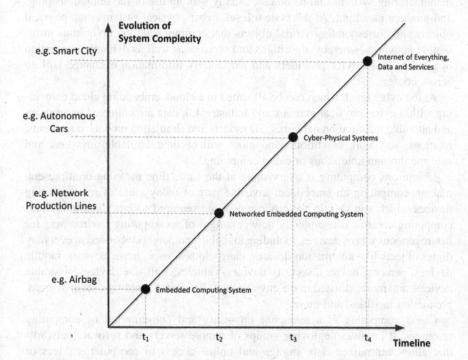

Fig. 3.1 Evolution of ECS into the Internet of Things (IoT), Data, and Services

technological and economical driver result in different international activities to transform the industry like:

- Industrie 4.0 (see Chap. 7): German initiative since April 2013 driven by the platform Industry 4.0.
- Advanced Manufacturing: US government initiative and funding program.
- Smart Manufacturing Leadership Coalition: nonprofit organization founded in 2012.
- Intelligent Manufacturing: national program driven by the Chinese Ministry of Science and Technology, released in March 2012.
- Similar programs have been initiated by the European Union in their Horizon 2020 program in the Industrial Technologies Program which focuses on key enabling technologies such as advanced manufacturing and processing (production technologies) and the future of manufacturing by the British government and others.

Hence, cyber-physical systems acquire the designated role as backbone of advanced manufacturing (Industry 4.0; see Chap. 7). Today's automation systems in industry will become intelligent, explorable, self-explanatory, self-aware, self-diagnostic, and interacting assets in manufacturing systems which will be the basis for context-sensitive decision-making and automatic optimization of manufacturing systems and resources, usually making use of the ubiquitous plug-and-produce paradigm. At the system level, cyber-physical systems map physical objects and corresponding virtual objects that communicate via ubiquitous information networks, whereby algorithms and services as well as dynamic integration of services and service providers and particularly information exchange will go across borders.

At the cyber level, which can be allocated in a cloud, embedding cloud computing within cyber-physical systems and Industry 4.0, data are collected in arbitrary and alterable information networks, 3D models and simulation models, documents, relations, and work conditions, and more will become available anywhere and anytime through ubiquitous or cloud computing.

Ubiquitous computing is everywhere at the same time meaning omnipresent, making computing an embedded, invisible part of today's life. Tiny computing devices which vanish into the environment are required to introduce ubiquitous computing creating a completely new paradigm of a computing environment for heterogeneous sets of devices, including invisible computers embedded in everyday things/objects like automation devices, cans, clothes, cups, home devices, mobile devices, personal devices, security devices, vehicles, wall-size devices, wearable devices, and more, situated in the environments, inhabited buildings, secured areas, production facilities, and more.

Cloud computing is a metaphor on utility and consumption of computing resources. It involves deploying groups of remote servers and software networks that allow centralized data storage and online access to computer services or resources.

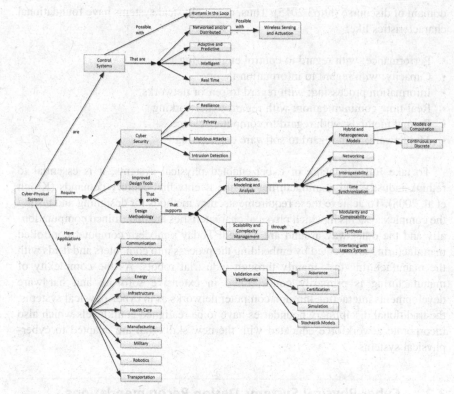

Fig. 3.2   Cyber-physical systems concept map (http3 2015)

The main consolidation of a typical cyber-physical system can be seen in integrating the dynamics of physical processes with those of the software and networking, providing abstractions and modeling, design, and analysis techniques for the integrated whole, as mentioned in http3 2015, which finally result in the so-called cyber-physical systems concept map, shown in Fig. 3.2.

From Fig. 3.2 it can be seen that cyber-physical systems are primarily allocated as an engineering discipline, focused on technology, with strong foundation in mathematical abstractions. Technical challenge is to conjoin abstractions that have evolved for modeling physical processes like differential equations, stochastic processes, and more with abstractions that have evolved in computer science with regard to algorithms and programs, which provide a procedural epistemology (Abelson and Sussman 1996). Former abstractions focus on the dynamic evolution of system state over time, whereas the latter focus on processes of transforming data. Computer science abstracts away core physical properties, particularly passage of times that are required to include the dynamics of the physical world in the domain of discourse (http3 2015). Thus, cyber-physical systems have foundational characteristics like:

- Performance: with regard to control engineering
- Capacity: with regard to information theory
- Information processing: with regard to sensor networks
- Real-time communication: with regard to networking
- Formal methods: with regard to computer science
- Middleware: with regard to software engineering

To take full advantage of cyber-enabled physical systems, it is essential to rethink industry and manufacturing for the twenty-first-century demands (Krogh et al. 2008). To achieve these requirements, new methods for designing and testing the complex systems in which physical characteristics are determined computationally and the other way around are needed. Today's modern computer-controlled manufacturing is executed by embedding the process into computers and deals with the manufacturing task mainly through industrial robots. As the complexity of manufacturing is primarily concentrated in extensive software and hardware developments interacting through computer networks as in cyber-physical systems, the traditional disciplinary boundaries have to be realigned at all levels which also incorporate a workforce educated with the new skills that are adapted to cyber-physical systems.

3.2   Cyber-Physical Systems Design Recommendations

From the current industrial experience, the limits in knowledge of how to combine computers and physical systems are known. Therefore, continuing in designing systems based on these limited knowledge (methods and tools) will not be efficient, and the risk of unsafe and unpredictable systems can be estimated. These shortcomings become extremely important in cyber-physical systems design, because these systems are heterogeneous comprising multiple types of physical systems and multiple models of computation and communication. Therefore, heterogeneity in cyber-physical systems design will result in system-/product-specific design flows which are inappropriate for design automation. This implies that the increasing design complexity and the lack of effective, specialized design automation tools can limit design productivity and increase time to market. This means that there is a need to realign the abstraction layers in the design flows and build a new infrastructure for agile design automation of cyber-physical systems.

Thus, the challenges in the design of cyber-physical systems result in the abstraction of levels which can be introduced as part of a stack-based process to abstract away the low-level architecture details and make the underlying system components more effective and transparent to the designer. Therefore, components at any level of abstraction should be made predictable and reliable so far as this is technologically feasible, because it is important to query the number and capabilities of the available system components, as well as for software development for code portability between the cyber and the physical part. If it is not technologically feasible, then the next level of abstraction must compensate by robustness as part of a robust principal component analysis. But abstractions do not directly encapsulate the essential characteristics which mean it is hard to predict whether the cyber part will meet requirements of the physical part or not (Pellizzoni 2015).

Let a successful systems design follow these principles assuming that it is technically feasible to build predictable and reliable components. Much harder is making wireless links predictable and reliable with regard to the increasing needs of interactive network traffic that may result in delays which raise the fundamental question on how to support delay guarantees over an unreliable medium like wireless. This is an important issue in the automotive industry because a premium car has around 75 sensors and a hundred switches that are connected by wiring of the order of 1,000 m making the wiring harness. This is very costly, very heavy, and very complicated from where much mechanical failure can occur in the harness. Therefore, automotive engineering tries to solve the fundamental question on how to support delay guarantees over the wireless medium. One possible option is compensating one level up, using robust coding and adaptive protocols

Another obvious fundamental question is whether it is technically feasible making software engineered systems predictable and reliable. At the foundations of computer architecture and programming languages, software is essentially perfectly predictable and reliable, when software was limited to refer to what is expressed in simple programming languages. With regard to imperative programming languages with no concurrency, like C, designers can count on a computer to perform exactly what is specified with essentially 100 % reliability.

A problem arises when scaling up from simple programs to software engineered systems and particularly to cyber-physical systems. The fact is that even the simplest C program is not predictable and reliable in the context of a cyber-physical system because the program does not express any aspect of the behavior that is essential to the cyber-physical system. It may execute perfectly, exactly matching its semantics, and still fail to deliver the behavior needed by the cyber-physical system. For example, it could miss timing deadlines. Since timing is not in the semantics of C, whether a program misses deadlines is in fact irrelevant to determining whether it has executed correctly. But it is very important to determining whether the system has performed correctly (Lee 2008). Thus, cyber-physical systems design requires an adopted systems and software engineering approach with regard to their intrinsic complexity, as shown in Fig. 3.3.

From Fig. 3.3 it can be seen that in cyber-physical systems design two essential engineering sectors are embedded, the systems and the software engineering areas. The systems engineering approach can be characterized as:

- Interdisciplinary approach to develop and implement complex technical systems in major projects in systems engineering.

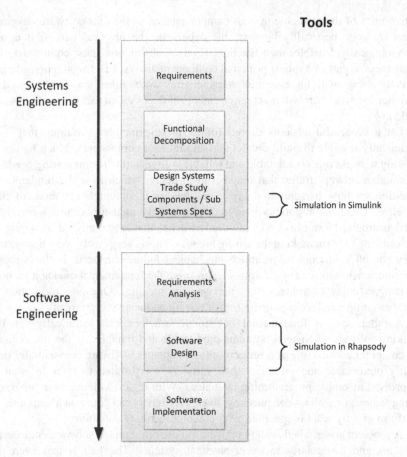

Fig. 3.3 Systems and software engineering approach in cyber-physical systems design modified after http4 2015

- Basic necessity, since in large complex projects, planning, coordinating, and implementing are more difficult to manage and can lead to massive problems in the execution of projects in systems engineering.
- Ongoing integration of software intensive embedded computing systems and global communication networks (GCN) in cyber-physical systems is considered to be the next big step in technological progress in ICT with a great deal of changing business potential and novel business models for integrated products and services.
- Mastering engineering of complex and trustworthy cyber-physical systems is crucial to planning, implementing, and sustaining business models.
- Current systems engineering frameworks, however, do not enable a conceptualization and design for the deep interdependencies among engineered systems and the natural world. Thus, there is a clear need for a new cyber-physical systems engineering framework (CPSEF) for the development, implementation,

and operation of highly efficient cyber-physical systems essential to rule the complexity of requirements of today's and tomorrow's cyber-physical systems in the manifold of application domains.

The software engineering approach can be introduced as:

- Engineering approach to systematic, quantifiable design, development, operation, and maintenance of software
- Engineering approach concerned with all aspects of software production
- The use of sound engineering principles in order to economically obtain software that is reliable and works efficiently on real machines

The IEEE Computer Society and ACM are the main US-based professional organizations of software engineering which have published guides to the profession of software engineering. The IEEE's *Guide to the Software Engineering Body of Knowledge* (*2004 Version*) defines the field and describes the knowledge IEEE expects a practicing software engineer should have.

Today cyber-physical systems present a range of challenges which call for better and more effective architectural design environments with regard to the:

- Complexity of the cyber-physical systems, which results from parallel and distributed development which usually use different tools and methods
- Electronics and software content that can be intricate
- Diverse sets of different algorithms with unique challenges
- Needs for integration and testing that can become costly and time-consuming
- Increased needs to manage requirement changes during the development cycle
- Desire to reuse existing and future intellectual properties

With regard to this specific constraints, SysML and UML (http5 2015) have been designed as an architectural framework and been validated across numerous industries with their separate views for functional, physical, and software architectures as well as their requirements capture and elicitation and have been expanded by Rhapsody which integrates a rich set of external components like:

- Code: C, C++, Java, or Ada
- Tools: Simulink, Statemate, SDL Suite

whose primary benefits are:

- Integration of various components that exist in the design environment into coherent, persistent design
- Reuse of existing IPs in new architectures/platforms
- Ease of the creation and maintenance of the design
- Provision of standard support to requirements traceability, change management, and other aspects of the engineering process
- Ability of co-execution to ensure correct interfacing and interaction

A very important intrinsic characteristic in cyber-physical systems design is the interface. The cyber-physical system interface inherits all the elements from the cyber and physical part and adds new elements that bridge the gap between computational and physical systems. To model the interactions between the cyber and the physical worlds, two directed connector types are essential, the physical-to-cyber (P2C) and the cyber-to-physical (C2P) connectors. Thus, simple sensors can be modeled as physical-to-cyber connectors and simple actuators can be modeled as cyber-to-physical connectors.

More complex interfaces between cyber and physical elements in cyber-physical systems require two directed transducer components which have ports to cyber elements on one side and ports to physical elements on the other side. These transducer components are the physical-to-cyber and the cyber-to-physical transducer types. If interfacing intelligent sensors, devices are modeled as P2C/C2P transducers, because they have to do more than a simple translation between cyber and physical nodes (Raihans et al. 2005). One of the major difficulties in providing tool support for architectural design and analysis is the need to tailor those capabilities to the application domain, which should be illustrated as an example for the architectural modeling of a temperature control system with two zones (rooms), as shown in Fig. 3.4. The zones have temperatures T_1 and T_2, respectively, and the temperature of the ambient environment is T_a. Let assume that the zones retain heat with thermal capacities C_1 and C_2, respectively, and lose heat to the ambient environment with thermal resistivities R_1 and R_2, respectively. Let Zone 2 be heated through a shared wall with thermal resistivity R_w with Zone 1. The thermostat is physically located in Zone 1, so it can only read the temperature in Zone 1. The heating can be powered on or off manually. When the heating is on, the thermostat in Zone 1 determines whether or not the heating should heat the room and turns a blower that forces hot air into the room on or off accordingly. The goal for the temperature control system is to maintain the measured temperature of Zone 1 close to the thermostat set point (Raihans et al. 2005).

The architecture of the temperature control system can be developed using the AcmeStudio environment (Schmerl and Garlan 2004). AcmeStudio is an

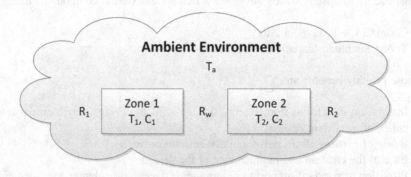

Fig. 3.4 Illustration of the temperature control system. For details, see text

architecture development environment, written as a plug-in to IBM's Eclipse IDE framework (http6 2003) that supports the Acme ADL (Garlan et al. 2000). Acme is a generally purpose ADL supporting component and connector architectures. In the AcmeStudio different architectural models and architectural style can be developed. At the one side of the architecture window is a type palette, which displays the available vocabulary for a particular domain. For the temperature control, for example, it has an architectural model in a pipe-filter style; the palette allows an architect to drag pipes, filters, data stores, and their associated interfaces into a diagram to create the model. The style used for this model is different from the other ones used for the temperature control. One shows selected elements from the models and styles in more detail, displaying their properties, rules, substructure, and typing which also allows the user to enter values for properties, define rules, etc.

With regard to the above remarks about AcmeStudio, the architectural modeling of a temperature control system can be executed, the result of which is shown in Fig. 3.5. The right-hand sector in Fig. 3.5 shows the cyber, physical, and the cyber-physical interfaces. Each zone is modeled as an energy storage component. The ambient temperature is modeled as a source component since it is independent of the room dynamics.

The top right corner of the canvas shows a bounding box that corresponds to the thermostat. The box contains a temperature sensor that senses the temperature of

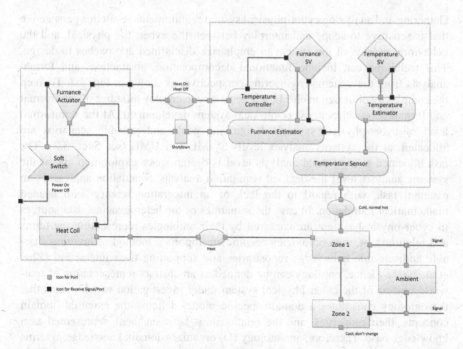

Fig. 3.5 Architectural modeling of a temperature control system using the cyber-physical system architecture design approach after Raihans et al. (2005)

Zone 1 and broadcasts it on a communication channel. A temperature estimator then reads the temperature from the communication channel and updates the temperature state variable. The furnace estimator reads the temperature and keeps updating the furnace state. The temperature controller is the part of the software that keeps track of when the furnace estimator changes the furnace state and communicates these changes to the furnace as commands to turn the heating on or off.

The furnace comprises a furnace actuator component that receives commands sent by the temperature controller in the thermostat and accordingly turns the corresponding heating element (heat coil) on or off. Finally, a transducer element called Soft Switch captures the manual on/off signal and communicates it to the furnace actuator software element to turn the heating element on or off.

Zone1 and Zone 2 are modeled as energy storage components and the ambient environment as a source. The connectors between ambient and the two zones are directional input-output connectors capturing the one-way coupling, while the connector between the two zones is a bidirectional shared-variable connector that captures the bidirectional coupling.

3.3   Cyber-Physical System Requirements

Gathering and analyzing cyber-physical system requirements result in a perspective that is sensitive to scope and interplay between the cyber, the physical, and the behavioral aspects of the system to emphasize disciplined approaches to design. This technique can include functional decomposition, abstraction, and formal analysis from the systems engineering perspective, as shown in Fig. 3.3. To keep the complexity of design in check, it is necessary to employ mixtures of semiformal and formal approaches to cyber-physical system development. At the semiformal level of cyber-physical systems design, the goals and possible scenarios are allocated, at the systems analysis levels SysML and UML (see Sect. 3.2). The task allocated at the formal analysis level is design space exploration and at the systems analysis level the detailed simulation analysis. Simulation analysis is an essential task, with regard to the lack of an integration science with needed mathematical foundation. In case the semantics of the heterogeneous data sources in cyber-physical system are captured by their ontologies representing the terms and relationships, then ontologies become an important method of building sharable and reusable knowledge repositories and supporting their interaction (Zhai et al. 2007). Hence, ontology can be defined as an abstract representation of real-world objects of the cyber-physical system under investigation which means that the ontology constitutes a domain-specific model defining the essential domain concepts, their properties, and the relationships between them, represented as a knowledge base. Therefore, an ontology (O) organizes domain knowledge in terms of concepts (C), properties (P), and relations (R). In other words, an ontology (O) is a triplet of the form (Möller 2014)

$$O = (C, P, R)$$

where C is a set of concepts essential for the domain, P is a set of concept properties essential for the domain, and R is a set of binary semantic relations defined between concepts in O. A set of basic relations is defined as $R_b = \{\approx, \uparrow, \nabla\}$ with the following interpretations (Zhai et al. 2007):

- For any two ontological concepts, $c_i, c_j \in C$, $\approx$ denotes the equivalent relation, meaning $c_i \approx c_j$. If two concepts, c_i and c_j, are declared equivalent in ontology, then instances of concept c_i can also be inferred as instances of c_j and vice versa.
- $\uparrow$ is the generalization notation. In cases where the ontology specifies $c_i \uparrow c_j$, then c_j inhibits all property descriptors associated with c_i; and these need not be repeated for c_j while specifying the ontology.
- $c_i \nabla c_j$ means c_i has part c_j. If a concept in ontology is specified as an aggregation of other concepts, it can be expressed by using ∇.

3.3.1 Requirements Engineering

Requirements engineering can be introduced as a process defining, documenting, and maintaining requirements, which are documented physical and functional needs that a particular design of a product or a process must be able to perform. The fields concerned with requirements engineering are systems and software engineering. The term comes into general use in the 1990s with the publication of an IEEE Computer Society tutorial (Thayer and Dorfmann 1997) and the establishment of a conference series on requirements engineering (http11 2015). Alan M. Davis maintains an extensive bibliography of requirements engineering (Davis 2011).

The activities involved in requirements engineering vary widely, depending on the type of system being developed and the specific style guide practices of the organization involved. These may include (http11 2015):

1. Requirements elicitation: practice of collecting requirements of a system from users, customers, and other stakeholders, sometimes also called requirements gathering.
2. Requirements identification.
3. Requirements analysis and negotiation: tasks determining the needs or conditions to meet for a new product, with regard to possibly conflicting requirements of the various stakeholders.
4. Requirements specification: documenting the requirements in a requirements document.
5. Systems modeling: developing models of the new system, often using a notation such as the Unified Modeling Language (UML); UML profiles are intensively used in modeling domain-specific distributed applications.

6. Requirements validation: checking that the documented requirements and models are consistent and meet stakeholder needs.
7. Requirements management: managing changes to the requirements as the system is developed and put into use.

These requirements are for some reasons presented as chronological stages; in practice there is considerable interleaving of these activities. For more details, see Chap. 6.

3.3.2   Interoperability

Interoperability describes the ability of systems working together, which means to interoperate. The term was initially defined for services in information and communication technology and systems engineering allowing a seamless information exchange. A more general definition refers to social, political, and organizational factors in regard of their impact to systems performance. From a more technical perspective, interoperability is the task building coherent services for systems when individual system components are technically different and managed by different software systems. Interoperability can be introduced as (http12 2015):

- Syntactic interoperability: necessary condition in case two or more systems are capable of communicating and exchanging data with regard to specified data formats and communication protocols. Extended Markup Language (XML) or Standard Query Language (SQL) standards are tools of syntactic interoperability.
- Semantic interoperability: ability to automatically interpret the information exchanged meaningfully and accurately in order to produce useful results as defined by the end users of the systems used which is of importance for the vertical integration means in manufacturing. To achieve semantic interoperability, the two or more systems must refer to a common information exchange reference model.

Interoperability must be distinguished from open standards. Open standards rely on a broadly consultative group including representatives from vendors, academicians, and others holding a stake in the development. That discusses and debates the technical and economic merits, demerits, and feasibility of a proposed common protocol. After the doubts and reservations of all members are addressed, the resulting common document is endorsed as a common standard. This document is subsequently released to the public and henceforth becomes an open standard. It is usually published and is available freely or at a nominal cost to any and all comers, with no further encumbrances. Various vendors and individuals can use the standards document to make products that implement the common protocol defined in the standard and are thus interoperable by design, with no specific liability or advantage for any customer for choosing one product over another on the basis of standardized features (http12 2015).

3.3.3 Real-Time Systems

A real-time system is required to complete its task and deliver its services on a timely basis which means that the time taken for the system to respond with an output from the associated input is within a sufficiently small acceptable timeline. Real-time systems include digital control, command and control, signal processing, and more. The *Oxford Dictionary of Computing* gives the following definition of a real-time system:

> *Any system in which the time at which output is produced is significant. This is usually because the input corresponds to some movement in the physical world, and the output has to relate to that same movement. The lag from input time to output time must be sufficiently small for acceptable timelines.*

The predictably dependable computer systems project (Randell et al. 1995) gives the following definition:

> *A real-time system is a system that is required to react to small stimuli from the environment (including the passage of physical time) within time intervals dictated by the environment.*

Fortunately, it is usually not a disaster if the system response is not forthcoming. These types of systems can be discerned from those where failure to respond can be considered just as bad as a wrong response. It is this aspect that distinguishes a real-time system from others, where response time is important but not crucial. Consequently, the correctness of a real-time system depends not only on the logical result of the computation but also on the time at which results are generated. Practitioners in the field of real-time computer systems design often differentiate between hard and soft real-time systems (Burns and Welllings 2001; Liu 2000):

- Hard real-time systems (HRTSs): systems where it is absolutely imperative that responses occur within the specified deadline.
- Soft real-time systems (SRTS): systems where response times are important but the system will still work correctly if deadlines are occasionally missed. Soft real-time systems can themselves be distinguished from interactive ones in which there are no explicit deadlines (Burns and Welllings 2001).

The use of the term soft does not imply a single type of requirement, but incorporates a number of different properties like:

- Deadline can be missed occasionally: typically with an upper limit of misses within a defined interval.

• Service can occasionally be delivered late: typically with an upper limit in tardiness.

As mentioned real-time embedded control systems are used for process control time- and/or life-critical application in manufacturing; for complex applications with regard to communication, command, and control in the military domain; as well as in the control of aircrafts, automobiles, autonomous robots, chemical plants, medical equipment, power distribution systems, and more.

The reliability requirement of real-time systems is to translate the need to meet critical task deadlines with a very high probability. Hence, the question has to be answered on how to schedule tasks that deadlines continue to be met despite processor permanent or transient or software failures (Chen et al. 2011a).

3.3.4   GPU Computing

Graphics processing units (GPUs) have evolved into general-purpose multiprocessors that are capable for highly parallel computation. The development was driven by the demand of high-definition real-time imagery. Graphics cards that accelerate the generation of three-dimensional real-time images became commercially available in the mid-1990s. The processors on these graphics cards mainly provided hardware-accelerated rasterization of textured triangles. In the late 2000s, the major graphics card manufacturers make their processors more accessible and provided tools to allow the use of more general-purpose programming languages. The CUDA computer architecture (http13 2015) developed by NVIDIA allows the software engineer to develop his GPU algorithms in a subset of the C programming language. Similarly the Stream SDK from AMD allows free programming using OpenCL, another extension of C.

CUDA provides a standard model for general computing on graphics processing units across the graphics card produced by NVIDIA. CUDA extends the C and C++ programming languages with new language primitives suitable for its parallel computation model. The CUDA developer kit includes cuBLAS, an implementation of the basic linear algebra subprograms. A good overview of the CUDA programming model and its application to numerous mathematical problems such as numerical linear algebra and fast Fourier transformation is given in Garland et al. (2008).

Graphics processing units are to hide memory latency by fast thread switching. On the G80 GPU, global memory access is not cached and involves a latency penalty of a hundreds of clock cycles. The G80 GPU is able to run 448 hardware threads in parallel. The threads are lightweight, and scheduling is very fast. In order to hide memory latency, software threads that are waiting for memory access to complete can be suspended with very low overhead, leaving the processor power to the active threads. Later GPUs like the Fermi GF100 GPU targeted a scientific computing by adding a L1 and a L2 cache in order to reduce latency. However, enabling the L1 cache reduces the amount of shared memory available to the program and thus potentially the number of threads that can be executed at the same time in parallel by the multiprocessor for a given program (Selke 2014).

3.4 Cyber-Physical Systems Applications

Cyber-physical systems cover an extremely wide range of application areas, which allow designing systems more economically by sharing abstract knowledge and design tools. This allows designing more dependable cyber-physical systems to get by applying best practices to the entire range of cyber-physical applications whereby the technological and economic drivers are creating an environment that enable and require a range of new capabilities. Progress during the past decade has produced examples of a new generation of systems that rely on cyber-physical technology such as:

- Advanced automotive systems
- Assisted living
- Avionics
- Critical infrastructure control (e.g., electric power, water resources, and communication systems)
- Defense systems
- Distributed robotics (telepresence, telemedicine)
- Energy conservation
- Environmental control
- Manufacturing
- Medical devices and systems
- Process control
- Smart structures
- Traffic control and safety
- Others

With regard to today's advanced automotive systems, the Defense Advanced Research Projects Agency (DARPA) has launched a grand challenge competition for American autonomous vehicles, funded by the US Department of Defense that bridges the gap between fundamental discoveries and military use. The initial DARPA Grand Challenge prize competition was created to stimulate the development of technologies needed to create the first fully autonomous ground vehicles capable of completing a substantial off-road course within a limited time. During the third competition in 2007, the urban area course involved a 96 km distance, approximately 60 miles, which has to be completed in less than 6 h, obeying all traffic regulations while negotiating with other traffic and obstacles and merging into traffic. The overall cyber-physical systems approach represents a transfer behavior in between input and output data which are processed in several specific units such as:

- Goal selection unit
- Global Positioning System (GPS) unit
- Intersection handling unit
- Lane driving unit
- Sensing unit
- Others

Eleven teams compete against each other representing major universities and large automotive and engineering companies. Teams were given maps sparsely charting the waypoints that defined the competition courses and the DARPA 2007 Basic Rules which have to be taken into account (http7 2015):

- Vehicle must stock or have a documented safety record.
- Vehicle must obey the California state driving laws.
- Vehicle must be entirely autonomous, using only the information it detects with its sensors and public signals such as GPS.
- DARPA will provide the route network 24 h before the race starts.
- Vehicles will complete the route by driving between specified checkpoints.
- DARPA will provide a file detailing the order the checkpoints must be driven to 5 min before the race start.
- Vehicles may "stop and start" for at most 10 s.
- Vehicles must operate in rain and fog, with GPS blocked.
- Vehicles must avoid collision with vehicles and other objects such as carts, bicycles, traffic barrels, and objects in the environment such as utility poles.
- Vehicles must be able to operate in parking areas and perform U-turns as required by the situation.

The six teams that successfully finished the entire course with their autonomous cars necessitate at the first place an average speed of 14 mph (22.53 km/h) throughout the course, at the second place an average speed of 13.7 mph (22.05 km/h) throughout the course, and at the third place an average speed of approximately 13 mph (20.92 km/h) throughout the course. These competitions represent the fundamentals of the smart mobility concept of cyber-physical systems, based on autonomous vehicles. The most recent challenge, the 2012 DARPA Robotics Challenge, focuses on autonomous emergency maintenance robots (http8 2015; http9 2015).

In 2010 Google announced their result developing autonomous driving cars based on Toyota Prius car models. These cars have traveled autonomously in total more than 1,000,000 miles on the roads and highways in California and Nevada. They are equipped with transmitter laser distance meters, radar and ultrasonic motion detectors, video cameras, and GPS receivers and can perceive its surroundings down to the smallest detail. The only accident was a pileup with five cars in the vicinity of the company's headquarters in Silicon Valley in the year 2011 as someone drove the car by hand (LeValley 2013). Autonomous automobiles have gone but still a long way ahead before they will drive us to work or shopping or kids to school. In this regard the car hacking problem has become an issue for discussing the potential hazards because attacker can very easily gain control over the electronic systems in cars (Pellizzoni 2015) such as:

- Start/stop/rev up/rev down engine.
- Brake/disable braking.
- Open doors.
- Determine your position through GPS.
- Listen to whatever the driver says in the car (without driver's knowledge).

The Infiniti Q50 is a steer-by-wire embedded vehicle. Attacker can remotely start and drive car from, e.g., a parking lot to his house without moving from it or can manipulate the engine management or can manipulate driver assistance systems.

Beside, the forgoing introduced advanced automotive cyber-physical systems will play an essential role in all relevant domains of science and engineering and will lead to breakthroughs, bringing a wide range of domains of action to emerge as shown in the cyber-physical concept map (http3 2015). This requires developing advanced modeling and integration frameworks because modeling techniques and comprehensive integrated toolchains for clearly defined case study design are an essential issue in cyber-physical systems applications. Major aspects include the holistic modeling of the system behavioral, computational, physical, and/or human constraints of the cyber-physical system and the seamless interoperability between the cyber-physical systems design tools.

Following the cyber-physical systems concept map shown in Fig. 3.2, the following application areas are of importance:

- Communication
- Consumer
- Energy
- Infrastructure
- Health care
- Manufacturing
- Military
- Robotics
- Transportation

They will be discussed more in detail.

3.4.1   Communication

Communication is simply the act of transferring a message from one place to another. In general communication means data transmission by the sender through a secure communication channel to a receiver, or to multiple receivers, whereby the sender encodes the data being transmitted into a form that is appropriate to the communication channel and the receiver then decodes the data transmitted to understand its meaning and significance. For communication a variety of technological solutions exist, the two most common being wireless technology and wired options—such as copper and fiber-optic cable. Wireless and wired systems communicating with other devices of the same ability have been one of the fastest-growing areas in many industries, because the challenges of wireless networks in automation are:

- Wireless networks differ significantly from traditional automation networks.
- Wireless networks are never 100 % reliable.
- Real-time communication underlies restrictions.

The opportunities of wireless networks in automation are:

- Mobility of systems and users
- Ad hoc communication
- Ad hoc configuration of field devices

Hence, wireless technology can improve data transmission much more than wired options because wireless enables distribution automation improvement, e.g., for the smart grid which can improve overall smart grid performance.

Using wireless communication significant progress has been made in many domains, such as (Chen at al. 2011a, b; Wan et al. 2012):

- Machine-to-machine (M2M) communication: refers to a technology that allows wireless and wired systems to communicate with other devices. M2M is considered as an integral part of the Internet of Things (see Chap. 4) and brings several benefits to industry and business. In general it offers a broad range of industrial applications such as automation, logistics, smart grid, smart cities, and more, mostly for monitoring and also for control purposes.
- Wireless sensor networks (WSNs): refer to spatially distributed autonomous sensors that monitor physical or environmental conditions and cooperatively pass their data through the network to a main location. In their bidirectional sensor network form, they also enable controlling the sensor.
- Wireless body area networks (WBANs): refer to a wireless network of wearable computing devices, which can be surface-mounted on the body in a fixed position. A WBAN system can also use wireless personal area network technologies as gateways for longer distances. Through gateway devices, it is possible to connect the wearable devices on the human body to the Internet.

The rationale behind M2M communications is based on two observations:

- Networked machines are more valuable than isolated ones.
- Multiple machines are effectively interconnected and more autonomous, and intelligent applications can be generated.

Thus, the development of machine-to-machine (M2M) communication enables new opportunities for the manufacturing and the information industry. Besides these application domains, various other M2M opportunities have already started to emerge, like assisted living, health care, smart homes, smart grids, and more. Form this few examples, cyber-physical systems would adopt the areas of M2M and **WSN**s because more and more sensor inputs and richer network connectivity are needed in the near future because WSN, M2M, and cyber-physical systems belong to the Internet of Things.

3.4.2 Consumer Interaction

The cyber-physical system enables connections with consumers with regard to their interaction. Thus, an integration of consumer action to automated urban life concepts is possible based on:

- RFID embedded solutions
- Infotainment
- Banking
- Electronic equipment intended for everyday use
- Energy
- Wearable computing devices
- Others

which follow the trend of convergence, combining the many elements of products, allowing consumers to face different decisions when purchasing. There is an ever-increasing need to keep product information updated and comparable, for the consumer to make an informed choice. Style, price, specification, and performance are all relevant which can be readable lodged in RFID tags. Hence, a gradual shift toward e-commerce Web storefront solutions can be declared which mostly are business-to-business (B2B) and business-to-customer (B2C) solutions that can, e.g., streamline the order to cash process throughout an organization through greater collaboration with customers via the Internet.

In this regard the Internet of Things (see Chap. 4) and nowadays the Internet of Everything find and serve consumer product-related information over the Internet to consumers in retail shopping environments, as well as at home and work and on the road to improve customer service and reduce cost while maintaining an online web site product catalog or developing a completely new look in the context of a smarter commerce advanced product catalog. This application includes Internet

information servers which store information pertaining to the Universal Product
Code Number (UPCN) preassigned to each consumer product registered with the
system, along with a list of Uniform Resource Locators (URLs) that point to the
location of one or more information resources on the Internet, e.g., World Wide
Web sites, which are related to such registered consumer products. Among the
capabilities and benefits provided by e-commerce Web storefronts are the following
(http10 2015):

- Enterprise software support
- Business-to-business (B2B) and business-to-consumer (B2C) support
- Real-time end-to-end integration of Web orders
- Real-time enrollment of new customers into the database with initial terms
  limited to credit card payment
- Real-time credit card authorization with integration to back-office functions
- Real-time inventory availability to the customer
- Real-time order status and shipment tracking with UPS, FedEx, etc.
- Customer-specific, real-time product pricing based on customer pricing structure
  derived from a database and advanced pricing catalog
- Call center for complete customer satisfaction

3.4.3 Energy

An important challenge for future energy systems in the context of cyber-physical
systems is the blend of the cyber and physical aspects of energy systems which
result in a new methodology that integrates the computation of the cyber and
physical energy processes through networks communication and/or embedded
real-time system components. Hence, physical components of the energy system
have cyber capability and cyber resources whereby the software is embedded in the
energy system subsystems or physical components, and the overall system
resources such as computing, communication network bandwidth, etc., are usually
limited. This results in enabling strategic approaches to achieving future cyber-
controlled energy infrastructures, including monitoring, control, and protection
solutions in the context of cyber-physical energy systems (CPESs). These systems
will adapt itself using predefined problem resolution in case a problem occur which
finally means it should be autonomous. Thus, the main cyber-physical energy
system research focuses on:

- Modeling energy systems
- Energy efficiency
- Energy resource management
- Energy monitoring and control

and advantages of cyber-physical energy systems in future energy systems, the
so-called smart grids. Smart grids use computers to control the grid energy

generation, load, and distribution assets. These computers itself interact over large-scale communication networks (Lemmon and Venkataramanan 2009). In smart grids, the distributed generated energy and the microgrids play a critical role in order to address the modernization of the electricity network. The microgrid concept assumes a cluster of loads and microsources operating as a single controllable system that provides both power and heat to its local area (Lasseter 2002). This concept provides a new paradigm for defining the operation of distributed energy generation. Therefore, in the smart grid concept, grid stability and reliability are enhanced through reconfigurable control (Lemmon and Venkataramanan 2009), and the energy systems must exhibit adaptive performance. To achieve this performance, cyber technologies capable of monitoring, communicating, and controlling the evolving physical system have to be embedded (Farhangi 2010). An initial strategy to obtain the adaptive performance of the energy system includes the installation of a monitoring sensor grid. In order to address these classes of issues, a model approach for cyber-physical energy systems is evident to use for smart grids, as introduced in Macana et al. (2011).

To understand and be able to determine the dynamic response of energy systems modeling is the best choice to reinforce an integrating view of the several entities representing the respective energy system. To be able to build models of such complex systems which successfully combine computational elements with elements from the physical world, it is important to use a new, integrated approach to the design like Modelica and Modelica-based tools which have good environments for such integrated model-based development. Modelica is a universal modeling concept (Tiller 2001) which has object-oriented facilities, and a large set of libraries in several physical domains allow for rapid prototyping of multidisciplinary applications. In Elsheikh et al. (2013), a Modelica-enabled rapid prototyping of a cyber-physical energy system via a functional mockup interface is introduced as a use case in modeling energy systems.

Looking at the energy system of the future, it is expected that this be composed of a large variety of technologies and applications. Hence, such systems can be introduced as physical-cyber network interconnections of many nonuniform components, such as diverse energy sources and different classes of energy users, equipped with their own local cyber (Ilic et al. 2008), which is qualitatively different from the currently used models that do not explicitly account for the effects of sensing and communications. However, not all physical components can usually be modeled because of the nonuniformity and the complexity of various classes of components. These components must be monitored and their models have to be identified. However, the diverse nature of the components to be considered, their interlinked topology, and the sheer size of the system lead to an unprecedented level of complexity which will result in severe problems in designing interoperable grid components, analyzing system stability, and improving efficiency. These require continuous time-based and discrete event-based models to describe these types of cyber-physical energy systems as introduced by Palensky et al. (2014).

Different modeling approaches referring to cyber-physical energy systems focusing on smart grid are published in Khaitan et al. (2015), which also contain a chapter on cyber attacks in the automatic generation control.

The monitoring and control strategies of the cyber-physical energy system can include networked-based control (Kottenestette et al. 2008), multiagent modeling and multiagent-based control (Lin et al. 2010), and online prediction (Zhang et al. 2009). All of these strategies consider that networks communication and/or computing have throughput limitations constraining what information may be communicated between the components. These approaches include two main aspects to address the current challenges in energy system issues (Macana et al. 2011):

- Communication and computing constraints: Networks communication constraints, such as time delays and throughput, can degrade a control energy systems performance and even cause system instability (Lemmon and Venkataramanan 2009; Yang 2006). The cyber-physical systems approach can consider both the physical constraints, as the network, and the computing constraints.
- Computing efficiency: Many optimization goals of current energy systems are based on reducing power consumption as an alternative to problems such as climatic changes and sustainable development. New cyber-physical tools offer an approach addressing computing efficiency objectives.

Today's power grid generators can be classified in three categories:

- Base load: run 24/7 to provide minimum demand level with good efficiency
- Intermediate load: run often to satisfy the average demand needs
- Peaking load: run sparingly to satisfy maximum load with generally poor efficiency

The forgoing mentioned mechanism does not work well when adding renewable energy sources in the mix. Many of the renewable sources are not flexible and the weather cannot be controlled in an appropriate way. Moreover, the modernization of the current power grid implies linking it with new concepts of communications, control, and computing which result in a new power grid generation called smart or intelligent grid (Farhangi 2010). The smart grid uses computers to control grid generation, load, and distribution assets, whereby the computers interact over large-scale communication networks (Lemmon and Venkataramanan 2009). In smart grids distributed generation and microgrids play a critical role to address the modernization of the electricity network as appropriate. The microgrid concept assumes a cluster of loads and microsources operating as a single controllable system that provides both power and heat to its local area (Lasseter 2002). This concept provides a new paradigm for defining the operation of distributed generation.

Several architectures have been proposed for the concept of microgrids showing an interconnection like distributed energy sources, such as microturbines, wind turbines, fuel cells, and photovoltaic panels integrated with storage devices, such as batteries, flywheels, and power capacitors on low-voltage distribution systems. This type of architectures highlights the presence of inverters direct current-to-alternate current (DC-AC) and alternate current-to-direct current (AC-DC) in the majority of connections of distributed generators. However, in smart grids grid stability and reliability are enhanced, as mentioned before, through reconfigurable control (Lemmon and Venkataramanan 2009), and the energy systems must exhibit adaptive performance (Macana et al. 2011).

3.4.4 Infrastructure

Cyber-physical systems often require a reliable and real-time wireless network infrastructure for sensing, communication, and actuation. Reliability can be achieved through mesh networking, channel hopping mechanism, and data link layer acknowledge. Real-time behavior requires at least a centralized data management and Time Division Multiple Access (TDMA) data link layer. In TDMA a base station coordinates access to one or more channels in its cell. However, if power is based on batteries, the short battery lifetime performance of wireless devices is becoming a critical factor. To prolong battery lifetime, one can increase the battery capacity, a problem which has no solution so far, or reduce energy expenditure. Since wireless devices energy expenditure may depend on the distance from its associated base station, the base station placement optimization is one of the most effective methods to address the battery lifetime issue. As shown in Liu et al. (2012) the optimization is a mixed-integer non-convex optimization problem which is NP hard meaning that no off-the-shelf optimization methods can be properly applied. To overcome this problem, the authors recommend an efficient algorithm called expansion-clustering-projection-contraction (ECPC).

From this example, it can be seen that focusing on control system aspects pervades critical infrastructures essential for reliable and real-time infrastructure services in cyber-physical systems, such as:

- Time Division Multiple Access: guarantee timely delivery
- Channel hopping and blacklisting: spread communication in all active physical channels and reduce interference to provide reliable communication
- Confidential and secure communication: use public and private keys to secure communication in joint processes and regular operations

Cybersecurity standards are essential, because sensitive information is stored on computers that have access to the Internet and many tasks are carried out by computers. Thus, in the 1990s, the Information Security Forum (ISF) published a comprehensive list of best practices for information security called Standard of Good Practice (SoGP). The ISF continues updating the SoGP; the latest version was published in June 2014 (http14 2015).

Today critical networking infrastructure and data management for cyber-physical systems are facilities for:

- Agriculture, food production, and distribution
- Electricity generation, transmission, and distribution
- Financial services (banking, clearing)
- Gas production, transport, and distribution
- Heating (e.g., natural gas, fuel oil, district heating)
- Oil and oil product production, transport, and distribution
- Public health (hospitals, ambulances)
- Security services (police, military)
- Telecommunication
- Transportation systems (fuel supply, railway network, airports, harbors, inland shipping)
- Others

Solutions for these topics can make use of infrastructure concepts such as near field communication (NFC) and IT code exchanges (QR codes).

Near field communication works using magnetic induction which means that a reader emits a small electric current which creates a magnetic field that in turn bridges the physical space between the devices which want to communicate. That field is received by a similar coil in the client device where it is turned back into electrical impulses to communicate data such as identification number status information or any other information. So-called passive near field communication tags use the energy from the reader to encode their response, while active or peer-to-peer tags have their own power source embedded and respond to the reader using their own electromagnetic fields.

QR is the abbreviation for quick response code, as the creator intended the code to allow its contents to be decoded at high speed. As an example: in 2006 the world's fastest SMS typer could manage 160 characters. An Internet address (URL) can be 250 characters long. To type an Internet address into the mobile browser would take about a minute for the fastest SMS typer at this time. Against this background, the market launch of the QR code becomes understandable, because the code replaces retyping of information. Instead of typing the information, the QR code is scanned. This requires a Java-enabled smart device or a smart device with a camera. The smart device camera photographs the code, and a specific applet in the smart device analyzes the code and decodes the information deposited on the screen of the smart device. In technical terms the QR code is a specific matrix barcode, readable by dedicated QR barcode readers and camera-based smart devices. The code consists of boxy modules (mostly black) arranged in a square pattern on a white background. The information encoded can be text, URL, or other data. Most of the time a QR code looks like this:

which is the QR code of the author and its university institution. QR codes are storing addresses and URLs may appear in magazines and on signs, buses, business cards, or just any object where users might need information. Hence, users with a camera-equipped smart device with the correct downloaded QR code reader application can scan the image of the QR code to display text and contact information, connect to a wireless network, or open a Web page in the smart device browser which gives reason for what the term QR code stands for: quick response code.

Recently much more advanced infrastructure concepts show up like transforming mobile devices into banking channels meaning bridging the existing gap for transactions between private individuals and the banking business which is not possible yet. For this purpose Apple is turning into a full-scale banking business and also entering another new realm—personal ID—because behind client's credit card and banking capacity, all elements needed for personal ID are available. ID cards are going to follow credit cards into smartphones. Thus, Apple announced the creation of a unique, potentially very lucrative financial infrastructure. By making a 0.15 % charge on each transaction, Apple will create a multibillion dollar business which is a value transfer away from fraud, which accounts for 5.5 billion US$ in the USA.

3.4.5 Health Care

The recent advances in sensors, low-power digital circuits, and wireless communications have enabled the design of low-cost, miniature, and lightweight sensor nodes which make cyber-physical systems an important candidate for health-care applications allowing inexpensive, noninvasive, continuous, in-hospital, and ambulatory as well as in-home health monitoring with almost real-time updates of medical records via the Internet (Milenković et al. 2006). These advances promise to provide cyber-physical systems the ability to observe patient conditions remotely and take actions regardless of the patient's location. Collected or cached sensor data and/or queries from other systems are sent to a gateway via the wireless communication medium. But security is a vital concern as patient data is confidential from legal and ethical perspectives. Hence, in designing cyber-physical system architectures for health-care applications, special attentions are needed to ensure data security which can be achieved by data integrity on dynamic-link library (DLL) and/or data confidentiality on network layer. There are also a number of other important issues to consider, for example, the requirement to store and manage the huge volume of data collected from many medical sensors. Thus, database management systems should be efficient and reliable. Moreover, medical data can provide useful insight to treatments necessary to save a patient's vital or life situation, and all data should be readily available and accessible to authorized medical personnel anytime from anywhere. In addition, health-care applications require huge computing resources for intelligent decision-making based on the big/smart data (patient), but the research on cyber-physical systems in health care is still in the early stages (Haque et al. 2014); hence, the number of cyber-physical systems proposed for health-care applications is negligible (Lee et al. 2012; Wu et al. 2011). Notable cyber-physical systems applications in health care reported in Haque et al. (2014) are:

- Electronic medical records (EMR): design of a cyber-physical system interface for automated vital sign readings (Mendez and Ren 2012).
- Medical cyber-physical systems and big data platform: big data processing framework for medical cyber-physical systems (Don and Dugki 2013).
- Fall detection system: detecting fall of elderly people by using an accelerometer on the head level identifying the fall via algorithms (Wang et al. 2008).
- HipGuard: analysis for the posture recovery period after hip replacement; detect position of the hip and applied load on it; alarm is raised and clinicians are notified if any harmful movement or load is applied to the operated hip (Iso-Ketola et al. 2008).
- MobiHealth: gathers data from the wearable sensor devices people carry all day (Konstantas and Herzog 2003).
- AlarmNet: wireless sensor network system prototype consisting of heart rate, oxygen saturation, and electrocardiogram (ECG) (Wood et al. 2008).
- Mobile ECG: system that uses smartphones as base station for ECG measurement and analysis (see Fig. 3.6) (Kailanto et al. 2008).

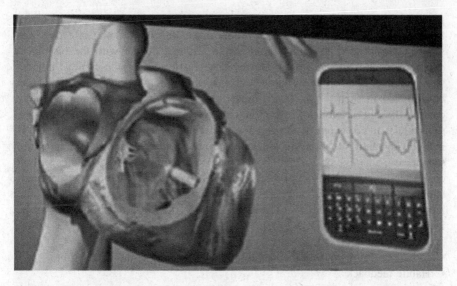

Fig. 3.6 Wireless pacemaker implanted directly into the heart chamber and extracorporeal monitoring by a nearby smartphone (http15 2015)

- iCabinet: utilizes RFID packing that can record the removed pill by breaking an electric flow into the RFID circuit (López-Nores et al. 2008).
- Secure and scalable cloud-based e-health architecture: for collecting and accessing large amount of data generated by medical sensor networks (Lounis et al. 2012).
- WSN cloud-based automated telemedicine: architecture of wireless sensor network (WSN), cloud integration mechanism, and dynamic collaboration between clouds to enable e-health care services (Perumal et al. 2012)
- CPS-MAS: cyber-physical medical systems modeling and analysis framework for safety verification (Banerjee et al.2011)
- Others

Nevertheless, computing technologies are the computational backbone of cyber-physical systems to improve the scalability of the system and to enable real-time data analysis in the manifold of medical applications integrating and managing the diverse medical device requirements and optimizing treatment delivery, to name a few. A rapidly growing field of cyber-physical research in health care focuses on environments that can exploit ubiquitous monitoring technologies to make assisted living and home care safer and more capable.

As reported in NITRD (2009), the direct consequence in prosthetic devices will depend upon sophisticated information and control technology. In devices such as artificial limbs, hands, or feet, the potential scope for intelligent prosthetics is just beginning to be explored, e.g., the possibility that a patient can control the prosthetic device through his own patterns of brain activity. If these technologies become used in many patients, the variety of challenges for safe, effective, and secure controlling of the interaction of the artificial device with the biological environment must be addressed. However, sophisticated prostheses must interact well with their users. If a prosthetic device has multiple functions, the system must be designed in such a way that the user stays aware of, and can appropriately control, the current mode of operation. Otherwise, mode confusion may cause the user to take an erroneous action and potentially induce harm because the patient is in highly dynamic situations, in which dangerous conditions can develop rapidly. In such a case, adaptive observer strategies need to be embedded to guarantee system controllability and observability (see Sect. 1.3). For many situations, visual feedback will not be possible, and haptic feedback will provide more information and increase system complexity. Therefore, some critical research areas can be identified including (NITRD 2009):

• Mode confusion causing the user to do the right thing at the wrong time
• Haptic feedback
• Actuating
• Intelligent prosthetics
• Data fusion and synchronization
• Human in the loop
• Adaptive control

Another important health-care domain is minimally invasive diagnostic and intervention technologies like less invasive methods for obtaining biometric data and performing surgical interventions, an important domain with regard to the risk of infection and collateral injury, as well as patient recovery time which can be lowered when large surgical wounds can be avoided. Precision microsurgery is already benefiting from the march of technology, with medical device miniaturization now reaching the scale of micro- and nanoelectromechanical systems, the so-called MEMS and NEMS (see Chaps. 4 and 5), with microelectromechanical systems (MEMS) ranging from millimeters through micrometers to nanometers (NEMS), like sensors that detect, for example, physical or chemical quantities or being integrated in smart clothes which is the integration into the cloth and of sensors, actuators, computing, power sources, etc., the whole being part of an interactive communication network.

Let us assume that such devices can be snaked through blood vessels to reach areas of the heart that control contraction rhythm. Let us also assume that such a device can perform microsurgery to induce a scar that reduces conductivity and alters the electrical signal to mitigate tachycardia, a common type of arrhythmia. Among the challenges for such devices is to avoid causing damage to the surrounding tissue and to offer precision control in the presence of varying natural motion and events in the complex human biological environment (NITRD 2009).

Hence, biological and medical advances can be expected opening new possibilities for interventions at all levels:

- System: cardiovascular, renal, respiratory, biomechanical, etc.
- Organ: heart, renal, liver, eye, ear, etc.
- Cellular: hormones, proteins, vitamins, amino acid lysine, etc.
- Molecular: B-1cell, IGM antibodies, etc.

The organ heart is composed of three major types of cardiac muscle: atrial muscle, ventricular muscle, and specialized excitatory and conductive muscle fibers. The atrial and ventricular types of muscles contract in a similar way as skeletal muscle fibers. The specialized excitatory and conductive fibers contract only feebly because they contain few contractile fibrils; instead they provide an excitatory system for the heart and a transmission system for rapid conduction of impulses throughout the heart. The period from the end of one heart contraction to the end of the next is called cardiac cycle. The cardiac cycle consists of a period of relaxation called diastole followed by a period of contraction called systole. Each cycle is limited by spontaneous generation of an action potential in the sino-atrial (S-A) node, a small crescent-shaped strip of specialized muscle cells in the top of the heart's right atrium. Hence, the S-A node can be introduced as pacemaker of the heart. Any action potential of the heart, which generates the electrical impulses that cause the heart to beat, begins in the S-A node. Therefore, a heart chamber contracts if an electrical impulse moves across it. To make the heart beat properly, the electrical signal travel down a specific path to reach the ventricles, the heart's lower chambers. In case of a disease, the heart can beat too fast (>80 beats per minute) called tachycardia, too slow (<60 beats per minute) called bradycardia, regularly (in between 60 and 80 beats per minute), or irregularly. Irregular heartbeats can be fast or slow and rhythmic or with no rhythm. Even irregular heartbeats can be regular in terms of a pattern, e.g., a beat that repeats itself every third beat, or chaotic rhythms that have no patterns at all, called irregularly regular. In case of bradyarrhythmias and tachyarrhythmias, the indication for a pacemaker is given. A pacemaker is a small device that is run by a battery. It helps the heart beat in a regular rhythm. Pacemakers can help pace the heart in cases of slow heart rate, fast and slow heart rate, or a blockage in the heart's electrical system. For that purpose the pacemaker system applies precisely timed electrical signals to induce heart muscle contraction and cause the heart to beat in a way very similar to a naturally heart rhythm.

Unlike traditional pacemakers, new pacemaker devices are completely self-contained and require no wires to connect it to the heart muscle. It's also implanted through a catheter, which bypasses the need for a chest incision. Medtronic launched a wireless pacemaker, revealed at TEDMED 2010 that can be implanted directly into the heart chamber via catheter and permanently latch itself into flesh with tiny claws. Then, medical doctors can wirelessly monitor and even control the functionality of the device through a nearby smartphone, as shown in Fig. 3.6.

3.4.6   Manufacturing

The purpose of manufacturing is to produce goods that can be sold to customers. Therefore, facilities are constructed to accomplish that goal. Capital, energy, human, information, and raw material resources are acquired, transported, and consumed in transforming the material into value-added products. The manufacturing system itself can be divided into five interrelated functions (Askin and Standridge 1993):

- Product design
- Process planning
- Production operation
- Material flow and facility layout
- Production planning and control

which include how machines are located and maintained, how parts are batched and dispatched, how performance is measured, and more. In this regard the first law of thermodynamics which states that energy is conserved in a system and the second law of thermodynamics which states that the entropy (disorder) of any system naturally increases through time can be used to describe the characteristics of a manufacturing system (see Chap. 7). Applying the first law of thermodynamics to manufacturing, it can be stated that the work in progress (WIP) is equal to the production rate (PR) multiplied by the throughput time (TI):

$$WIP = PR \cdot TI$$

Increasing the work in progress level by dispatching more material to the shop floor will increase the production rate as well as the throughout time. If the increased production rate reaches the capacity limit of the used machine of the shop floor, meaning the machine is fully utilized, work in progress will increase too resulting in a longer throughput time. The second thermodynamic law applied to manufacturing focuses on matter which is conserved, because manufacturing systems process materials from their raw to finished product state. This means in the long term that a stable manufacturing system cannot have any inventory accumulation; input must equal output (Askin and Standridge 1993).

Meanwhile, tremendous progress has been made in advancing cyber-physical systems technology over the last years in manufacturing lines, where machines can perform many work processes by communicating with work in progress, machine utility, components, and more. For this purpose cyber-physical systems use real-time capable and context-sensitive integration of sensors, and actors, and may incorporate distributed artificial cognition in interaction with human cognition.

In this regard the adoption of the Internet of Things (see Chap. 4) is an important element for smarter manufacturing because manufacturing will benefit through collecting data from wireless connected sensors and communicating that data after processing to used machines of the shop floor, factory floor workers, plant managers, and many other operational aspects of the supply chain. This result in building smarter manufacturing systems with the help of the Internet of Things, in the context of cyber-physical systems, the new paradigm Digital Manufacturing also called Industry 4.0 (see Chap. 7). In this context cyber-physical manufacturing systems integrate intelligent networks along the entire value chain that communicate and control each other autonomously with significantly lower intervention by plant operators. The following features of Digital Manufacturing/Industry 4.0 are essential and should be taken into account when implementing smarter manufacturing systems:

- Horizontal integration through value networks refers to topics that make business strategies and new value networks sustainable being supported by using cyber-physical systems.
- End-to-end digital integration of cyber-physical systems and software engineering across the entire value chain require appropriate IT systems to provide end-to-end support to the entire value chain, from product development to manufacturing systems engineering, production, and service.
- Vertical integration and networked manufacturing systems require integration of cyber-physical system to create flexible and reconfigurable manufacturing systems which mean that configuration rules will be defined to be used on a case-by-case basis to automatically build a structure of the machines of the shop floor for every situation, including all associated requirements in terms of models, data, communication, and algorithms.

An example for introducing vertical integration and networked manufacturing system can be seen in custom manufacturing and how individual customer's requirements can be met. Today's automotive industry is characterized by static production lines with predefined sequences which are hard to reconfigure to make new product variants. Software-supported manufacturing execution systems (MESs) are normally designed with narrowly defined functionality based on the production line's hardware and are therefore equally static. Let's assume tomorrow vehicles become smart products that move autonomously through the assembly shop from one cyber-physical system-enabled processing module to another. The dynamic reconfiguration of production lines makes it possible to mix and match equipment with which vehicles are fitted; furthermore, individual variations, e.g., fitting a seat from another vehicle series, can be implemented at any time in response to logistical issues such as bottlenecks (Industrie 4.0 2015).

In the textile manufacturing era, the textile process chains in high-wage countries are mostly described along the production chain. To adopt these textile process chains on the Digital Manufacturing level (Industry 4.0), information that flows through all levels of an enterprise needs to be connected to other entities of the textile process which enables a flexible and fast production, feasible to deal with an order of a lot size. Thus, in the cyber-physical systems world, machine communicates to each other and the plant operation can be realized. They inform about their status and upcoming problems such as maintenance. In this case, the factory will reconfigure itself in order to fulfill the customer's production order. Textile machines with open interfaces will be highly flexible and able to independently adapt status based on an overall information platform. Can, core, and warp beam and fabric will become carriers of information which lead to autonomic textile process chains (Gloy and Schwarz 2015).

A main aspect of production in the cyber-physical systems world is the human-machine interaction. The use of smart personal devices, such as smartphones, tablets, or head-mounted displays, offers a huge potential for innovation. Smart personal devices can be used to make production more transparent by providing relevant production key parameters in a sophisticated way. In addition, guidance programs can lead to optimized production or faster act in case of machine breakdowns. Also aspects of tele-maintenance, such as repair of machine supported by the machine produces, are possibly easier (Gloy and Schwarz 2015).

Self-optimization of the warp tension of textile machines is one of the innovative digital technologies in Digital Manufacturing. The aim of this method is to enable the loom to set the warp tension automatically on a minimum level without reducing the process stability. Therefore, as part of the self-optimization, the model of the process has to be set up to enable the loom to set the warp tension automatically on a minimum level without reducing the process stability and implemented in the weaving process. So far the loom is able to create its process model for a given process domain independently, than the machine run an experimental design, and automatically determined at respective test points the warp yarn tension. The operating point can be determined with the aid of quality criterion, such as that the warp tension becomes minimal (Gloy 2013).

In addition, further sensors are embedded in the weaving process. A system for automatic in-line flaw detection in industrial woven fabrics is discussed. This system operate on low-resolved ($\approx$200 ppi) image data, and the new cyber-physical system describes process flow to segment single yarns in high-resolved ($\approx$1000 ppi) textile images. This work is partitioned into two parts: First, mechanics, machine integration, vibration canceling, and illumination scenarios are discussed, based on the integration into a real loom. Second, the software framework for high-precision fabric defect detection is presented. The system is evaluated on a database of 54 industrial fabric images, achieving a detection rate of 100 % with minimal false alarm rate and very high defect segmentation quality (Schneider et al. 2012).

Some of the most relevant research topics on advanced manufacturing systems are:

- Advances in cloud manufacturing and supply chain
- Advanced modeling and coordinated control for industrial processes
- CPS for networked manufacturing processes
- CPS for networked manufacturing systems
- Intelligent sensing technology and instrumentations
- Internet of Things in smarter manufacturing

- Real-time production planning and adaptive control
- Smart factory and smart supply chain

For more details, see Chap. 7.

3.4.7 Military

Embedded computing systems (see Chap. 2) form a ubiquitous, networked, computing systems class. Such systems range from large supervisory control and data acquisition (SCADA) systems that manage physical infrastructure to military devices such as unmanned vehicles, to communication devices such as cell phones, to special vehicles such as satellites. Such devices have been networked for a variety of reasons, including the ability to conveniently access systems health information, perform software updates, provide innovative features, lower costs, and improve ease of use (http16 2015). Hackers have shown that these kinds of networked embedded computing systems are vulnerable to remote attack, and such attacks can cause physical damage while hiding the effects from monitors.

Therefore, the research focuses on a technology for the construction of high-assurance cyber-physical military systems (HACMS), where high assurance is defined to mean functionally correct and satisfying appropriate safety and security properties. Achieving this goal requires a fundamentally different approach from what the software community has taken to date. Consequently, this new HACMS approach adopts a clean-slate, formal methods-based approach to enable semiautomated code synthesis from executable, formal specifications. In addition to generating code, HACMS seeks a synthesizer capable of producing a machine-checkable proof that the generated code satisfies functional specifications as well as security and safety policies. A key technical challenge is the development of techniques to ensure that such proofs is composable, allowing the construction of high-assurance systems out of high-assurance components (http16 2015).

Key HACMS technologies include interactive software synthesis systems, verification tools such as theorem provers and model checkers, and specification languages. Recent fundamental advances in the formal methods community, including advances in satisfiability (SAT) and satisfiability modulo theories (SMT) solvers, separation logic, theorem provers, model checkers, domain-specific languages, and code synthesis engines, suggest that this approach is feasible. If successful, HACMS will produce a set of publicly available tools integrated into a high-assurance software workbench, which will be widely distributed for use in both the commercial and defense software sectors. HACMS intends to use these tools to (1) generate open-source, high-assurance, and operating system and control system components and (2) use these components to construct high-assurance military vehicles. HACMS will likely transition its technology to both the defense and commercial communities. For the defense sector, HACMS will enable high-assurance military systems ranging from unmanned vehicles (e.g., UAVs, UGVs, and UUVs) to weapon systems, satellites, and command and control devices.

With regard to the forgoing mentioned topics, the research challenges in the military domain are:

- Synthesis of attack-resilient control systems
- Synthesis of operating systems code
- Specification languages: function, environment, hardware, resources
- Composition
- Proof engineering
- Scaling
- Attack/fault response
- Verification and validation of complete system
- Managing time: synchrony, asynchrony, concurrency (Schneider et al. 2012)

For flight control systems, this results in:

- Synthesized flight control systems and proofs from embedded computing systems domain-specific languages (EDSL)
- Verified integrity of control flow
- Verified simple architecture extended to monitor security properties (Fischer 2013)

3.4.8   Robotics

Robotics, in different forms, are very powerful elements of today's industry. They are capable of performing many different tasks and operations precisely and do not require common safety and comfort elements humans need. However, it takes much effort and many resources to make robot function properly working. The motions of the robot are controlled through a controller that is under the supervision of a computer, which itself is running a program which contains the robot functionality. Thus, if the program is changed, the actions of the robot will change accordingly. The intention using robots usually is to use a device that can perform many different tasks and thus is very flexible in what it can do, without having to redesign the device. Hence, the robot is designed being able to perform any task that can be programmed simply be changing the program itself. Robots can be classified in different ways according to the Japanese Industrial Robot Association (JIRA), the Robotics Institute of America (RIA), and more. The advantages and disadvantages of robots can be introduced as follows (Niku 2001):

- Robots can, in many situations of automation, increase productivity, safety, efficiency, quality, and consistency of products.
- Robots are able to work in hazardous environments without the need for life support, comfort, or concern about safety.

- Robots don't need environmental comfort.
- Robots can work 24/7 without experiencing fatigue or being bored; they don't need medical insuration or a vacation.
- Robots have repeatable precision 24/7, unless something happens to them like a wear-out failure or a wear-out failure period.
- Robots work much more accurate than humans. New wafer-handling robots have microinch accuracies.
- Robots, although superior in certain senses, have limited capabilities in:
  - Actuators
  - Controller
  - Degrees of freedom
  - Fine motor skills with reference to dexterity of the effector
  - Real-time response
  - Sensors
  - Vision systems
- Robots are costly with regard to:
  - Initial cost of equipment
  - Installation costs
  - Need for peripherals
  - Need for training
  - Need for programming

However, robotic systems are an important category of cyber-physical systems. The ability of robots to interact intelligently with the environment rests upon:

- Embedded computation and communication
- Real-time control
- Perception of the ambient world around them

More advanced robotic systems will realize the vision of cyber-physical systems which include increasingly intelligent robotic systems like robotic surgery systems, robots for assisted living in smart homes, robot teams for exploration and emergency and rescue response, as well as autonomous driving vehicles which belong to intelligent ambient robotic systems. All such types of robots are telerobots and get their intelligence wireless and the functionality through networking. Sensors, actuators, microprocessors and/or microcontrollers, databases, cloud computing, and control software will work together without the need to be collaborated.

Considering the advantages and disadvantages, capabilities and limitations, the design methodology for cyber-physical robotic systems should support:

- Development activities such as:
  - Modeling
  - Simulation
  - Verification/testing

- Methodologies such as:
  - Simulation, which contains model test, model-in-the-loop, and rapid prototyping
  - Prototyping, which contains hardware-in-the-loop and software-in-the-loop
  - (Pre-)production, which contains robotic system test
- Supported tools
- Supported libraries
- Integrated toolchain
- Hard and soft real-time constraints

A pick-and-place smart manufacturing robot provides services of moving components either one or two positions over. The objective is to move a stack of components from one location to the next such that the final stack is ordered according to different components. The ordering is an emerging behavior based on local control for each of the different component. Stereoscopic vision of a video stream of the virtual world enables the robot to first locate where the stack of components is. Feedforward control allows a quick pick-and-place action. Feedback control is used to move the components between locations.

3.4.9 Transportation

The transportation systems sector is a vast, open, interdependent network moving millions of tons of freight and millions of passengers. Every day, the transportation systems network connects cities, manufacturers, and retailers by moving large volumes of freight and passengers through a complex network of roads and highways, railways and train stations, sea ports and dry ports, and airports and hubs (Möller 2014; Sammon and Caverly 2007). Thus, the transportation systems sector is the most important component of any modern economy's infrastructure in the globalized world. It is also a core component of daily human life with all of its essential interdependencies, such as demands for travel within a given area and freight transportation in metropolitan areas, which require a comprehensive framework in which to integrate all aspects of the target system. The transportation systems sector also has significant interdependencies with other important infrastructure sectors (e.g., energy sector). Transportation and energy are directly dependent on each other for the movement of vast quantities of fuel to a broad range of customers, thereby supplying fuel for all types of transportation. Moreover, cross-sector interdependencies and supply chain implications are among the various sectors and modalities in transportation that must be considered (Sammon and Caverly 2007).

The transportation systems sector consists of physical and organizational objects interacting with each other to enable intelligent transportation. These objects include information and communication technology (ICT), the required infrastructure, vehicles and drivers, interfaces for the multiple modes of transportation, and more (Torin 2007). Advanced transportation systems are essential to the provision of innovative services via multiple modes of transportation interacting and affecting each other in a complex manner, which cannot be captured by a single existing model of transportation systems traffic and mobility management.

Transportation systems models enable transportation managers to run their daily businesses safely and more effectively through a smarter use of transportation networks. But the transportation systems sector in today's open, interdependent network encompassing urban and metropolitan areas requires optimization of all operating conditions. This can be successfully achieved if the interactions between transportation modes, the economy, land use, and the impact on natural resources are included in transportation systems planning strategies.

The proposed future of multimodal transportation systems cannot be measured through planning alone. Mathematical models of transportation systems and mobility management, incorporating both real and hypothetical scenarios, should be embedded in transportation systems analysis, including the evaluation and/or design of traffic flows, determining the most reliable mode of operation of physical (e.g., a new road) and organizational (e.g., a new destination) objects, and the interaction between the objects and their impact on the environment. These mathematical models are fundamental to the analysis, planning, and evaluation of small-, medium-, and large-scale multimodal transportation systems (Cascetta 2009). The success of model-based scenario analysis can be evaluated by the resulting forecast or prediction of the transportation system response. An ideal design or operational methodology for a transportation system can be achieved using model-based analysis in conjunction with backcasting or backtracking. Thus, modeling and simulation can play a central role in planning, developing, and evaluating multimodal transportation systems, improving transportation efficiency and keeping pace with the rising demands for optimizing multimodal transportation systems.

Various simulation models capture different aspects of a transportation system enabling evaluation of complex simulation scenarios where each one represents a certain aspect of a transportation system or a certain operational strategy. Multimodal transportation systems models can be classified as (Möller 2014):

- *Supply models*: representing the multimodal transportation systems sector services used to travel between different operating points within a given area
- *Demand models*: predicting the relevant aspects of travel demand as a function of system activity and level of service provided by the transportation system
- *Assignment models*: using the objects of the multimodal transportation system assignments

As forgoing mentioned, personal mobility is the key to the success and prosperity of every country's economy. But the growing population in the world's largest megacities and the increasing amount of traffic is leading to paralysis. In the world's 30 biggest megacities, paralyzed traffic flows generate annual costs of more than 266 billion US$. The answer to how to get a grip on the problem of increasing passenger transportation lies within the paradigm of networked mobility by intelligently linking transportation data and modes, so people can quickly and easily use different mobility models as needed to get where they are going (Möller 2014). Integrated offers and a comprehensive management function will play a central role by bundling various options and offering services from an integrated transportation platform. These are the key findings of a recent study entitled *Connected Mobility 2025* by Roland Berger Strategy Consultants (Berger 2013).

Thus, tools are needed for the manual and the automated analysis of the various properties of transportation. Taking into account a vehicle model in transportation, it may be necessary to verify that:

- Automatic transmission control does not oscillate when the car is driving on a fixed-gradient road.
- Cruise control can stay within a certain error of the desired speed.
- Transportation expends energy only within reasonable bounds of what is expected to be needed under such circumstance.

This requires methods and techniques to the application of information to improve traffic control systems performance guaranteeing real-time properties when designing cyber-physical systems in the transportation systems sector. Hence, the integration of information and transportation processes in the context of control instruction information will result in:

- Information flow
- Control command
- Behavioral control utility of the traffic control system which is a need in cyber-physical systems-based traffic control
- Information flow among computer systems
- Traffic light and traffic sign systems (Fidêncio and Cota, 2014; Jehle 2014)
- Travelers in the traffic control system
- Others

Thus, integration of information and transportation processes can be realized as part of the traffic control process. Hence, future intelligent transportation systems are focusing on guaranteeing crash avoidance of autonomous cars and efficient algorithms for collision avoidance at traffic intersections and real-time motion planning for autonomous urban mobility, an important issue to realize smart cities (see Sect. 3.5). With regard to these topics, it has to be stated that the cyber-physical systems approach in transportation is not only about the application of methods of advanced information and communication theory, but it is also about the theory building new systems by integration of information and communication processes, physical processes, and the cyber which can be introduced as intelligent transportation systems. Thus, the development of future cyber-physical transportation control systems has to start from its reanalysis which requires an approach as introduced in Jianjun et al. (2013). In this regard cyber-physical transportation systems can make transportation control not only facing to the operation of the transportation system but also facing to people and their behavior in the transportation system directly.

Thus, for cyber-physical transportation systems, all information involved in the process to achieve the transportation control objective, whatever it takes, system, materials, or energy, rely on and in whichever ways they are presented, obtained, stored, transmitted, and processed. Hence, intelligent transportation systems (Weiland and Purser 2000) are envisioned to address the numerous challenges faced by the transportation sector. One category of solutions envisioned in intelligent transportation systems pertains to the real-time and reliable delivery of traffic-related information to drivers both for safety-critical applications (such as blind spot warnings during lane changing) and for applications that improve driving experience and help the environment (such as notification of congestion and rerouting advise that can help to alleviate traffic congestion and lost productivity) (Gokhale et al. 2010, Zhao and Cao 2008). Supporting these applications requires a thorough understanding of the intelligent transportation systems problem with its types of communication networks:

- Wireless network among vehicles for vehicle-to-vehicle (V2V) communication
- Wireless network that involves vehicles communicating with the road-side infrastructure (V2I)
- Predominantly wireline network that connects multiple infrastructure elements

Real-timeliness and reliability of information dissemination via vehicle-to-vehicle and road-side infrastructure communication is a hard problem due to multiple challenges as described in Gokhale et al. (2010). Hence, some challenges are imposed by the physics of the system including the wireless radio transceiver power, shared nature of the wireless channel, mobility of the vehicles, and density of the vehicles. Other challenges arise from the vagaries of the cyber infrastructure including behavior of protocols like IEEE 802.11 Media Access Control (MAC), Address Resolution Protocol (ARP), Internet Protocol (IP) addressing and routing, and the Transmission Control Protocol (TCP) retransmission and congestion control.

3.5 Smart Cities and the Internet of Everything

Access to digital technology and digital operating options is becoming more and more essential if considering the importance of smartphones, tablets, gadgets, and other smart devices. Every smartphone today contains more computing performance than the first Apollo rocket that flew to the moon in the 1960s. In today's new technological age, it is easier to access to information in much better quality with regard to speed and quantity. The reason for that is the Internet of Things (see Chap. 4), because it is increasing the connectedness of people and things on a scale that once was unimaginable. It is expected that by 2017, 3.5 billion people will be connected to the Internet, 64 % of them via mobile devices, and people and connected things will generate massive amounts of data estimated to 40 trillion gigabytes (http17 2015). Hence, connected devices outnumber the world's population today by 1.5–1. Therefore, the pace of Internet of Things market adoption is accelerating because of:

- Growth in analysis and cloud computing
- Increasing interconnectivity of machines and personal smart devices
- Proliferation of applications connecting supply chains, partners, and customers

An advanced form of the Internet of Things (IoT) (see Chap. 4) is the Internet of Everything (IoE) which is the networked connection of people, data, processes, things, and services which create the forgoing mentioned vast amounts of data and allow access to a manifold of innovative services which never has been thinkable and available before. When these data are analyzed and intelligently merged with the new innovative services, the possibilities of the Internet of Everything seem endless. In a white paper published by Cisco (http17 2015), three major concerns for the Internet of Everything have been identified:

- Internet of Everything will automate connections of systems and services: meaning the types of services that are offered will completely change. Hence, people must not for longer be proactive connected to the network. For example, elderly people can manage their health care from home rather than a hospital or nursing home getting automatic reminders to take medicine or sent automatic information to an emergency station if an elderly has struggled and fell down, which has been measured by a velocity sensor in the elderly's smart watch (Wang et al. 2008).
- Internet of Everything will enable personal communications and decision-making through intelligence embedded within sensors and devices: meaning the types of services that are offered by the Internet of Everything will completely change as well as the way they are delivered to citizens. For example, intelligence embedded within sensors will allow filtering out relevant information and even applying analytics to come up with an alert or warning to people with regard to the actual situation in a chemical factory.
- Internet of Everything will uncover new information and services through the deployment of sensors and other information-gathering devices: applying sensor fusion and multi-sensor fusion methods allows to identify information hidden so far.

Thus, the Internet of Everything will be the backbone for smart cities because they rely on connections and information to transform the quality of life of citizens. It is the vision of smart cities creating communities that become the places where people may want to live, to learn, to raise their kids, and more and where business seeks to invest. These require making up of many technology transitions, including the Internet of Things. In this context smart cities use information and communication technology; networks communication including the Internet and its most advanced form, the Internet of Everything; and sensors to automate routine processes and provide rapid and intelligent decision-making which will allow beside cost-saving much more efficiency with regard to the offered functions and services.

Smart cities are based on digital strategies which introduce the way on how to build more and efficient infrastructures and services, the intrinsic key factors of which, such as (http18 2015):

- Digitally enabled administration—e-government: ensuring that all appropriate public services are available online or over the smartphone to reduce unnecessary travel
- Digitally enabled information and communication center: ubiquitous broadband infrastructure of wired and wireless networks that generate a platform for fixed and mobile communication, digital infrastructure and systems throughout the city, and Internet connectivity in all spheres of smart cities activities
- Digitally enabled education: enables access to educational resources and instructors and improves collaboration between students and instructors. Remote access labs allow lab-based training minimizing the need to travel through integration of educational institutions within a smart digital transformed education platform.
- Digitally enabled grids and utilities: usually usage of real-time metering and control systems for smart grid electricity networks and gas and water utilities allows improved usage reporting and more effective alignment of supply and demand to reduce overall natural resource consumption
- Digitally enabled environment: ensuring that all appropriate activities are undertaken reducing CO_2 emission and addressing impact of climate change on Web portals that help people better understand
- Digitally enabled health care: use of integrated digital health records, medical consulting systems via remote diagnosis and telehealth care, in-home monitoring, and networked technologies to leverage medical professional's expertise and adoption to demographic developments through digital prevention systems and portals
- Digitally enabled transport and urban mobility: improving efficiency, reliability, and safety of transportation modes. For the road mode, this will include intelligent parking services, e.g., the closer a parking spot is to the center of the smart city, the higher the hourly costs, and smart traffic lights that adapt to the demand with regard to early morning and late afternoon commuting in and out a metropolitan area to shape and manage demands controlling traffic lights in response to real-time traffic flow data. Moreover vehicle tracking is an important issue in digitally enabled transport. Vehicle tracking systems are commonly used by fleet operators for fleet management functions such as fleet tracking, fleet routing, fleet dispatching, onboard information, and security. Along with commercial fleet operators, urban transportation agencies also use this technology for a number of purposes, including monitoring the schedule adherence of buses in service, triggering changes of buses' destination sign displays at the end of the line (or other set location along a bus route), and triggering prerecorded announcements for passengers. For the air mode, this will include intelligent runway incursion avoidance systems (Schönefeld and Möller 2012) and for the sea mode vessel traffic seaborne systems. Although portals and apps will help

people planning journeys, and/or to decide to take public instead of private transportation systems as well as supporting use of e-cars and e-bikes

- Digitally enabled industries: ensuring that construction industry minimizes the use of primary resources and prolongs lifetime of products, minimizes waste and emission during manufacturing, and uses more energy-efficient automation systems. Rethinking the design from its end, which means product design, is not only oriented along cost and functionality, but it also has to take into account which and how materials and components can be recycled for re-usage in new products.
- Digitally enabled crowdsourcing: open publishing of data sets by private and public agencies to promote transparency and stimulating crowdsourcing of innovation and coproduction of services.
- Others.

From a more general statement, smart cities are expected to be the future concept for municipalities around the globe, using the power of ubiquitous communication networks, highly distributed wireless sensor technology, and Semantic Web (inclusion of semantic content in Web pages), to create current and future challenges and intelligent new proactive services which require to communicate data to the information and communication centers to manage requests on demand. Hence, smart city technologies have to integrate and analyze the massive amounts of data to anticipate, mitigate, and even help to prevent serious problems like crime hot spots and others. Moreover, smart cities connect citizens at work or to local government and encourage more direct participation, interaction, and collaboration based on the opportunities offered by the Internet of Everything, the backbone of a smart city. Hence, the key factors that will shape and build the smart city reality are:

- Networks of interconnected computers to networks of interconnected things and/or objects which finally result in the ubiquitous connectivity which represents the essential infrastructure of the twenty-first century. Ubiquitous connectivity allows access to high-bandwidth, Internet, and mobile network connectivity anytime and anywhere.
- Things:
  - Have their own Internet Protocol addresses
  - Embedded in simple and/or complex components or systems
  - Use sensors to obtain information from the ambient and/or environment, e.g., food products that record temperature along supply chain
  - Use actuators to interact with, e.g., air conditioning valves that react to the presence of people in a meeting room
  - Others
- Refers to networked interconnection of everyday things.
- Described as a self-configuring wireless network of sensors whose purpose is to interconnect everything.

At a more practical level, a good resource highlighting smart cities exemplars is the Intelligent Community Forum (http19 2015).

3.6 Case Study: Cyber-Physical Vehicle Tracking System

Analyzing and designing cyber-physical systems such as a vehicle tracking system require an a priori knowledge of whether the system being analyzed or designed can be assumed being controllable, observable, and/or identifiable. Controllability, observability, and identifiability are important properties of systems (see Chap. 1). With regard to the analysis of linear systems, it can be said that a linear system is state controllable when the system input u can be used to transfer the system from any initial state to any arbitrary state in finite time. Moreover, a linear system can be said to be observable if the initial state $x(t_0)$ can be determined uniquely for a given output $y(t)$ for $t_0 \le t \le t_1$ for any $t_1 > t_0$. If a mathematical model of a system can be written in the state notation, the method of controllability, observability, and identifiability analyses can be used for model predictions.

Road traffic systems are an important part of the overall transportation system, which includes not only a large number of human-made infrastructures, such as large bridges across lakes or rivers, long and big tunnels, urban elevated bridges, etc., but also a huge variety of vehicles, people, and goods in the complex transportation road environment. Especially, in urban traffic control systems, a large number of digital devices and information systems are available, as well as complex management and control systems. This allows developing road infrastructure cyber-physical systems, vehicle-road coordinated cyber-physical systems, traffic control cyber-physical systems, and cyber-physical vehicle tracking systems, respectively. The functions of these applications are shown in Table 3.1 (Jianjun et al. 2013).

3.6.1 Vehicle Tracking System

To achieve efficient and safe road transportation is one of the motivations to carry out the research on cyber-physical transportation systems as it possesses information and physical and other features with regard to the essential needs of a cyber-physical vehicle tracking system which is an important issue due to the growing numbers in vehicle volume in recent years in the public and the private sector. Public and private transportation are faced with the problem to transport increasing volumes of passengers and freight. Within this process, freight must be identified several times. Currently in many applications, barcode systems are in use to identify the freight and the respective vehicle to which the freight belongs. But these barcode systems have some weakness and repeatedly failed and the freight may end up in the wrong truck. Therefore, transporting the wrong freight with a truck means wasting resources. Thus, the convergence of ubiquitous computing with embedded computing systems like onboard units in trucks is an important milestone enabling large-scale distributed cyber-physical computing systems which are strongly coupled with their physical environment. Hence, radio frequency identification (RFID; see Chap. 4), as a component for wireless communication, becomes of great interest in transportation and logistics in the global economy business, i.e.,

Table 3.1 Function and constraints of cyber-physical transportation systems

	Physical traffic process	Information technology process	Functions
Road infrastructure cyber-physical system	Mechanics changing process of key transportation infrastructure like bridge, culvert, tunnel, subgrade, slope, roadside, etc.	Ubiquitous sensing in a wide range of reliable interconnected depth perception, forecast, warning, and monitoring	Real-time monitoring road facilities and transportation meteorological environment detection
Vehicle road coordinated cyber-physical system	Relation between synergic relationship process of car to car and car to road which are running in the road and communication process	Wireless, high speed, high reliability, security communications, automatic driving	High-speed information exchange to guarantee safety of vehicles in efficient access
Traffic control cyber-physical system	Road traffic system process and traffic control process	Traffic control systems model description, traffic system control, and traffic behavior control instruction optimization calculation	More secure and efficient dynamic road traffic control
Vehicle tracking cyber-physical system	Relation between car to truck and truck to road running in the road, communication, and traffic control process	Wireless, high speed, high reliability, security, communications of depth perception	High-speed information exchange to guarantee real-time monitoring of vehicles in efficient access

in process optimization, in freight transportation, in the transportation, and in the logistics domain.

With the emergence of the recently released 6LoWPAN (Mulligan 2007), the convergence between cyber-physical systems and the Internet of Things becomes a reality because it enables using the Internet as supportive infrastructure to sensor networks, similarly to its integration with RFID systems. This also allow beside tracking that freight is transported with the appending truck tracking the position of the truck with regard to delivery on demand at the right destination of the truck, and more. Tracking in this sense means that RFID readers are used to monitor RFID-tagged vehicle movements. With regard to the term vehicle, any mobile item used to carry freight or passengers is meant. Thus, various kinds of pallets, forklifts, and other put-away and load units fall under this category, as well as various passenger cars and cargo trucks.

RFID tracking application in transportation and logistics in general is implemented in order to gather up-to-date information of tagged freight and their movements, facilitating effective and in-time management. For this reason, we would remind the problems of stolen or lost freight, as well as freight delivered incorrectly or with significant delays. This problem should not be underestimated because it has a huge impact on time and money spent for developing a performing stand-alone system for tracking freight for successful business operation. Therefore, RFID might be a stepping stone to achieving success in this field.

Without demonstrating other technologies, it will be stated that RFID does not require having a line of sight being established. Furthermore, RFID tags are resistant against environment impact, such as physical interaction with other items. Moreover, RFID supports multi-object recognition, so that several tags can be read simultaneously. However, it is possible to extend the list of RFID advantages, but there are still some potential drawbacks that have to be kept in mind when using this technology for vehicle tracking.

In order to mitigate the risk of unsuccessful RFID implementation, there should be a comprehensive requirements analysis performed beforehand.

3.6.2   RFID-Based Vehicle Tracking System

Vehicle tracking systems are commonly used by fleet operators for fleet management functions such as fleet tracking, fleet routing, fleet dispatching, onboard information, and security. Along with commercial fleet operators, urban transportation agencies also use this technology for a number of purposes, including monitoring the schedule adherence of buses in service, triggering changes of buses' destination sign displays at the end of the line (or other set location along a bus route), and triggering prerecorded announcements for passengers.

With regard to the forgoing mentioned, vehicle tracking systems can also be understood as an integrated part of a layered approach to vehicle protection, recommended by the National Insurance Crime Bureau (NICB) to prevent motor vehicle theft. This approach, which is not considered here, requires four security layers which are based on the risk factors pertaining to a specific vehicle. Vehicle tracking system is one such layer and is described by the NICB as very effective in helping police recover stolen vehicles.

In order to investigate the requirement for RFID-based vehicle tracking, several RFID vehicle tracking application use cases were identified for further consideration. These are:

- Tracking tagged load units as a part of logistics and supply chain management
- Implementing RFID vehicle tracking in road systems

To illustrate both cases, there is one simplified figure provided in Fig. 3.7, showing common RFID system structure and its interaction with other system components (Deriyenko 2012).

The first tracking application belongs to freight in logistics and supply chains. Here freight items are usually tracked at several stages as they pass through the business workflow process. For better transparency, it is possible to perform tracking every time the freight approaches and leaves each stage.

There are more several ways using RFID systems by integrating them in road systems. The first example was implementing RFID wagon tracking by Finnish

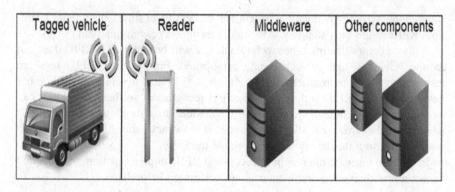

Fig. 3.7 RFID-based vehicle tracking system after Deriyenko (2012)

railroad operator (Wessel 2011). Setting up readers along railroads allows generating more precise information about the train location at particular moment of time. Another way of using RFID systems on road systems is the system of automatic payment collection on toll roads, which makes it possible to overcome such severe problem as traffic jams at toll points and to reduce labor costs (Xiao et al. 2008). The system consists of onboard units based on RFID tags usually fixed on windshields or bumpers of moving vehicles and RFID readers located at the toll stations. To assure effective work of the system, each tag should be associated with corresponding payment account. In a sunny-day scenario, as tagged vehicle enters the toll area and passed by the reader, its tag is requested to provide information for identification. Once the information is read by the reader, money is charged from the linked account.

Other examples to illustrate the usage of RFID in road systems are managing parking lots and tank stations (Pala and Inan 2007; Mathis 2012). However, these approaches are really close to the one with toll collection.

3.6.3 Requirements Analysis

The requirements analysis for an RFID-based vehicle tracking system is essential to identify at the one hand the most relevant system constraints and on the other hand the essential knowledge required conducting a systems design including RFID readers, RFID tags, and RFID middleware. But the RFID-based vehicle tracking system should not only gather data, but it also needs to preprocess them according to specific business operational rules. Against this background, the requirements analysis is based on analyzing available research projects and research papers. Out of them the following requirements are identified (Deriyenko 2012):

- *Integration*: The main goal of the RFID-based vehicle tracking system is providing the stakeholders in transportation and logistics with valuable, complete, and reliable information on time and in a convenient form. Creating a system combined of RFID tags and readers for vehicle tracking results in generating certain amount of data, in some cases quite significant one. Gathering this data and storing them as stand-alone does not make any sense; it has to be processed and presented to the user. This results in an obvious requirement that the RFID-based vehicle tracking system should be integrated with other enterprise informational system components with the help of a middleware in order to provide them with data for further business operational use.

- *Data cleanup*: Gathered data in its pure form most likely are not user-friendly and would not be of a great value for the user. Due to this reason in most cases, a preprocessing according to particular business needs is required. Therefore, the middleware (see Fig. 3.7) should not only be part of an interconnection bus but has also to perform such pivotal functions as data cleanup, deleting duplicates, ordering, arranging data against selected granularity level, and carrying out other preprocessing operations, preparing the data for the respective business usage.

- *High throughput*: Implementation of RFID components is initially aimed on increasing system throughput ability; therefore, successful operating with high amount of vehicles tracked within certain period of time is one of the most central requirements for vehicle tracking. Obviously, the importance of this requirement and concrete indicators directly depend on particular business operational needs and constraints that have to be taken into account, in particular available budget resources.

- *Real-time operation*: Along with ability to let high amount of items quickly pass through, the RFID-based vehicle tracking system should have feasible ways to provide the data about them to the user. In most cases, retrospective data gathered by tracking vehicle has a certain value for the business operation, but its importance cannot be compared with regard to real-time operational data. Updating information received from readers within short bounded time is a vital requirement for vehicle tracking. However, the update rate may fluctuate according to business operational specifics. Considering the use-case toll collection: Obviously, there should be no significant delays with proceeding vehicle data in order to perform payments. The same can be said for logistics and supply chain management activities: Users should be able to access the most up-to-date information about vehicle movements; otherwise, the whole system loses its advantage. Therefore, the whole RFID-based vehicle tracking system has to assure short response time. With regard to Fig. 3.7, this requirement, however, requires that other enterprise system components, which means more than the one considered within this use case, have to be adapted to this needs.

- *Reliability*: Reliability of RFID systems depends on a number of influencing factors including radiofrequency interference, technical infrastructure, reader and tag configuration and placement, etc. In general problems can be arising while tracking can be roughly divided into two groups: false positives and false negatives. False positives owe their name to their origin: These are the situations when the system treats items as present, while in reality it is absent or should not be taken into consideration. Thus, positive situation with item presence in the system is actually false. This can happen due to several reasons. On the one hand, it is possible being confronted with the situation that one item can be scanned two times either by the same or by different readers. One of the solutions for that problem, mentioned in the literature, is to force tags to respond only in case when their first digits match the digits requested by the reader. However, this solution makes the whole system more overwhelmed. On the other hand, it is important to avoid readers signal collision as a result if their reading range is overlapping. Another situation that can be identified is scanning a tag that is supposed to be located beyond reader's reading range. Therefore, tags and reader positions should be controlled properly with technical indicators of both devices, as well environment specifics.

Nevertheless, some of the problems mentioned can be solved at a certain level of data preprocessing. But in general the RFID-based vehicle tracking system should embed a feature reducing the amount of false positives for real-time broadcasting by using effective anticollision solutions and requesting algorithms and other adequate approaches.

One of the most frequently mentioned problems that fall into false-negative category is the presence of metal or water, affecting tag readability which can be a problem for both cases. The reason for that is very simple because freight-tagged load may contain pallets carrying, for example, bottles with water. Also mistracking of vehicles can cause numerous inconveniences and result in additional business operational costs.

To overcome these problems, there might be several reasonable solutions applied, for example, using metal as antenna or changing antenna impedance. In any case, regardless of the used approach, the RFID-based vehicle tracking system should be able to overcome obstacles, such as metal and water, preventing tags from being read.

Moreover, the RFID-based vehicle tracking system must be able to detect if some of its components is down. This means that it has to have appropriate user notification algorithms. As mentioned earlier, the reader can perform a request using first digits of the tag identification number. Theoretically, such approach can help reveal necessary tag absence or failure. However, this solution is applicable not for all cases of RFID-based vehicle tracking, since there should be all tags' identifications stored in the system.

For example, it can help to detect if some load pallet is missing or its tag is not readable, but for obvious reasons, this is unfeasible for the toll collection use case. However, in the latter situation, missing tag functionality is not necessary, since the tag absence or breakdown will be identified anyway due to the car disability to pass the barrier gate without it. Anyway, the requirement mentioned above can be optional and refers not only to tags monitoring but to readers and middleware as well.

3.6.4 Further Research

The primary goal of this case study was to introduce to cyber-physical vehicle tracking and summarize the requirements for RFID-based vehicle tracking with regard to cyber-physical and ubiquitous computing systems approaches (Moeller et al. 2015). The requirements identified are allocated at high level and are suitable for further investigation and a deeper level of detail. At the same time, as RFID technology continues to increase its market share and getting into new fields of applications. Thus, the requirements analysis cannot be a one-time action; it has to take a continuous character. Thus, a great prospective for further research activities is available in this field.

3.7 Exercises

What is meant by the term *cyber-physical system*?
Describe the characteristics of a cyber-physical.
What is meant by the term *Internet of Things*?
Describe the characteristics of the Internet of Things.
What is meant by the term *physical world*?
Describe the characteristics of the physical world.
What is meant by the term *virtual world*?
Describe the characteristics of the virtual world.
What is meant by the term *systems engineering*?
Describe the characteristics of systems engineering.
What is meant by the term *software engineering*?
Describe the characteristics of software engineering.
What is meant by the term *machine-to-machine communication*?
Describe the characteristics of machine-to-machine communication.
What is meant by the term *wireless network*?
Describe the characteristics of wireless network.
What is meant by the term *wireless sensor network*?
Describe the characteristics of wireless sensor network.
What is meant by the tem *Internet Protocol version 6*?
Describe the opportunities of this manifold of addresses.
What is meant by the term *wireless body area networks*?
Describe the characteristics of wireless body area networks.
What is meant by the term *microelectromechanical systems*?
Describe the characteristics of microelectromechanical systems.
What is meant by *horizontal integration through value networks*?
Describe the characteristics of horizontal integration through value networks.
What is meant by *vertical integration in manufacturing systems*?
Describe the characteristics of vertical integration in manufacturing systems.
What is meant by the term *smart city*?
Describe the characteristics of the smart city.

What is meant by the term *Internet of Everything*?
Describe the characteristics of the Internet of Everything.
What is meant by the term *digitally enabled transport*?
Describe the characteristics of the digitally enabled transport.
What is meant by the term *digitally enabled health care*?
Describe the characteristics of the digitally enabled health care.
What is meant by the term *digitally enabled grids and utilities*?
Describe the characteristics of the digitally enabled grids and utilities.
What is meant by the term *digitally enabled industry*?
Describe the characteristics of the digitally enabled industry.
What is meant by the term *digitally enabled crowdsourcing*?
Describe the characteristics of the digitally enabled crowdsourcing.
What is meant by the term *vehicle tracking*?
Describe the characteristics of the cyber-physical vehicle tracking system.

References

(Abelson and Sussman 1996) Abelson, H., Sussman, G. J.: Structure and Interpretation of Computer Programs. MIT Press, 1996

(Askin and Standridge 1993) Askin, R. G., Standridge, C. R.: Modeling and Analysis of Manufacturing Systems, John Wiley Publ. 1993

(Banerjee et al. 2011) Banerjee, A., Gupta, S. K. S., Fainekos, G., Varsamopoulos, G.: Towards modeling analysis of cyber-physical medical systems. In: Proceed. 4th International Symposium on Applied Sciences in Biomedical Communication technologie (ISABEL), pp. 154–158, 2011

(Berger 2013) Berger 2013. Connected Mobility 2025 by Roland Berger Consultants, 01/2013

(Burns and Wellings 2001) Burns, A., Wellings, A.: Real-Time Systems and Programming Languages. Addison Wesley Publ., 2001

(Cascetta 2009) Cascetta, E.: Transportation Systems Analysis, Springer Publ. New York, 2009 (Manheim 1979)

(Chen et al. 2011a) Chen, M., Gonzalez, S., Vasilakos, A., Cao, H., Leung, V.: Body Area Networks: A Survey. ACM/Springer Mobile Networks and Applications, Vol. 16, No. 2, pp.171–193, 2011

(Chen et al. 2011b) Chen, M., Leung, V., Huang, X., Balasingham, I., Li, M.: Recent Advances in Sensor Integration. International Journal of Sensor Networks, Vol. 9, No. 1, pp.1–2, 2011

(Davis 2011) Davis, A., M.: Riquirements Bibliography; http://www.reqbib.com/

(Elsheikh et al. 2013) Elsheikh, A., Awais, M. U., Widl, E., Palensky, P.: Modelica-enabled rapid prototyping of cyber-physical energy systems via the functional mockup interface. In: IEEE Workshop on Modeling and Simulation of Cyber-Physical Energy Systems. 2013, DOI: 10.1109/MSCPES.2013.6623315

(Deriyenko 2012) Deriyenko, T.: RFID Application in Vehicle Tracking, project work in ITIS class Internet of Things, TU Clausthal, 2012

(Don and Dugki 2013) Don, S., Dugki, M.: Medical cyber-physical systems and bigdata platforms. In: Proceed. Medical Cyber-Physical Systems Workshop, 2013

(Farhangi 2010) Farhangi, H.: The path of the smart grid. IEEE Power and Energy Magazine, Vol. 8, pp. 18–28, 2010

(Fidêncio and Cota 2014) Fidêncio, A. X., Cota, E.: Smart Traffic Light: A Simulation Model, Student Project Work, TU Clausthal, Germany, 2014

(Fischer 2013) Fischer, K.: High Assurance Cyber Military Systems (HACMS), DARPA Report 5/20/13

(Garlan et al. 2000) Garlan, D., Monroe, R. T, Wile, D.: Acme: Architectural Description of Component-Based Systems – Foundations of Component-Based Systems. Cambridge University Press, 2000

(Geisberger and Broy 2012) Geisberger, E., Broy, M.: Integrated Research Agenda Cyber-Physical Systems (in German), Springer Publ. 2012

Garland, M., Le Grand, S., Nickolls, J., Anderson, J., Hardwick, J., Morton, S., Phillips, E., Zhang, Y., Volkov, V.: Parallel Computing Experiences with CUDA. Micro IEEE, Vol. 28, No. 4, pp. 13–27, 2008

Gloy, Y.-S.: Modellbasierte Selbstoptimierung des Webprozesses, PhD Thesis, RWTH Aachen 2012, Published by Shaker Verlag 2013

(Gloy and Schwarz 2015) Gloy, Y. -S., Schwarz, A.: Cyber-Physical Systems in Textile production – the next industrial revolution? http://www.textile-future.com/textile-manufacturing.php?read_article=1829

(Gokhale et al. 2010) Gokhale, A., McDonald M. P., Drager, S., McKeever, W.: A Cyber Physical System Perspective on the Real-time and Reliable Dissemination of Information in Intelligent Transportation Systems. In: Network, Protocols and Algorithms, Vol. 2, No. 3, pp. 116–136, 2008

(Haque et al. 2014)Haque, A. A., Aziz, S. M., Rahmann, M. : Review of Cyber-Physical Systems in Helathcare. In: International Journal of Distributed Sensor Networks, Vol. 2014, Article ID 217415, 20 pages, http://dx.doi.org/10.1155/2014/217415

(Ilic et al. 2008) Ilic, M. D., Le, X., Khan, U. A., Moura, J. M. F.: Modeling Future Cyber-Physical Energy Systems. In: IEEE Power and Energy Society General Meeting – Conversion and Delivery of Electrical Energy in the 21st Century, pp. 1–9, 2008, and DOI:10.1109/PES.2008.4596708

(Industrie 4.0 2015) Industrie 4.0 2015. Secretariat of the Platform Industry, Recommendations for implementing the strategic initiative INDUSTRIE 4.0, Final report of the Industry4.0 Working Group, 2013

(Iso-Ketola 2008) Iso-Ketola, P., Karinsalo, T., Vanhala, J.: HipGuard: a wearable measurement system for patients recovering from a hip operation. In: Proceed. 2nd Internat. Conference on Pervasice Computing Technologie in Healhcare, pp. 196–199, 2008

(Jehle 2014) Simulation of a Traffic Light Junction. Student Project Work, TU Clausthal, 2014

(Jianjun et al. 2013) Jianjun, S., Xu, W., Jizhen, G., Yangzhou, C.: The analysis of traffic control cyber-physical systems. In: Procedia – Social and Behavioral Sciences, Vol. 96, pp. 2487–2496, 2013

(Kailanto et al. 2008) Kailanto, H., Hyvärinen, E., Hyttinen, J. : Mobile ECG measurement and analysis system using mobile phone as the basis station. In: Proceed. 2nd International Conference on Pervasive Computing Technologies for Healtcare, pp. 12–14, 2008

(Khaitan, et al. 2015) Khaitan, S. K., McCalley, J. D., Liu, C-C., Eds.: Cyber-Physical Systems – Approach to Smart Electric Power Grid, Springer Publ. 2015

(Konstantas and Herzog 2003) Konstantas, D., Herzog, R.: Continuous Monitoring of vital constans for mobile users: the MobiHealth approch. In. Proceed. 25th Annual International Conference of the IEEE Engineering in Medicine and Biology Society, pp.3728–3731, 2003

(Kottenstette et al. 2008) Kottenstette, N., Koutsoukos, X., Hall, J., Sztipanovits, J., Antsaklis, P.: Passivity-based design of wireless networked control systems for robustness to time-varying delays. In: Proc. Real-Time Systems Symposium, pp. 15–24, 2008.

(Krogh et al. 2008) Krogh, B. H., Lee, E., Lee, I., Mok, A., Pappas, G., Rajkumar, R., Sha, L. R., Vincentelli, A. S., Shin, K., Stankovic, J., Sztipanovits, J., Wolf, W., Zhao, W.: Cyber-Physical Systems: Executive Summary, 2008

(Lasseter 2002) Lasseter, R.: Microgrids. In: IEEE Power Engineering Society Winter Meeting, Vol. 1, pp. 305–308, 2002

(Lee 2008) Lee, E. A.: Cyber-Physical Systems: Design Challenges. Techncal Report No. UCB/EECS-2008-8, 2008

(Lee et al. 2012) Lee, I., Sokolsky, O., Chen, S.: Challenges and research directions in medical cyber-physical systems, In: Proc. IEEE, Vol. 100, No. 1, pp. 75–90, 2012

(Lemmon and Venkataramanan 2009) Lemmon, P. C. M. D., Venkataramanan, G.: Position paper – using microgrids as a path towards smart grids. In: New Research Directions for Future Cyber-Physical Energy Systems, 2009.

(LeValley 2013) LeValley, D.: Autonomous Vehicle Liability – Application of Common Carrier Liability. Seattke Univ. Law Report, Vol 36, pp.5–26, 2013

(Li et al. 2011) Li, W., Jagtap, P., Zavala, L., Joshi, A., Finin, T. : CARE-CPS : Context-Aware Trust Evaluation for Wireless Networks in Cyber-Physical System using Policies. In: Proceed. IEEE International Symposium on Policies for Distributed Systems and Networks (POLICY), pp. 171–172, 2011

(Lin et al. 2010) Lin, J., Sedigh, S., and Miller, A.: Modeling cyber-physical systems with semantic agents. In: Proc. IEEE 34th Annual Computer Software and Applications Conference Workshops (COMPSACW), pp. 13–18, 2010

(Liu 2000) Liu, J. W. S.: Real-Time Systems. Prentice Hall Publ., 2000

(Liu et al. 2012) Liu, J. Kou, T., Chen, Q., Sherali, H. D.: On Wirleless Network Infrastructure Optimization for Cyber-Physical Systems in Smart Buildings. In: Wang, X. L., Zheng, R., Jing, T., Xing, K. (Eds.): Lecture Notes Computer Science 7405, pp. 607–618, Springer Publ. 2012

(López-Nores et al. 2008) López-Nores, M., Pazos-Arias, J. J., Garcia-Duque, J., Blanco-Fernández, Y. : Monitoring medicine intake in the networked home : the iCabiNet solution. In: Proceed. 2nd International Conference on Pervasive Computing Technologies for Healtcare, pp. 116–117, 2008

(Lounis et al. 2012) Lounis, A., Hadjidj, A., Bouabdallah, A., Challal, Y. : Secure and scalable cloud-based architecture for e-health Wireless sensor networks. In: Proceed. International Conference on Computer Communication Networks (ICCCN), pp. 1–8, 2012

(Macana et al. 2011) Macana, C. A., Quijano, N., Mojica-Nava, E.: A Survey on Cyber Physical Energy Systems and their Applications on Smart Grids, IEEE, 2011, DOI: 10.1109/ISGT-LA. 2011.6083194 Conference: ISGT-LA 2011

(Mathis 2012) Mathis, R.: Neste Oil launches automated vehicle identification at fueling stations, 2012. [Online]. http://secureidnews.com/news-item/neste-oil-launches-automatedvehicle-iden tification-at-fueling-stations/

(Mendez and Ren 2012) Mendez, E. O., Ren, S.: Design of cyber-physical interface for automated vital signs reading in electronic medical record systems. In. Proceed. IEEE International Conference on Electro/Information Techonolgy (EIT), 2012

(Milenković et al. 2006) Milenković, A., Otto, C., Jovanov, E.: Wireless sensor networks for personal health monitoring: Issues and an implementation. In: Computer Communications, Vol. 29, No. 13–14, pp. 2521–2533, 2006.

(Möller 2013) Internet of Things, Lecture ITIS Study Program TUC, 2013, http://video.tu-clausthal.de/vorlesung/408.html

(Moeller et al. 2015) Möller, D. P. F.: Introduction to Transportation Analysis, Modeling and Simulation. Springer Publ. 2014

(Moeller et al. 2015) Moeller, D. P. F., Deriyenko, T., Vakilzadian, H.: Cyber Physical Vehicle Tracking System: Requirements for using Radio Frequency Identification Technique. In. Proceed. IEEE EIT, 2015

(Mulligan 2007) Mulligan, G.: The 6LoWPAN architecture, In: EmNets Proceed. 4th Workshop on Embedded Networked Sensors, pp. 78–82, ACM, 2007

(Niku 2001) Niku, S. B.: Introduction to Robotics, Prentice Hall Publ. 2001

(Ning 2013) Ning, H.: Unit and Ubiquitous Internet of Things, CRC Press, 2013

(NITRD 2009) NITRD 2009. High Confidence Medical Devices: Cyber-Physical Systems for 21st Century Health Care. NITRD published by National Science Foundation 2006

(Pala and Inanc 207) Pala, Z., Inanc, N.: Smart Parking Applications Using RFID Technology, In RFID Eurasia, 1st Annual Confereance, 2007; DOI: 10.1109/RFIDEURASIA.2007.4368108

(Palensky et al. 2014) Palensky, P., Widl, E., Elsheikh, A.: Simulating-cyber physical energy systems: challenges, tools, and methods. In: IEEE Transactions on Systems Man and Cybernetics Part C (Applications and Reviews), Vol. 44, No. 3, pp. 318–326, 2014, and DOI: 10.1109/TSMCC.2013.2265739

(PCAST 2007) PCAST 2007. PCAST: Leadership under Challenge: Information Technology R&D in a Competitive World, PCAST by Executive Order 13226, Published 2007

(Pellizzoni 2015) Pellizzoni, R.: Cyber-Physical Systems: www.engineering. waterloo.ca; accessed January 2015

(Perumal et al. 2012) Perumal, B., Rajasekaran, P., Ramalingan, H. M : WSN integrated Cloud for Automated Telemedicine (ATM) based e-healthcare applications. In: Proceedings of the 4th International Conference on Bioinformatics and Biomedical Technology (IPCBEE), Vol. 29, pp. 166–170, 2012

(Raihans et al. 2005) Raihans A., Cheng, S.-W., Schmerl, B., Garlan, D., Krogh, B. H., Agbi, C., Bhave, A.: An Architectural Approach to the Design and Analysis of Cyber-Physical Systems. Electronic Communications of the EEAST, Vol. 21, pp.1–10, 2009

(Randell et al. 1995) Randell, B., Laprie, J.C., Kopetz, H., Littewoods, E., Eds.: Predictably Dependent Computing Systems, Springer Publ. 1995

(Sammon and Caverly 2007) Sammon, J. P. and Caverly, R. J.: Transportation Systems, US Department of Homeland Security, USA, 2007

(Sammon and Caverly 2007) Schmerl, B., Garlan, D.: AcmeStudio: Supporting Style-Centered Architecture Development. In: Proceedings of the 26th International Conference on Software Engineering, Scotland, 2004

(Schneider et al. 2012) Schneider, D., Holtermann, T., Neumann, F., Hehl, A., Aach, T., Gries, T.: In: Proceed. 7th IEEE Conference on Industrial Electronics and Applications, pp. 1494–1499, 2012. DOI: 10.1109/ICIEA.2012.6360960

(Schönefeld and Möller 2012) Schönefeld, J., Möller, D. P. F.: Runway incursion prevention systems: A review of runway incursion avoidance and alerting system approaches. In: Progress in Aerospace Sciences, Vol. 51, pp. 31–49, 2012

(Seitz et al. 2007) Seitz, N., Kameas, A., Mavrommati, I., Eds.: The Disappearing Computer, Springer Publ., 2007

(Selke 2014) Selke, G.: Design and Development of a GPU-accelerated Micromagnetic Simulator. PhD Thesis, University of Hamburg, 2014

(Tiller 2001) Tiller, M. M.: Introduction to Physical Modeling with Modelica, Kluwer Academic Publ., 2001

(Torin 2007) Torin, M.: "War Rooms" of the Street: Surveillance Practices in Transportation Control Centers. Communication Review 10 (4): 367–389 2007

(Wan et al. 2012) Wan, J., Li, D., Zou, C., Zhou, K.: M2M Communications for Smart City: An Event-based Architecture. In: Proc. 12th IEEE Internat. Conf. on Computer and Information Technology, pp. 895–900, 2012

(Wan et al. 2013) Wan, J., Chen, M., Xia, F., Li, Di, Zhou, Keliang: From Machine-to-Machine Communications towards Cyber-Physical Systems, ComSIS Vol. 10, No. 3, pp. 1105–1128, 2013

(Wang et al. 2008) Wang, C. C., Chiang, C. Y., Lin, P. Y.: Development of a fall detecting systems fort he elderly residents. In: Proceed. 2nd International Conference on Bioinformatics and Biomedical Engineering, pp. 1359–1362, 2008

(Weiland and Purser 2000) Weiland, R. J., Purser, L. B.: Intelligent Transportation Systems. In: Transportation Research Board, 2000

(Wessel 2011) Wessel, R.: Finnish railroad streamlines operations, 2011. http://www.rfidjournal.com/articles/view?8594

(Wood et al. 2008) Wood, A. D., Stankovic, J. A., Virone, G.: Context-aware wireless sensor networks for assisted living and residential monitoring. In: IEEE Network, Vo. 22, No. 4, pp.26–33, 2008

(Wu et al. 2011) Wu, F. J., Kao, Y. F., Tseng, Y, S.: From wireless sensor networks towards cyber physical systems," In: Pervasive and Mobile Computing, Vol. 7, No. 4, pp. 397–413, 2011

(Xiao et al. 2008) Xiao, Z., Guan, Q., Zheng, Z.: The Research and Development of the Highways Electronic Toll Collection System, In: Proceed. 1st International Workshop on Knowledge Discovery and Data Mining, pp. 359–362, 2008

(Yang 2006) Yang, T.: Networked control system: a brief survey. Control Theory and Applications, IEE Proceedings, Vol. 153, no. 4, pp. 403–412, 2006

(Zhai 2007) Zhai, J., Zhou, Z., Shi, Z., Shen, L.: An Integrated Information Platform for Intelligent Transportation Systems Based on Ontology, In: IFIP Vol. 254, Research and Practical Issues of Enterprise Information Systems, pp. 787–796, Eds. I. Xu, A. Tjoa, S. Chaudhary, Springer. Pub. 2007

(Zhang et al. 2009) Zhang, F., Shi, Z., Wolf, W.: A dynamic battery model for codesign in cyber-physical systems. In: Proc. 29th IEEE International Conference on Distributed Computing Systems Workshops, ICDCS, pp. 51–56, 2009

(Zhao and Cao 2008) Zhao, J., Cao, G.: VADD: Vehicle-Assisted Data Delivery in Vehicular Ad Hoc Networks. In: IEEE Transactions on Vehicular Technology, Vol. 57, No. 3, pp. 1910–1922, 2008

Links

(http1 2015) http://www.heise.de/netze/rfc/rfcs/rfc2460.shtml; accessed January 31st 2015
(http2 2015) http://tools.ietf.org/html/rfc3513; accessed January 31st 2015
http://cyberphysicalsystems.org/; accessed January 31st 2015
(http4 2015) http://www.research.ibm.com/haifa/conferences/hvc2012/papers/HVC2012Eldad_Palachi.pdf; accessed January 31st 2015
(http5 2015) http://www.sysml.org/; accessed January 31st 2015
(http6 2003) http://www.eclipse.org/whitepapers/eclipse-overview.pdf, 2003; accessed January 31st 2015
(http7 2015) http://en.wikipedia.org/wiki/DARPA_Grand_Challenge_%282007%29; accessed January 31st 2015
(http8 2015) http://www.darpa.mil/newsevents/releases/2012/04/10.aspx; accessed January 31st 2015
(http9 2015) https://www.youtube.com/watch?v=AQXvM6Am6BQ; accessed January 31st 2015
(http10 2015) https://www.premierway.com/products/smartercommerce/e-com-merce-and-mobile-sales; accessed January 31st 2015
(http11 2015) http://en.wikipedia.org/wiki/Requirements_engineering#cite_note-5; accessed January 31st 2015
(http12 2015) http://en.wikipedia.org/wiki/Interoperability; accessed January 31st 2015
(http13 2015) http://docs.nvidia.com/cuda/index.html; accessed January 31st 2015
(http14 2015) https://www.securityforum.org/shop/p-71-173; accessed January 31st 2015
(http15 2015) http://www.engadget.com/2010/10/28/medtronic-debuts-tiny-lead-less-pacemaker-at-tedmed-2010/; accessed January 31st 2015
(http16 2015) http://www.darpa.mil/Our_Work/I2O/Programs/High-Assurance_Cyber _ Military_Systems_%28HACMS%29.aspx; accessed January 31st 2015

(http17 2015) http://www.cisco.com/web/solutions/trends/iot/overvies.html; accessed January 31st 2015

(http18 2015) http://www.cisco.com/web/strategy/docs/Is_your_city_smart_ enough-Ovum_Analyst_ Insights.pdf; accessed January 31st 2015

(http19 2015) http://www.intelligentcommunita.org; accessed January 31st 2015; accessed January 31st 2015

Introduction to the Internet of Things

4

This chapter begins with a brief introduction to the Internet of Things in Sect. 4.1, which identifies the enabling technologies for its use. Section 4.2 introduces radio frequency identification (RFID), a wireless automatic identification technology whereby an object with an attached RFID tag is identified by an RFID reader. Section 4.3 introduces the principal concept of wireless sensor network technology which has important applications, such as remote environmental monitoring and target tracking. This technology has been enabled by the availability of sensors that are smaller, cheaper, and intelligent. Section 4.4 introduces powerline communication technology that enables sending data over existing power cables. With just power cables running to an electronic device, it can be powered up, and, at the same time, data can be controlled/retrieved from it in a half-duplex manner. Section 4.5 refers to RFID applications. Section 4.6 introduces a case study on the concept of the Internet of Things. Section 4.7 contains comprehensive questions from this chapter, followed by references and suggestions for further reading.

4.1 The Internet of Things

The Internet is a global system of interconnected computer networks that use the standard Internet Protocol Suite (TCP/IP) to serve billions of users worldwide. It is a network of millions of private, public, academic, business, and government networks, from local to global in scope. Originating from the Advanced Research Projects Agency Network (ARPANET) around 1970, it was available in the 1980s and became popular circa 1990. The Internet of Things (IoT), also known as the Internet of Objects, refers to the networked interconnection of everyday objects. Today, the Internet of Things has become a leading path to the smart world of ubiquitous computing and networking. It is described as a self-configuring wireless network of sensors whose purpose is to interconnect all things. The concept was originally introduced by the Auto-ID Labs, founded in 1999 at Massachusetts Institute of Technology (MIT) where an important effort was made to uniquely

© Springer International Publishing Switzerland 2016   141
D.P.F. Möller, *Guide to Computing Fundamentals in Cyber-Physical Systems*,
Computer Communications and Networks, DOI 10.1007/978-3-319-25178-3_4

identify products. Auto-ID Labs, originally founded by Kevin Ashton, David Brock, and Sanjay Sarma, helped to develop the Electronic Product Code (EPC), a global RFID-based item identification system intended to replace the Universal Product Code (UPC) barcode (Auto-ID 2013). Today, Auto-ID Labs are the leading global network of academic research laboratories in the field of networked RFID. They are comprised of seven of the world's most renowned research universities located on four different continents. These institutions were chosen by the Auto-ID Center, together with EPCglobal, to architect the Internet of Things (IOT 2013).

A thing, object, or entity is any possible item in the real world that joins the communication chain. Therefore, the initial main objective of the Internet of Things was to combine communication capabilities characterized by data transmission. The main object in the IoT was RFID. Thus, the Internet of Things can be thought of as the building of a global infrastructure for RFID tags: a wireless layer on top of the Internet. A network of interconnected computers communicates with a network of interconnected objects constantly tracking and accounting for millions of things, from razor blades to banknotes to car tires. These objects sometimes have their own Internet Protocol (IP) addresses, are embedded in complex systems, and use sensors to obtain information from their environment, e.g., food products that record the temperature along the supply chain and/or use actuators to interact with it, e.g., air conditioning valves that react to the presence of people.

The growth in the forms of information and communication networks is evident by the widespread use of mobile devices. The number of connected mobile devices worldwide surpassed $2*10^9$ in mid-2005 and is approximately $25*10^9$ in 2015, as shown in Table 4.1.

From Table 4.1, it can be seen that the Internet of Things (IoT) represents the point in time when more devices are connected to each other than people are connected with/and/or to devices. This has an impact on today's world and will change everything, including each of our lives because staying connected has become an integral and intimate part of the 24/7 paradigm of everyday life for many millions of people. The IoT has become an important concept in the global economy because wireless technology is making it possible to interact with the IoT anywhere and everywhere at any time. This has opened the opportunity for new ubiquity-based products and services with a high degree of innovation and a major impact on society and business. In a process as revolutionary as the IoT, the participation of all groups in society is crucial, particularly in value creation by connecting new places, such as manufacturing floors, energy grids, health-care facilities, and transportation systems, to the Internet.

Table 4.1 Connected devices in relation to world population in the third wave of computing

Year	2003	2010	2015	2020	Increase
World population	$6.3*10^9$	$6.8*10^9$	$7.2*10^9$	$7.6*10^9$	+20.6349 %
Connected devices	$500*10^6$	$12.5*10^9$	$25*10^9$	$50*10^9$	$+10^2$
Connected devices per person	0.0793 %	1.8382 %	3.4722 %	6.5789 %	+82.9621 %

The availability of the Internet and advances in software and telecommunication services with the ability to connect every object and/or thing, with any object and/or thing, at any time and in any media, have accelerated the worldwide penetration of the IoT paradigm. In particular, the basic idea that every object and/or thing can also be part of a tiny computer and/or microchip that is connected to the Internet has outperformed any forecast. The enabling technologies of the IoT are:

- RFID
- Sensor and actuator
- Miniaturization
- Nanotechnology
- Smart entities

In addition, the increasing processing power available in the smallest of packages or devices in networked computing is the substructure for the IoT paradigm. RFID and sensors, among other technologies, have been increasingly deployed and allow the real-world environment to be embedded into the IoT networked services. Entity-to-entity-oriented IoT applications are monitored in real time, depending on their actual status, while the IoT automatically reacts. This has finally resulted in smart objects or things which can act smarter than objects or things which have not been tagged with a unique visual or invisible identification code or equipped with sensors or actuators. These new smart objects will obviously raise many questions on topics such as Chaouchi (2010):

- Addressing, identifying, and naming
- Choice of the transport model
- Communication model of these connected objects or things
- Connecting technology of the objects or things
- Economic impact and the telecommunication value chain evolution
- Interoperability between objects or things
- Possible interaction with existing models, such as the Internet
- Security and privacy
- And more

Most of the Internet services were designed to satisfy person-to-person interaction. In contrast, IoT services rely on easy location and tracking of connected entities which means a new dimension has been added to the world of *anytime*, *anyplace* connectivity for *anyone* . . ., that is, connectivity for *anything*. Therefore, the relevant characteristics of the IoT are:

- *Connectivity*: generating and processing data traffic on the IoT. Connecting entities can be wireless, as with RFID (see Sect. 4.2), or wired, as with powerline communication (see Sect. 4.5). The IoT also allows the connection of heterogeneous entities.

- *Connections*: multiplying and creating entirely new dynamic networks of networks and IoT. The IoT is neither science fiction nor industry hype but is based on solid technological advances and visions of network ubiquity that are zealously being realized.
- *Embedding*: short-range mobile transceivers in a wide array of additional gadgets and everyday items, such as smartphones, enabling new communication forms between people and things and/or entities and between things and entities themselves.

As stated in a 2005 UN report, "*Today, in the 2000s, we are heading into a new era of ubiquity, where the users of the internet will be counted in billions and where humans may become the minority as generators and receivers of traffic*" (Biddlecombe 2005). The roadmap of the IoT is shown in Fig. 4.1 (Source: http://en.wikipedia.org/wiki/Internet_of_Things).

Typical views of the IoT are the following application domains (see Sect. 4.5) in conjunction with cyber-physical systems:

- Automation
- Smart cities
- Smart grids
- Smart health care
- Smart lighting
- And more

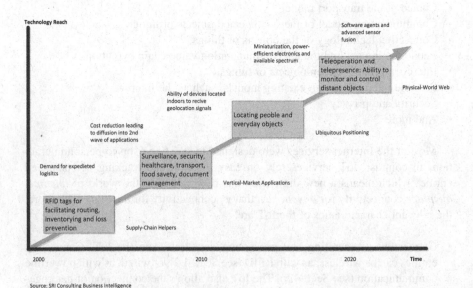

Fig. 4.1 Roadmap of the Internet of Things

In today's IoT paradigm, many things or objects will be part of the network in one form or another. This is where RFID and wireless sensor network (WSN) technologies will meet this new approach, as the information and communication systems used are invisibly embedded in the environment. RFID technology allows for contactless communication between powered reading devices, so-called active transponders, and passive identification tags, or transponders. This results in the generation of data which has to be stored, processed, and presented in a seamless, efficient, and easily interpretable form. Therefore, the IoT consists of services that are commodities and delivered in a manner similar to traditional commodities, and it is partly inspired by the success of RFID technology. RFID is now widely used for tracking things or objects or people or animals. Hence, the RFID system architecture is marked by a sharp dichotomy of simple RFID tags and an infrastructure of wireless networked RFID readers. This architecture optimally supports the tracking of physical things or objects within well-defined confines but limits the sensing capabilities and deployment flexibility that more challenging application scenarios require. The architectural model for the IoT is a decentralized system of autonomous physical things or objects with sensing, processing, and network capabilities based on RFID tags as wireless sensors that are uniquely identifiable and information related.

Internet search trends during the last 10 years, using applications such as Google to search for the terms Internet of Things, wireless sensor networks, and ubiquitous computing, are shown in Fig. 4.2 (Gubbi et al. 2013).

4.2   Radio Frequency Identification Technology

From a historical point of view, RFID dates back to the 1940s when the British Air Force used RFID-like technology in World War II to distinguish between enemy and friendly aircraft. The theory of RFID was first explained in 1948 in a conference paper (Stockmann 1948) with the title, *Communication by Means of Reflected Power*, published by the Institute of Radio Engineers. In 1973, M.W. Cardullo received the first US patent for an active RFID transponder with rewritable memory on January 23. An RFID transponder is a microchip attached to an antenna that is packaged such that it can be applied to an object. It is called active if the transponder has a power source. Also in 1973, C. Walton received a patent for a passive transponder used to unlock a door without a key. The key was replaced by a card with an embedded transponder communicating a signal to a reader near the door. When the reader detected a valid identity number stored within the RFID tag, the reader unlocked the door. Such a transponder is called passive when it has no power source and cannot actively broadcast a signal. The US government was also working on RFID systems. In the 1970s, a group of scientists at Los Alamos National Laboratory came up with the concept of putting a transponder in a truck and readers at the gates of secure facilities. The gate antenna woke up the transponder in the truck, which responded with an ID and, potentially, other data, such as the driver's ID. This system was commercialized in the mid-1980s when the Los

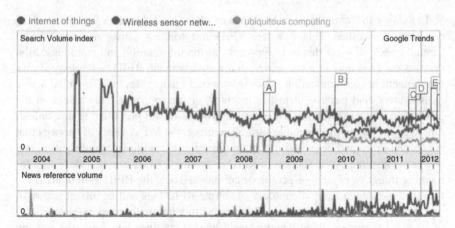

Fig. 4.2 Google search trends since 2004: Internet of Things, wireless sensor networks, and ubiquitous computing. SPOT points are listed (Gubbi et al. 2013):
A. Algorithms and protocols for wireless sensor networks provides you with a comprehensive resource, *MarketWatch*, Nov. 18, 2008
B. Internet of Things: from vision to reality, *MarketWatch*, Apr. 14, 2010
C. CCID consulting: China's Internet of Things industry sees a landscape characterized by clustering in four regions, *MarketWatch*, Oct. 4, 2011
D. China hi-tech fair highlights Internet of Things, *MarketWatch*, Nov. 21, 2011
E. ARM unveils low-power chip for the Internet of Things, Reuters UK, Mar. 13, 2012
F. Web connected objects get a voice on the Internet of Things, *Winnipeg Free Press*, Apr. 25, 2012

Alamos scientists who worked on the project left to form a company to develop automated toll payment systems. These systems have become widely used on roads, bridges, and tunnels around the world (Roberti 2005).

RFID technology can help to provide operational efficiencies and improve handling transparency in the logistics of on-demand distribution. RFID systems incorporate microelectronic devices, called transponders, and reading units. Transponders are more commonly known as tags, and they are attached to the things or objects to be identified. Tags are available in a large variety of forms and functional characteristics and are classified into active and passive tags:

- *Passive tags*: read/write range is shorter than most active tags; they do not possess an onboard source of power to broadcast a signal.
- *Active tags*: read/write range is longer than most passive tags and has its own power source to broadcast a signal.

Passive tags, such as the one shown in Fig. 4.3, are relatively inexpensive. They cost anywhere from 20 cents to several dollars because they do not contain a power source. They draw power from a reader's radio signals which induce a current in the tag's antenna using either inductive coupling or electromagnetic capture. This power is used both for chip operation and broadcasting. These tags essentially

Fig. 4.3   Passive tag (Cisco 2008)

Chip

Antenna

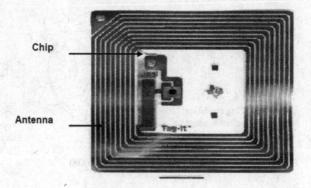

reflect back the radio waves from the reader in order to broadcast, a phenomenon sometimes known as backscatter. However, their signal range is very low, usually less than 10 ft. Semipassive tags fall somewhere between the two; they use a battery for a chip's standby operation but draw energy from the reader during active broadcasting (Cisco 2008).

Low-cost tags, applicable to the grocery industry, cost from 20 cents to 35 cents, and the latest tag developments promise tags that will cost just around 5 cents (Kärkkäinen 2003). However, tags can also cost several dollars depending on many factors such as:

• Data capacity
• Form
• Operating frequency
• Range
• Performance requirements
• Presence or absence of a microchip
• Read/write memory

Passive RFID tags vary in how they broadcast to RFID readers and how they receive power from the RFID reader's inductive or electromagnetic field. This is commonly performed by two basic methods:

• *Load modulation and inductive coupling in the near field* as shown in Fig. 4.4. The RFID reader provides a short-range alternating current magnetic field that the passive RFID tag uses for both power and broadcasting. Through inductive (near field) coupling, the magnetic field induces a voltage in the antenna coil of the RFID tag, which powers the tag. The tag broadcasts its information to the RFID reader. Each time the tag draws energy from the RFID reader's magnetic field, the RFID reader itself detects a corresponding voltage drop across its antenna leads. Thus, the tag can communicate binary information to the reader by switching a load resistor on and off to perform the load modulation. When the tag performs load modulation, the RFID reader detects this action as amplitude

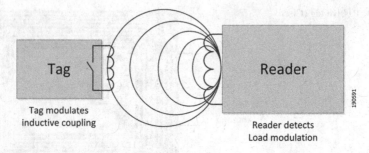

Fig. 4.4 Passive tag load modulation (Cisco 2008)

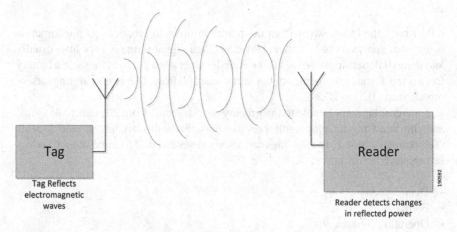

Fig. 4.5 Passive tag backscatter modulation (Cisco 2008)

modulation of the signal voltage at the reader's antenna. Load modulation and inductive coupling are available for passive RFID tags using frequencies from 125 to 135 kHz and 13.56 MHz. Limitations that exist with regard to the use of such low frequencies include the necessity to use larger antennas, low data rate and bandwidth, and a decay in the strength of the electromagnetic field of $1/r^6$, where r represents the distance between a low-frequency interrogator and a passive RFID tag (Cisco 2008).

- *Backscatter modulation and electromagnetic coupling in the far field* as shown in Fig. 4.5. Again, the RFID reader provides a medium-range electromagnetic field that the passive RFID tag uses for both power and broadcasting. Through electromagnetic (far-field) coupling, the passive RFID tag draws energy from the electromagnetic field of the RFID reader. However, the energy contained in the incoming electromagnetic field is partially reflected back to the RFID reader by the passive tag antenna. The precise characteristics of this reflection depend on the load connected to the antenna. The tag varies the size of the load that is placed in parallel with the antenna in order to apply amplitude modulation to the reflected electromagnetic waves, thereby enabling it to broadcast information payloads back to the RFID reader via backscatter modulation. Tags using backscatter modulation and electromagnetic coupling typically broadcast over a longer range than inductively coupled tags. Passive RFID tags operate at 868 MHz and higher frequencies. Limitations that exist with regard to the use of far-field coupled tags are due to a much slower rate of attenuation associated with the electromagnetic far field. Antennas used for tags employing far-field coupling are typically smaller than their inductively coupled counterparts (Cisco 2008).

The price of RFID readers varies depending on the type of reader. RFID readers communicate wirelessly with the RFID tags through electromagnetic waves and send a signal to someone for further elaboration or processing.

Most tags are resistant to environmental temperature and other external factors and can be read and reprogrammed at least 300.000 times before replacement. Therefore, when utilized in recyclable transportation containers, the same tags can be used many times (Kärkkäinen 2003).

Active tags are typically used in real-time tracking of high-value assets in closed-loop systems, which usually justify the higher cost of the active tag. Active RFID tags are physically larger than passive RFID tags. They contain RAM, which enables the active tag to store information from attached assets. This memory also makes active RFID preferable to passive RFID. Active RFID is available at operating frequencies of 303, 315, 418, 433, 868, 915, and 2400 MHz with read ranges of 60–300 ft. Active RFID tag technology typically displays very high read rates and read reliability because of the higher transmitter output, optimized antenna, and reliable source of onboard power. The cost of active RFID tags varies significantly depending on the amount of memory, the battery life required, and whether the tag includes added-value features, such as onboard temperature sensors, motion detection, or telemetry interfaces, and more. The durability of the tag housing also affects price, with the more durable or specialized housings required for specific tag applications available at higher costs. As with most electronic components of this nature, prices for active tags can be expected to decline as technological advances, production efficiencies, and product commoditization all exert a downward influence on market pricing (Cisco 2008).

Table 4.2 provides approximate values for the characteristics of high- and low-frequency tags. The exact values depend upon a combination of factors, such as tag type (active or passive), the presence of radio noise or radio-absorbing materials in the environment, the size and power gain of the antenna, and the type of reader.

In general, the relative values for various characteristics of tags operating at different frequencies can be summarized as shown in Table 4.3.

RFID tags are available in many forms, including glass capsules, disks, cylindrical tags, wedge-shaped tags, smart cards, and key-chain fobs, and can range from a few square millimeters to up to a few inches long. Different form factors are suitable for different applications. Small glass capsules from 2 mm to 1 cm can

Table 4.2 Characteristics of active and passive tags (Zaheeruddin and Mandviwalla 2005)

Tag frequency	General tag type	Range	Transmission rates	Power consumption
Low	Passive	<1.0 m	1–2 kb/s	20 µW
High		1.5 m	10–20 kb/s	200 µW
Ultra high	Active	10–30 m	40–120 kb/s	0.25–1.0 W
		20–100 m[a]		

[a]With battery-powered tags

Table 4.3 Comparison of high- and low-frequency tags (Zaheeruddin and Mandviwalla 2005)

Tag frequency	Relative range	Transmission rates	Power consumption	Relative cost	Environmental susceptibility
Low	Shorter	Lower	Lower	Lower	Lower
High	Longer	Higher	Higher	Higher	Higher

be injected directly under the skin through large-gauge hypodermic needles to tag cattle, as glass is nonreactive and nonbiodegradable.

Similarly, different operating frequencies are suitable for different purposes, and there is no ideal frequency for all applications. For instance, while higher frequencies may be needed in the shipping industry for longer ranges, low frequencies may be more suitable for access control purposes.

The components of an RFID system include:

- Transponders (tags) that allow items to be identified
- Readers that allow tags to be interrogated and to respond
- Software (RFID middleware) to control the RFID equipment and manage the data and interfaces with enterprise applications

The reader receives identity information broadcast from and stored in the tag. An RFID system can be built using several readers and tags. The reader is able to read the identity of each tag. The reader is also capable of storing information into a tag as well as altering the state of the tag. The information collected by the reader is not really useful unless it is connected to a network server. Therefore, two more components are required for an RFID system: a server and a network. An RFID reader is a complex device which consists of the following elements:

- Amplifier
- Carrier and carrier cancelation
- Demodulator block
- Modulator block
- Network interface to the interface
- Receiving antenna
- Transmitting antenna

Operating the readers and tags in the different frequency bands and using the available protocols for the applications accessing the tags through readers can be performed by the RFID system software, also called middleware. In general, middleware architecture consists of three components (Harish 2010):

- *Device interface*: provides the necessary functionality to establish a connection between the core processing interface and the RFID hardware to enable the RFID system to discover, manage, and control readers and tags
- *Core processing interface*: the decision-making component that manages and manipulates the large amount of raw RFID data before passing it to the application interface
- *Application interface*: responsible for delivering RFID data to and from the specific application

RFID applications are being used today in an ever-increasing number of industries for purposes such as:

- Access control
- Baggage handling
- Fraud prevention
- Inventory management
- Package tracking
- And more

These applications can be classified according to the major purpose for deploying RFID:

- *Authentication*: authentication applications, such as smart cards with no contact, are simple, used for automatic payments of small amounts of money, and usually assume the tag holder to be a person rather than a thing and/or object.
- *Automatic data acquisition*: the major thrust of supply chain applications is automatic data acquisition. In most automatic data acquisition applications, things and/or objects, such as items produced, cases, and pallets, are tracked automatically, and the captured data is used to derive enterprise applications, such as supply chain management systems, customer relation management systems, and enterprise resource planning systems.
- *Identification*: RFID as main identification platform in conjunction with sensor fusion.
- *Location tracking*: people and objects in open space.

4.3   Wireless Sensor Networks Technology

Intelligent sensors not only observe processes and the environment, they also process the data measured and transfer the results obtained. This allows monitoring and/or controlling of complex systems concentrating on specific areas, which can serve as an early warning system. An example of such an application is intelligent vibration sensing in wind turbines in regenerative energy systems which switch off the system if the tower vibrations reach dangerous high amplitudes. With the use of wireless communication, intelligent sensors can be directly connected to sensor networks, e.g., to monitor the wind turbines of the regenerative energy system. Hence, wireless interconnected sensors have a large advantage in difficult environmental conditions, where wired sensor connections can be avoided.

With advancements in microelectronic components and the related miniaturization of intelligent functions, wireless sensors can be implemented in decentralized locations where they are needed. This is usually accomplished by using embedded computing systems (see Chap. 2), which have been developed to assist in many of the functions of daily life and industry. Particularly important in this context is communication with the outside of the immediate network. It should be noted that communication often takes place through the user interface of the device itself, which calls for more advanced technologies. This has been achieved in recent years by the development of networked sensors, the so-called wireless sensor networks. Intelligent sensor nodes are wirelessly linked to computer networks. Current and planned applications of WSN range from early warning systems in production control to so-called smart dust.

Smart dust belongs to one of the three forms of devices for the ubiquitous computing paradigm proposed by Marc Weiser (1991) and can be considered as useful ubiquitous devices as introduced by Poslad (2009). Thus, smart dust is composed of systems of many tiny microelectromechanical systems (MEMS), ranging from millimeters to micrometers to nanometers, such as sensors that detect physical or chemical quantities or are integrated into smart clothes (the integration of sensors, actuators, computers, power sources, etc. into the cloth, the whole being part of an interactive communication network). Smart dust is usually wirelessly operated on a computer network and distributed over a specific area to perform tasks, such as using RFID to sense a smart dust component introduced through the IoT paradigm. The size of an antenna for a tiny smart dust communication device ranges from a few millimeters to centimeters, and it may be vulnerable to electromagnetic disablement and destruction by microwave exposure.

4.3.1   Sensor Technology

Sensors produce electrical signals (usually voltages) which are analogous to the physical quantity to be measured. Often, the voltage is proportional to the measurand. Then, the sensor voltage can be described by:

$$V_{\text{sensor}} = K \cdot m$$

where V_{sensor} is the voltage produced by the sensor, K is the sensitivity constant of the sensor, and m is the measurand. The sensitivity of a sensor indicates how much the sensor's output changes when the input quantity being measured changes. It is basically the slope $\Delta y/\Delta x$ assuming a linear characteristic.

Sensors can be used to instrument and monitor environments, track assets through time and space, detect changes in the environment which have been defined to be important, control a system with regard to being in a closed vicinity within a defined range of change, and adapt services to improve their utility. Sensors are also used in everyday applications, such as touch-sensitive elevator buttons, lamps which dim or brighten by touching the base, and more. With advances in mechatronics and easy-to-use microcontroller platforms, the use of sensors has been expanded beyond the more traditional fields of flow, pressure, or temperature measurement. Analog sensors, such as potentiometers and force-sensing resistors, are still widely used. Applications include manufacturing and machinery, aeronautics, automotive, medical, and robotics.

Sensors need to be designed with little effect on the physical quantity measured, which requires that the sensor be fabricated much smaller to reduce measurement error. With the advent of new technology, sensors are now manufactured on a microscopic scale, such as microsensors that use the so-called MEMS technology. They range in size from nanometers to micrometers to millimeters and are fabricated as discrete devices or large arrays (Berlin and Gabriel 1997). MEMS perform two different types of functions: as a sensor and as an actuator. Both sensors and actuators act as transducers, converting one signal to another, as discussed above. Of specific interest are transducers that convert environmental information into digital signals and vice versa. MEMS sensors can convert environmental information, such as temperature, humidity, and pressure, into an electrical signal. MEMS actuators work in reverse to sensors; they convert an electrical signal into physical information to move or control devices, such as motors, hydraulic pistons, and relays. These MEMS components have high resonant frequencies leading to higher operating frequencies (Poslad 2009).

Collections of millions of cooperating sensing, actuation, and locomotion mechanisms can be introduced as a form of programmable matter because they can self-assemble into arbitrary three-dimensional shapes (Goldstein et al. 2005), forming the basis for much more fluid, flexible computers and human-computer interfaces (Poslad 2009).

MEMS design differs from that of the equivalent macroscale devices which are comprised of mechanical and discrete electronic components on silicon-based integrated circuits, consisting of several layers of doped silicon added to a substrate.

MEMS sensors incorporate the requisite sensing, communication, and computing hardware, along with a power supply, into a volume no more than a cubic millimeter, achieving the usual performance in terms of sensor functionality and communication capability (Kahn et al. 2000). These millimeter-scale nodes are called smart dust and are the size of motes.

MEMS actuators are engineered in a different way than microscale devices. They usually use integrated circuit design and nanotechnology.

4.3.2   Sensor Networks

Sensor networks usually contain a large number of sensors and nodes. A sensor node can be a component of a larger network of sensors. Each sensor node in the sensor network is capable of performing processing, gathering sensor information, and communicating with other connected nodes in the network.

The main components of a sensor node are the microcontroller, the transceiver, the external memory, the power supply, and one or more sensors, which results in the typical architecture shown in Fig. 4.6.

The microcontroller component of the sensor node, shown in Fig. 4.6, performs specific tasks, processes data, and controls the functionality of other components in the sensor node. Microcontrollers are usually used because of their low cost, flexibility in connecting to other devices, easy programming, and low-power consumption.

With regard to wireless communication, sensor nodes usually make use of the industrial scientific and medical (ISM) radio band, which allows broadcasting, spectrum allocation, and is available globally. Despite the intent of the original allocations in recent years, the fastest-growing uses of the ISM band have been for short-range, low-power communication systems, such as Bluetooth devices, near field communication (NFC) devices, sensor networks, and wireless computer networks, which all use frequencies allocated to low-power communications as well as ISM.

Transceivers of sensor nodes represent a combination of a transmitter unit and a receiver unit into a single device. The operational states of transmitters are transmitting, receiving, idle, or sleep, which refers to their realization as state machines that perform some operations automatically. A state machine is a mathematical model of computation used to design both computer programs and sequential logic circuits, acting on a set of inputs and computing a set of outputs. Thus, a finite state machine has a finite number of states to represent its state of processing. Its actions depend upon its internal state, and any inputs adhere to a specific syntax.

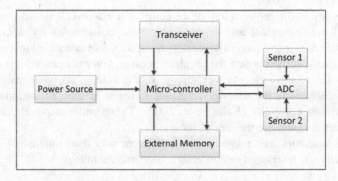

Fig. 4.6 Sensor node architecture

The memory requirements of sensor nodes depend on the application. There are two categories of memories usually used in sensor nodes: user memory to store application-related or personal data and program memory to program the device. The program memory can also contain the device's identification data.

An important issue in the development of a wireless sensor node is ensuring that adequate energy is available to power the system because the sensor node requires power for sensing, communicating, and data processing. More of the required energy is used for data communication than for any other process. For example, the energy cost of transmitting 1 Kb a distance of 100 m (330 ft) is approximately the same as that used for the execution of 3 million instructions by 100 million instructions per second/W processor (http1 2015). Power is stored either in batteries or capacitors. Batteries, both rechargeable and nonrechargeable, are the main source of power supply for sensor nodes. Wireless sensor nodes are typically very small electronic devices. They can only be equipped with a limited power source of less than 0.5–2 ampere-hour and 1.2–3.7 V (http1 2015).

The energy efficiency for communication in a sensor network can be increased when using a multi-hop topology (Zhao and Guibas 2004). In an N-hop network, overall transmission distance is Nd, where d is the average one-hop distance. The minimum receiving power is P_r, and the power at the transmission node is P_t. Thus, the power advantage P_A of an N-hop transmission versus a single-hop transmission over the same distance can be described as follows (Poslad 2009):

$$P_A = \frac{P_t Nd}{N.P_t Nd} = \frac{(Nd)^2 P_r}{(N.d^2 P_r)} = N^{(n-1)}$$

Sensors are hardware devices that range in scale from nanosensors to macrosensors. They act as data generators and data pre- or post-processors of the data to be monitored. In the case of RFID sensors, the data processing is less complex compared with other sensor types. The continuously produced analog signal x of the sensor is digitized into a proportional digital quantity by an analog-to-digital converter and sent to the microcontroller for further processing. Sensors are traditionally classified into the following categories:

- *Passive sensors*: are self-powered and sense data without actually manipulating the environment by active probing
- *Omnidirectional sensors*: have no notion of direction involved in their measurements
- *Narrow-beam sensors*: have a well-defined notion of direction of measurement
- *Active sensors*: require continuous energy from a power source and sense data by actively probing the environment

Sensor nodes are used in:

- Air traffic control
- Battlefield surveillance

- Environmental monitoring
- Industrial automation
- Robotic landmine detection
- Target tracking
- Wildfire detection
- And more

Sensors and sensor nodes also perform many activities in society with potential civil, industrial, medical, and military applications. In the medical application domain, sensor nodes can be deployed to:

- Assist disabled patients
- Monitor disabled patients
- Monitor ambulatory patients
- Track disabled patients
- And more

Sensor nodes also play an important role in other domains, such as:

- Control systems
- Data acquisition (SCADA) systems
- Supervisory systems
- And more

Sensors can also be deployed in crisis and emergency response systems, such as those that respond to crises and unexpected events (Mehrotra et al. 2004).

For digital recording of analog quantities, an analog-to-digital converter (ADC) is required. The task of the ADC is to convert analog input variable X into a proportional output number. In many cases, time-dependent signals are digitized. For this purpose, the input quantity to be converted has to be sampled at a certain time and held. This task is performed by sample-and-hold (S+H) circuits. Very often, nonelectrical signals should be digitally processed. Then, prior to the actual analog-to-digital conversion, the nonelectrical quantity has to be converted into an electrical voltage.

The sensors used to detect nonelectrical quantities map an electrical voltage as output to the nonelectrical input. In Fig. 4.7, the block structure of analog inputs to a digital data converting system is shown. The control of the analog-to-digital conversion is efficient for the following reasons:

- Several input channels are used; switching to a channel is controlled by using an analog multiplexer.
- After achieving the settling time, the sample-and-hold circuit switch is placed on hold; thus, the conversion of a stable analog signal is possible. This does not affect the integration of the converter.

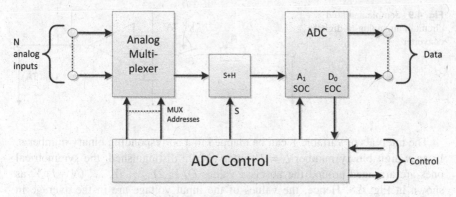

Fig. 4.7 Components of an analog-to-digital conversion system

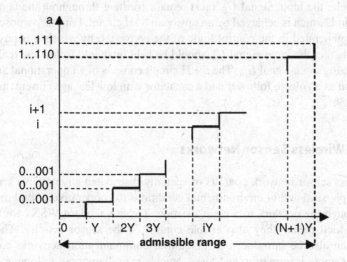

Fig. 4.8 Analog-to-digital converter resolution characteristic

- Analog-to-digital conversion is started by the start of conversion (SOC) mode.
- After completion of the conversion, the analog-to-digital converter activates the end of conversion (EOC) mode to the controller (ADC control).
- ADC control transfers the converted (digitized) measurements to the following data processing unit (not shown in Fig. 4.7).

Analog-to-digital converters have a characteristic transmission curve in common, as shown in Fig. 4.8, with respect to the following:

- Continuous abscissa pool y
- Discrete ordinate pool a

Fig. 4.9 Sample and hold
circuit of an analog-to-digital
converter

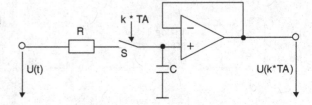

The intervals of variable Y can be mapped to a corresponding binary number a. In an n-digit binary number $N = 2^n$, intervals are distinguished; the symmetrical ones are arranged around the abscissa values O, Y, $2Y$, ... iY, ... $(N-1)$ Y, as shown in Fig. 4.8. Hence, the values of the input voltage are in the average in accordance with the converted binary number.

Since the conversion process of analog-to-digital converters takes more than one clock cycle, the input signal U_{in} must remain constant throughout the conversion time period, which is achieved by an upstream S+H circuit. For this purpose, a hold signal is generated in the control logic of the converter, by which it is possible to define whether the input signal U_{in} should be held or should follow the real course of the analog input signal U_{in}. The S + H circuit consists of an operational amplifier connected as a voltage follower and a capacitor with low leakage current, as shown in Fig. 4.9.

4.3.3 Wireless Sensor Networks

A wireless sensor network consists of spatially distributed autonomous sensors to monitor physical and/or environmental conditions to cooperatively send measured data through the network to a main location. Today's modern WSNs are bidirectional, which means they also enable control of the sensor activity. Therefore, WSNs can also be introduced as ubiquitous communication networks to access relevant remote information and tasks, anywhere and anytime, following the IoT paradigm. Wireless sensor networks are used in many industrial and consumer applications, such as industrial process monitoring and control, machine health monitoring, and more. Against the background of the manifold applications, different combinations of network functions and services are required, which results in a wide variety of wireless sensor networks based on the infrastructure, network range, frequency range used, bandwidth, and power consumption. Despite the disparity in the objectives of sensor applications, the main task of wireless sensor nodes is to sense and collect data from a target domain, process the data, and transmit the information back to specific sites where the underlying application resides. Conducting this task efficiently requires the development of an energy-efficient routing protocol to set up paths between sensor nodes and the data sink.

Wireless sensor networks are built of nodes, from a few to several hundreds or even thousands, where each node is connected to one or sometimes several sensors. Each such sensor network node typically has several parts, as shown in Fig. 4.6. The topology of the wireless sensor network can vary from a simple star-based network topology for monitoring and security applications to an advanced multi-hop wireless mesh network, where the propagation technique between the hops of the network can be routing or flooding (Dargie and Poellabauer 2010; Sohraby et al. 2007).

With regard to the aforementioned as described in Sohraby et al. (2007), which is taken as the basis of the design for routing protocols for wireless sensor networks, the following must be considered: the power and resource limitations of the network nodes, the time-varying quality of the wireless channel, and the possibility of packet loss and delay. To address these design requirements, several routing strategies for wireless sensor networks are available.

- *Flat network architecture*: has several advantages, including minimal overhead to maintain the infrastructure and the potential for the discovery of multiple routes between communicating nodes for fault tolerance.
- *Network structure*: imposes a structure on the network to achieve energy efficiency, stability, and scalability. Here network nodes are organized into clusters in which a node with higher residual energy, for example, assumes the role of a cluster head. The cluster head is responsible for coordinating activities within the cluster and forwarding information between clusters. Clustering has the potential to reduce energy consumption and extend the lifetime of the network.
- *Data-centric approach*: disseminates interest within the network. The approach uses attribute-based naming, whereby a source node queries an attribute for the phenomenon rather than an individual sensor node. The interest dissemination is achieved by assigning tasks to sensor nodes and expressing queries relative to specific attributes. Different strategies can be used to communicate interests to the sensor nodes, including broadcasting, attribute-based multicasting, geocasting, and anycasting.
- *Location to address sensor node*: location-based routing is useful in applications where the position of the node within the geographical coverage of the network is relevant to the query issued by the source node. Such a query may specify a specific area where a phenomenon of interest may occur or the vicinity to a specific point in the network environment.

In Sohraby et al. (2007), several routing algorithms that have been proposed for data dissemination in wireless sensor networks are described. The design trade-offs and performance of these algorithms are also discussed.

In general, routing algorithms are based on various network analyses and graph-theoretic concepts or in operations research, including shortest route, maximum flow, and minimum span problems. Routing is closely associated with dynamic programming and the optimal control problem in feedback control theory.

The shortest path routing schemes find the shortest path from a given node to the destination node. If the cost, instead of the link length, is associated with each link, these algorithms can also compute minimum cost routes. Algorithms are centralized (find the shortest path from a given node to all other nodes) or decentralized (find the shortest path from all nodes to a given node). There are certain well-defined algorithms for shortest path routing, including:

- *Dijkstra algorithm*: which has polynomial complexity
- *Bellman-Ford algorithm*: finds the path with the least number of hops

Routing schemes based on competitive game theory notions have also been developed (Lewis 2004).

Large-scale communication networks contain cycles (circular paths) of nodes. Moreover, each node is a shared resource that can handle multiple messages flowing along different paths. Therefore, communication nets are susceptible to deadlock, wherein all nodes in a specific cycle have full buffers and are waiting for each other. Then, no node can transmit because no node can get free buffer space, so all transmission in that cycle comes to a hold.

Livelock is the condition wherein a message is continually transmitted around the network and never reaches its destination. Live lock is a deficiency of some routing schemes that route messages to alternate links when desired links are congested without taking into account that the message should be routed closer to its final destination.

Many routing schemes are available for routing with deadlock and livelock avoidance (Lewis 2004). Flooding is a common technique frequently used for path discovery and information dissemination in wired and wireless ad hoc networks, as described in Sohraby et al. (2007). The routing strategy is simple and does not rely on costly network topology maintenance and complex route discovery algorithms. Flooding uses a reactive approach whereby each node receiving a data or control packet sends the packet to all of its neighbors. After transmission, a packet follows all possible paths. Unless the network is disconnected, the packet will eventually reach its destination. Furthermore, as the network topology changes, the packet transmitted follows the new routes. Figure 4.10 illustrates the concept of flooding in a data communications network. As shown in the figure, flooding in its simplest form may cause packets to be replicated indefinitely by network nodes.

To prevent a packet from circulating indefinitely in the network, a hop count field is usually included in the packet. Initially, the hop count is set to approximate the diameter of the network. As the packet travels across the network, the hop count is decremented by one for each hop that it traverses. When the hop count reaches zero, the packet is simply discarded. A similar effect can be achieved using a time-to-live field, which records the number of time units that a packet is allowed to live within the network. At the expiration of this time, the packet is no longer forwarded. Flooding can be further enhanced by identifying data packets uniquely, forcing each network node to drop all of the packets that it has already forwarded. However, such a strategy requires maintaining at least a recent history of the traffic to keep track of which data packets have already been forwarded.

The application domains of wireless sensor networks are manifold, as mentioned. Some examples are:

Fig. 4.10 Flooding in a data
communication network
model

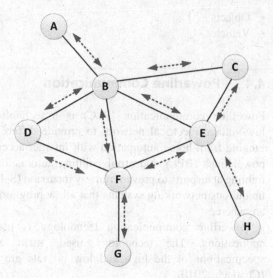

- *Smart road monitoring*: enables all cars to interact with each other which can
  help prevent:
  - Car accidents
  - Being stuck in traffic congestion
  - Driving too fast in school or hospital zones
- *Sport events monitoring*: allows:
  - Monitoring vital parameters of runners
  - Supervising a target line
  - Tracking the ball in soccer games
- *Waste managementmonitoring*: monitoring the decomposition process of a
  waste dump with regard to the temperature inside, chemical reactions, etc., in
  order to react in case a critical level is reached

More general monitoring applications include (Yick et al. 2008):

- Indoor/outdoor environmental monitoring
- Health and wellness monitoring
- Power monitoring
- Inventory location monitoring
- Factory and process automation monitoring
- Seismic and structural events monitoring

In addition to monitoring tracking applications in wireless sensor networks,
tracking can also include:

- Animals
- Humans

- Objects
- Vehicles

4.4   Powerline Communication

Powerline communication (PLC) is a technology that uses a medium and low-voltage electrical network to provide different telecommunication services, ranging from home automation with Internet access, often called broadband over power lines (BPL), industrial building automation, data transmission in airfield lighting at airports to prevent runway incursion (Schönefeld and Moeller 2012), and ubiquitous networking systems that allow program applications to follow the user anywhere.

Powerline communication technology is used in high and low bit-rate applications. The technology used, main characteristics, and technical specifications of the high and low bit rate are shown in Tables 4.4 and 4.5 (Chaouchi 2010).

In Table 4.6, several PLC technologies (HomePlug, UPA, and CEPCA) are shown which have been developed for the emerging worldwide market for standard equipment. They are not interoperable and are indicative of why it is important for the PLC market to establish an international standard (Chaouchi 2010).

The main technique used to transmit PLC signals over these electric media is to add a modulated signal of low amplitude to the low-voltage electrical signal around a center of carrier frequencies (Moeller and Vakilzadian 2014a).

In the late 1990s, the deregulation of the energy and telecommunication markets and the recent availability of PLC technologies encouraged this concept. To provide an optimum speed on each PLC link, HomePlug technologies, a

Table 4.4 Low bit-rate technologies, characteristics, and specifications

PLC type	Technology	Characteristics	Technical specs
Low bit rate	X10	Protocol for communication among electronic devices used for home automation; popular in the home environment with millions of units in use worldwide	Throughput <1 Kbits/s
	HomePlug CC	Specification for advanced command and control technology; serves as the basis for enabling a new era of convenience, safety, and security in the home	Throughput <50 Kbits/s
			Indoor and outdoor
			Specific MAC layer
	Echelon	Software controlling collection and distribution of civilian telecommunications traffic conveyed using communication satellites	Throughput <10 Kbits/s
			Commonly used in the home

Table 4.5 High bit-rate technologies, characteristics, and specifications

PLC type	Technology	Characteristics	Technical specs
High bit rate	HomePlug AV	Standard enabling devices to communicate with each other and the internet over existing home electrical wiring; provides sufficient bandwidth for applications such as high-definition TV (HDTV) and voice over IP (VoIP)	Throughput 200 Mbits/s
			MAX TCP throughput 60 Mbits/s
			Outdoor usage
	UPA	Universal Powerline Association chipsets for home streaming of HD video content, audio files, and other formats, such as photographs	Throughput 200 Mbits/s
			MAX TCP throughput 60 Mbits/s
			Outdoor usage
	CEPCA	Consumer Electronics Powerline Communication Alliance; promotes advanced high-speed PLC to utilize and implement a new generation of consumer electronics products through rapid, broad, and open industry adoption.	Throughput 220 Mbits/s
			MAX TCP throughput 70 Mbits/s
			In-home usage in Japan

Table 4.6 Technologies/standards for PLC indoor environments

Technologies/ standards	Industrial consortium	Technologies
HomePlug	Consortium HomePlug (USA) Leader: Intellon	HomePlug 1.0, Turbo (throughput of 14 and 85 M) HomePlug AV (throughput 200 M)
		Technology: OFDM, CSMA/CA
UPA	Consortium UPA (EU)	UPA (throughput 45 M
	Leader: DS2	UPAHD (throughput 200 M)
		Technology: OFDM, CSMA/CA
CEPCA	Consortium CEPCA (Japan)	HD-PLC (throughput 220 M)
	Leader: Panasonic	Technology: wavelets, TDMA
IEEE	IEEE P1901 WG	Draft standard based on HomePlug AV

manufacturing consortium comprised of industry leaders at each level of the value chain, from technology to services and content, has successfully launched over 195 certified products. The products developed by HomePlug members offer different modes of modulation on each subband. The Universal Powerline Association (UPA) has launched high-speed powerline network specifications (Digital Home Standards (DHS)) and provides modems capable of transmitting high data rates in the Mb/s range. Standards have been developed for PLC applications which follow IEEE P1901, a standard for high-speed (up to 500 Mbit/s at the physical layer) communication devices via electric power lines, often called broadband over power lines (BPL), and IEEE P1775, a standard for PLC equipment and electromagnetic compatibility requirements testing and measurement methods. Therefore, the availability of power lines enables private and public users to utilize these lines for broadband communication without any additional infrastructure.

Electrical companies make use of PLC for low bit-rate data transfer (<50 Kbit/s) signal transmission in the 3–148 kHz frequency band to monitor some home automation products, such as smart meters. This monitoring feature makes PLC an enabler for sensing, control, and automation in large systems comprised of a large number of components spread over a relatively wide area in industrial applications, e.g., large-scale control in automation and manufacturing systems serving as a useful common communication network connecting a large number of devices (Bumiller at al. 2010) and energy control networking technology for smart grids, smart cities, smart buildings, and smart mobility. These few examples show that there is no need for additional wires to power devices, e.g., smart meters in buildings, to communicate with the neighborhood data concentrator. Powerline communication, on the other hand, can traverse the power lines to reach the data concentrator. For these reasons, most utilities around the world have chosen PLC for their smart grid projects; and most cities have chosen ubiquitous computing and PLC for their smart streetlight projects (Echelon 2015).

Another point to be considered in a PLC network is that changes within the network happen often. For example, changes in a medium-voltage energy system will affect the network transfer function; physical removal; or, in addition to the network, a repair or network improvement can be expected. Thus, changes in topology or signal transmission affect the communication, but the PLC network would be able to purge these changes, making PLC different from other types of networks.

4.4.1   Internet of Things and Powerline Communication

Services over the Internet of Things (IoT) evolved based on the need identified for person to person, person to thing, and other interactions, such as thing to person, person to machine, and thing to thing or machine to machine, based on ubiquitous and pervasive computing. Such systems are based on RFID technology. New services for the IoT (Moeller and Vakilzadian 2014b), the meeting point of the real and the virtual worlds, are built using RFID combined with other technologies, such as sensor technology, mobile technology, and smart metering.

The IoT and PLC are important areas of concentration with regard to the home networking application. This application is based on the idea of building a smart home environment by connecting things to things, things to a human, or a human to things at home. This concept emerged before IoT and RFID became popular. Within the IoT paradigm, services have been developed for home networking as part of the IoT services, but they do not have the same connectivity issues as RFID or sensors, tiny devices with limited resources, mainly battery powered (Chaouchi 2010).

With regard to smart cities, PLC and IoT can help to reduce the worldwide use of energy. About two-thirds of the world's energy and 60 % of water are consumed by cities which generate 70 % of greenhouse gases. Today's awareness of climate change is forcing the current infrastructure to become more efficient. With innovative technologies, our cities can be made more environmentally friendly and, at the same time, offer a higher quality of life and reduce costs. Besides the traffic sector, buildings are targeted because they are responsible for 40 % of energy consumption and produce about 21 % of greenhouse gases (CO_2). Using efficient technologies allows power consumption and CO_2 emissions from buildings to be reduced without compromising on comfort. The concept of a smart environment is based on IoT and PLC research to create smart cities, smart homes, smart buildings, and smart mobility.

A green or smart building requires innovative solutions to build an infrastructure that makes use of ubiquitous computing, the IoT, and PLC. This infrastructure includes:

- Access control systems
- Building automation
- Electrical installation
- Fire protection and evacuation
- Heating, ventilation, and air conditioning
- Identification systems
- Intrusion detection and video surveillance
- Water supply and water consumption

Using IoT and PLC in conjunction with sensors and RFID for fire protection and evacuation enables emergency personnel to instantly recognize who is/was in the building, who is still in the building, and where and what makes an evacuation and firefighting operations more efficient. In the case of electrical installation technology, energy-saving lamps can be replaced with light-emitting diodes (LEDs) which save tons of CO_2 and thousands of dollars on the electricity bill. Moreover, lights will be shut off when a room is unoccupied. In addition, the room temperature can be reduced when the resident leave for work; and it can be increased to a comfortable level in advance of the occupants arriving back home from work. The home system can also be controlled from outside of the home, for example, to determine if windows are closed or if the coffee maker has been shut off. The energy provider can read the energy consumption by day, week, or month and by application area. Thus the operation of buildings and/or homes will be more simple, safe, reliable, environmentally friendly, and cost effective by using smart devices in conjunction with IoT and PLC.

These smart devices can collect data through interconnected sensors and actuators and react in real time at the request of events. Thus, having a reliable communication network infrastructure, such as a ubiquitous networking system, that allows a smart device program to follow the user is a major concern. This infrastructure needs to be inexpensive, easy to install and maintain, and reliable.

Traditional networks with high performance and reliability are expensive and difficult to manage in a home environment. These factors make ubiquitous networks and PLC good candidates for network infrastructure for smart home environments because of their ready availability and almost no installation cost for PLC.

4.4.2 Smart Grid

A smart grid is an electrical grid that uses analog or digital information and communications technology (ICT) to gather and act on information, such as information about the behaviors of suppliers and consumers, in an automated fashion. This improves efficiency, reliability, economics, and sustainability of the production and distribution of electricity while decreasing the cost with regard to power distribution and consumption (Niyato et al. 2011).

Smart grids contain three major functionalities:

- Power generation
- Power distribution
- Power consumption

Smart grid policy is organized in Europe as the Smart Grids European Technology Platform. Smart grid policy in the USA is described in 42 U.S.C. ch. 152, subchapter IX§17381. The rollout of smart grid technology also implies a fundamental reengineering of the electricity services industry, although typical usage of the term is focused on the technical infrastructure (Torriti 2012).

Thus, energy management as part of the smart grid approach only deals with power consumption in which different appliances can be monitored regularly with the help of smart meters and controlled by a central controller to balance the energy consumption at different times. This is an important application in the smart home environment. A smart meter is an electrical meter that records consumption of electric energy in intervals of an hour or less and communicates that information, on at least a daily basis, back to the electric power company for monitoring and billing. Besides the automatic meter reading features, smart meters are also used for two-way communication to reduce load, connect/reconnect remotely, and interface to gas and water meters. Operational field tests have been performed.

Energy consumption can also be displayed on customer computers. For this reason, the smart meter software offers numerous features, such as displaying consumption changes and switching off individual devices. The communications package allows the smart meter to be integrated as a single solution in the remote readout (www.open.pr).

In a recently published survey, it was found that consumers are interested in real-time energy pricing plans in exchange for allowing their utility to remotely control in-home devices. Using a smart meter, real-time pricing plans provide the optimum in energy and peak load savings when combined with home energy equipment.

These smart home solutions are enabled by ubiquitous computing in conjunction with PLC and IoT which can be offered as cloud computing service by the respective energy suppliers and can revolutionize energy savings and can pave the way for a new generation of demand response programs. These programs include access to smart thermostats as most popular home energy management devices followed by lighting controls and other smart appliances.

For the digital natives, a smart phone/tablet app-based home energy management is the most important consideration.

From the foregoing mentioned, it can be seen that smart home energy management systems offer the essential services to track power consumption. As in energy management systems, data about the energy consumption need to be communicated to a central place for analysis which requires a reliable network. For this purpose the availability of PLC is a good candidate as a home energy monitoring communication medium because, as with an increase in demand for real-time pricing of energy, there is a need to balance the energy consumption and peak time load. Smart meters and real-time communication with a central energy management station are the basic requirement for this real pricing application. PLC will serve as a communication medium between this central energy management station and the smart meter at home.

A more visionary case of energy management is the control and data acquisition whereby the grid load is monitored regularly and reported to the central energy management control center from where commands are sent to manage the generation and consumption. This needs to be executed in real time to keep the grids stable. Therefore, smart homes will be the major help to build up an energy management system, with an automated control system (consisting of a smart home network based on PLC, a rule-based context management system and the IoT access) to monitor the energy consumption at different times regularly and alert the owner or manage the energy supply itself by controlling the energy supply of the manifold and different devices at home. Doing so smart grids require real-time information online to balance the load and do it with the respective smart meters at home. Up-to-date information is required from this smart meter to enable the user to be updated remotely on the energy consumption.

Decoding PLC signals will take some effort to figure out how to read them. Important details on how to decode one commonly used PLC smart meter system are already available on the web, which may result in privacy problems. Once someone has figured out how to decode a PLC system, he will boast about it. There are websites dedicated to this sort of information. Therefore, the US Congressional Research Service cites two studies showing that a smart meter reading every 15 min can access sufficient details to find out what goes on inside a private home. This includes information obtained through measures with embedded statistical methods, e.g.:

• When people cook food
• When people go to bed and wake up
• When people shower

- What is stored in the refrigerator; what is consumed
- When the residents are not at home

This kind of information can be used in various ways, including by would-be burglars and abductors, tabloid magazines, government agencies, the insurance industry, and so forth. Think about the following refrigerator information scenario: IoT and RFID in conjunction with PLC store the information what a person took out of the refrigerator, on which day and at what time, e.g., every day late at night, a person consumes three cans of cola, three bowls of chocolate ice cream, and two big bars of chocolate. This is the person's secret, but unfortunately, someone hacked into the person's smart home environment and posted his secret on the Internet with his name and address. This is a scenario which is not amusing. Hence, security and privacy of smart grids and smart metering networks are important to their rollout and acceptance by the public. Therefore, research in this area is ongoing, and smart meter users need to be reassured that their data is secured.

In Efthymiou and Kalogrids (2010), a method for securely anonymizing frequent, e.g., every few minutes, electrical metering data sent by a smart meter is discussed. Although such frequent metering data may be required by a utility or electrical energy distribution network for operational reasons, this data may not necessarily need to be attributable to a specific smart meter or consumer. It does, however, need to be securely attributable to a specific location, e.g., a group of houses or apartments, within the electricity distribution network.

In Kaplantzis (2012), the smart meter is introduced as a gateway to the household, with the ability to constantly monitor attached devices but, more importantly, the ability to switch them on and off, making security a pressing issue. Hence, a hacker who successfully dissimulates a smart meter can access confidential information, change control commands, and deny access to legitimate systems. Therefore, it becomes apparent that threats, such as repudiation, masquerading, and unauthorized access, need to be addressed to avoid hacking of the communication medium used by the smart devices (wireless and Bluetooth), cracking the device's smart card, attacking the IT infrastructure of the electrical grid itself, and even intercepting IP streams of devices that choose to connect via the Internet.

As described in Kaplantzis (2012), a secure smart grid has to uphold the following requirements while managing data:

- *Confidentiality*: requires that only the sender and the intended receiver be able to understand the contents of a message. Confidentiality of information generated and transmitted by the smart grid is paramount to customer privacy and grid success.
- *Integrity*: requires that the sender and receiver ensure the message is authentic and has not been altered during transit without detection. In smart grids, integrity means preventing changes to data measured and control commands by not allowing fraudulent messages to be transmitted through the network.
- *Availability*: requires that all data is accessible and available to all legitimate users. Since the smart grid is not only communicating usage information but also

controlling messages and pricing information, the availability of this information is crucial for successful operation and maintenance of the grid.

- *Nonrepudiation*: requires that sender and receiver cannot deny they were the parties involved in the transmission and reception of a message. Accountability of the members of a data transaction is critical when it comes to financial interactions. However, data generated by the network can be owned by different entities (customer, data management services, billing systems, utilities) at different times of the data life cycle.

4.4.3 Smart Home Energy Management

The increased demand for energy has resulted in an efficient system for energy suppliers to manage energy consumption at the household level. There are different load control programs for individual appliances, such as water heating systems, cooling units, etc. However, the model proposed for Smart Home Environment Energy Management Systems (SHE^2MS) monitors and controls appliances according to a well-defined set of energy usage limits and comfort levels of the homeowners. Because different appliances are running in a house at different times, these appliances may not be needed all of the time, they may not need to be run at the same time, or some of them can be turned off at times of peak energy consumption to balance the load or to reduce expenses by shifting the load to some other time slots, if this does not interfere with the comfort level of the homeowners. Definitions of these different time spans can use different factors, such as appliance usage patterns, comfort level of the residents in case an appliance can be turned off without affecting the home-hold operation/needs, energy prices at different times, or availability of the energy load at a specific time span.

To define the priority for an appliance's operation, SHE^2MS will turn some devices off to balance the load or to reduce the expenses at peak times. Devices with the lowest priority will be turned off first, such as an e-car which needs to be recharged by 8 or 9 a.m. Charging the e-car by the end of the night will have the high priority if it has not already been done. Heating or cooling appliances can have the lowest priority during the daytime when nobody is at home. Usage of other appliances, such as refrigerators and dishwashers can be prioritized depending on the needs of the user.

A preference for a comfort level over energy saving must be defined, as there will be times when a resident does not want a specific appliance to be turned off even if its use is causing an excess in load. Therefore, the performance of a SHE^2 MS is based on the potential energy limit. Thus, for all operating appliances, the condition demand limit needs to be met. Exploiting the availability limits will result in distraction in the appliance operation. As the demand for energy from operating appliances exceeds the available limits set by the SHE^2MS, devices will be turned off, starting with the lowest-priority devices. Hence, the workflow of the SHE^2MS can be summarized in the following steps, as described in a project by Ahmed (2013) as part of the Internet of Things class at TUC (2013):

- Analyze operating conditions of appliances.
- Calculate energy consumption.
- Check available energy limits.
- Check the priority of operating devices.
- Check the status of all appliances.
- If appliances are operating at the defined comfort level, turn those devices on.
- If the energy load is exceeding the defined limits, turn the low-priority devices off.
- Resume operation of the devices turned off when the minimum time required by the high-priority devices has been completed or the appliances are no longer operating.

This project shows that energy-efficient buildings require real-time monitoring, measurement, and facility management through implementation of different energy management control techniques and algorithms so that energy consumption in buildings can be reduced.

4.5   RFID Applications

RFID technology is one of the leading enabling technologies in the IoT. In terms of cost, passive RFID tags range from 0.25 US$ up to 10 US$, depending on functionality, packaging, and application. Prices of passive tags are highly dependent on the volume of tags ordered. Low volumes generally lead to higher prices per tag than higher volumes (millions). There is the potential for even lower prices for simple tags as standards solidify and as larger numbers of tags are used. Thus, passive tags with read-only serial numbers will approach 0.05 US$ today and 0.01 US$ in the near future. Table 4.7 shows the prices and field of operation for different tags (Miles et al. 2008).

The MIT Auto-ID Center introduced a tag classification system which is also used by EPCglobal. Tags are grouped into passive, semipassive, and active, as shown in Table 4.8 after Botero and Chaouchi (2010):

- *Class 0*: read-only tags with simple ID numbers of 64 or 96 bits; can be EPC; cannot be modified.

Table 4.7 Price and field of operation for different tags

Price per tag [$]		Wide area tracking		Active (UWB WiFi)
500	Passive tags			Semipassive tags
50				
5	Local area tracking UHF			
0.50	Close area tracking			
0.05	HF/LF			
Distance [m]	1	10	100	1,000

Table 4.8 Tag classes.
For details see text

Tag class	Type	Capabilities
Class 0	Passive	Read only
Class 1		Read, write once
Class 2		Read/write
Class 3	Semipassive	Increased range
Class 4	Active	Tag communication
Class 5		Reader capabilities

- *Class 1*: read/write passive tags; can only be written once, either by manufacturer or user.
- *Class 2*: read/write passive tags; can be written several times; additional functionalities like data logging and/or cryptography may be included.
- *Class 3*: semipassive tags with extra energy tags can increase reading distance range and provide new functionalities, e.g., sensors.
- *Class 4*: provide communication functionalities with other active tags and have features as in Class 3.
- *Class 5*: have reader capabilities enabling the tag to communicate with all types of tags.

Other areas for RFID applications are:

- Access control, tracking, and tracing of individuals and goods
- Animal monitoring
- Data warehousing and analysis
- Homeland security
- Household
- Location tracking
- Inbound and outbound logistics and supply chain
- Logistics and supply chain monitoring
- Loyalty, membership, and payment
- Luggage tracking
- Medical and pharmaceutical
- Passenger tracking
- Production, monitoring, and maintenance
- Product safety, quality, and information
- Public transportation
- RFID security
- Smart city
- Smart energy
- Smart home
- Smart mobility
- Smart ports
- Vehicle tracking
- And more

As a more detailed introduction to RFID applications, the medical application will be examined in more detail, where RFID enables tracking and tracing of objects and/or things. In the case of patient tracking, the following activities are essential (http4 2015):

- *RFID-enabled patient identification and location assistance*: often needed to ensure patient safety when urgent medical action is needed. Patient tags with RFID can meet this need.
- *RFID-enabled wristbands*: used by hospitals to identify, locate, and obtain status updates on patients in the hospital to better manage patient care. They also provide accurate identification to maintain the safety of newborn infants.
- *RFID readers installed at doors*: used by facilities housing Alzheimer's disease patients to help determine the location of patients and maintain their safety in the event they attempt to enter unauthorized areas.
- *RFID-enabled prescription containers*: blind patients and visually impaired individuals can use devices placed on prescription containers to "read" the RFID-tagged prescriptions and learn dosage amounts.

In addition to patient tracking, product and device tracking is another important task that needs to be accomplished safely and securely in a hospital. In this case, the following activities (and much more) are required:

- *Tracking radioactive isotopes* throughout a facility, from storage to transport and from administration to disposal. RFID tags and readers can automate these tasks, thereby saving time and resources. Active RFID tags with read/write capabilities can be used to detect seal integrity for containers and individual packages. Tags can record time and duration of seal loss, allowing even problems that occur midshipment to be detected.
- *Tracking the location, operating time, and maintenance of medical devices*, such as respirators, anesthesia machines, blood pumps, infusion pumps, and more. RFID tags and readers can automate these tasks, thereby saving time and resources. Active RFID tags with read/write capabilities can be used to detect the start of surgery in the operating theater, determine the medical devices used, and identify the right anesthesiologist to run the respective anesthesia device. Tags can record start, stop, and elapsed time of the surgery.

Another important task in medical aid is in aviation, as described in Sect. 6 of the *IATA Medical Manual* (IATA 2013). The average healthy passenger tolerates air travel very well; however, the cabin environment may present significant challenges to those with medical problems. More people are traveling, including the elderly and those with medical problems, because of the changes in demography and attitudes toward air travel. Therefore, every airline should have a medical clearance procedure. The International Air Transport Association (IATA) believes that medical guidelines should be reasonably consistent and based on accepted physiological principles for the benefit and protection of the passenger and the

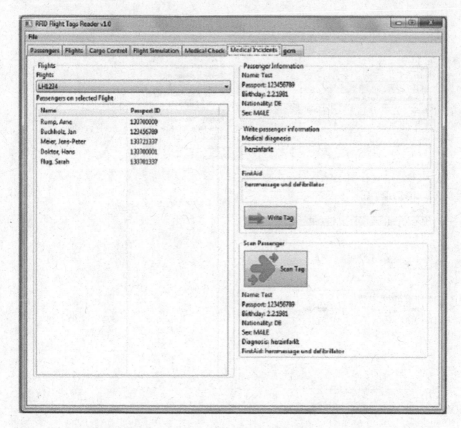

Fig. 4.11   Passenger-related medical information for an airline flight

safety of the flight. In Sect. 6.1.2, the general guidelines for medical clearance are introduced. With regard to this, a generic passenger-specific database with graphical user interface has been developed for a trial. In Fig. 4.11, the initial graphical user interface is shown in which additional passenger information with regard to medical information is recorded.

In the event of a medical problem on board an airplane, the crew asks if there is a medical doctor (MD) on the flight. This can also be embedded within the expanded passenger data list, as shown in the graphical user interface in Fig. 4.12.

This information can be merged with the passenger seat reservations. Both types of information are also attached to the electronic boarding pass which contains an RFID. These RFID tags can be read by onboard readers used by the crew to automate these tasks, thereby saving time and resources because the crew can immediately contact the medical doctor (MD) and request his/her assistance with a medical emergency on board. Moreover, the MD can cyber-physically consult with the airline's medical center via the Internet, getting advice for medical aid on board or medical care after landing.

Fig. 4.12 MD-passenger-related information for an airline flight

4.6   Case Study: Luggage Tracking System

RFID is used to automatically identify and/or locate individual physical objects, such as luggage in the aviation business. The Société Internationale de Télécommunication Aéronautique (SITA) reported in 2007 that ≈42 million pieces of luggage worldwide have been lost. This requires additional work for airlines.

- 2005: 61 % of luggage was misrouted; 16 % could not be loaded on board; 13 % was misrouted due to ticketing failure and security checks; 7 % was misrouted during loading and unloading failures; and 3 % was misrouted due to incorrect luggage tags, etc.
- 2008: 20 % less pieces of passenger luggage were misdirected compared to 2007.
- 2009: 40 % less pieces of passenger luggage were misdirected compared to 2007, which means 25 million pieces of luggage were misdirected (http2 2014).

- SITA reports that 52 % of misrouting of luggage happens during transfer of luggage from one plane to another.
- Lost and found activities are costly for airlines due to financial compensation. SITA reports about 3.9 billion US$ is spent by the airlines worldwide.
- Reducing the loss of luggage further cannot be achieved with barcode labels.
- An alternative solution is to use RFID transponders.
- SITA reported that the number of misdirected luggage dropped when RFID was used.

Besides reducing the number of lost or misdirected pieces of luggage by using RFID, distinct luggage identification and fully automated luggage transport from check-in to boarding can be accomplished.

Using RFID-tagged luggage or freight results in faster loading and unloading of aircrafts. Another positive side effect of RFID-tagged luggage is the faster identification of luggage with no owner on board when unloading. With regard to the increased number of passengers, the following assumptions can be made for the case study:

- Decreased dispatch time.
- Misrouted luggage is expensive.
- More pieces of luggage.

RFID allows, in contrast to barcodes:

- Fast scanning
- High memory capacity
- Minimum number of read errors

Assume the following specifications for the case study:

- Trays (plastic bowls) run at high velocity on conveyor belts.
- Tilt trays for sorting.
- RFID-chip arrays.
- Considerably faster than barcodes.
  - Approx. 10 m/s
- System
  - Frequency: 13.56 MHz
  - Passive transponder
  - Memory
    8 kByte FRAM
    20 kByte EEPROM

Luggage is loaded into the aircraft via a mobile conveyer belt through the cargo area door, where an RFID reader can be positioned to scan the tags. This requires

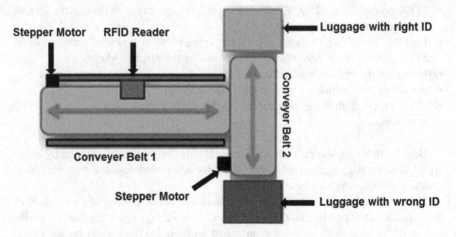

Fig. 4.13 Model of the RFID-based luggage transportation system

standardization of the reader system in relation to the door width, depending on the aircraft.

For luggage transportation in the luggage cargo hold, a model was built to test two scenarios based on the physical model shown in Fig. 4.13 (Waldmann 2013):

Scenario 1:

This scenario consists of an inductive coupled system with a service area of 1 m and an operating frequency of 13.56 MHz. To guarantee that pieces of luggage can be identified in any position, several readers are necessary. The readers must be positioned at angles of 0°, 45°, and 90° on the wall and at the ceiling to ensure safe detection of the tags. Passive smart label tags with a memory capacity of 128 bytes are appropriate.

Scenario 2:

A geared reader mounted at the ceiling has a narrow cone for a service area with which the whole inner area of the luggage hold can be observed. The operating frequency is 868 MHz. Passive smart label tags with a memory capacity of 128 bytes are used.

To run the RFID Flight Tag software prototype, the classes, MasterControlProgram and CheckInTab, have been defined as well as the classes, Engine and Manager. The relationships between these classes are shown in the following excerpt of the class diagram. The MasterControlProgram class generates an object of class Manager which generates an object of class Engine, as shown in Fig. 4.14 (Waldmann 2013).

The case study begins with the MasterControlProgram class, which results in the window in Fig. 4.15. There are four tabs from which the user can select an operation. First, the Passengers tab is activated, which is grouped into three arrays

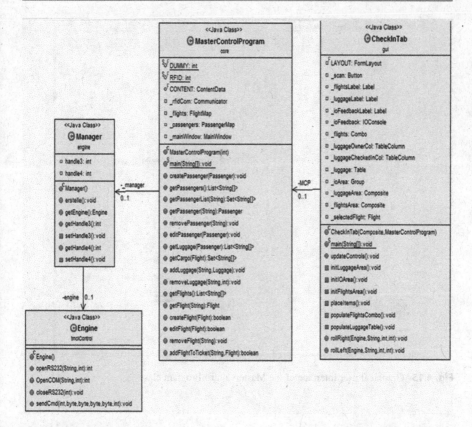

Fig. 4.14 Class diagram

with three tables. The left side shows the passengers. A passenger can be generated by clicking the *Plus* button below in the screenshot of Fig. 4.15 (Waldmann 2013).

After clicking the Passenger button, a pop-up window opens which contains passenger-specific information, as shown in Fig. 4.16. NB: Passport-Number consists of a 9- or 11-digit number; the nationality is given by capital letter.

After generating a passenger, clicking the *Plus* button in the upper right area of the window generates a piece of luggage which is linked to the passenger. Afterward, a transition is required to the Flights tab to generate a flight. Clicking the *Plus* button results in a pop-up window in which the flight data can be entered. NB: the Flight-ID is based on two capital letters and subsequent four digits (Fig. 4.17).

After flight generation, the user can access the Passengers tab again by clicking the *Plus* button at the bottom right to allocate the passenger to the right flight. Once this has been done, the message "Ticket invalid" changes to "Ticket valid" to indicate the assignment was successful. Now the data set containing passenger, luggage, and booked flight is ready to write to the tag. This is done by clicking the Write Tag button.

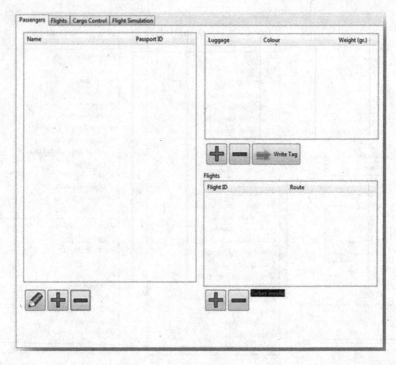

Fig. 4.15 Graphical user interface of the MasterControlProgram class

Fig. 4.16 Graphic user
interface pop-up for
passenger-specific
information

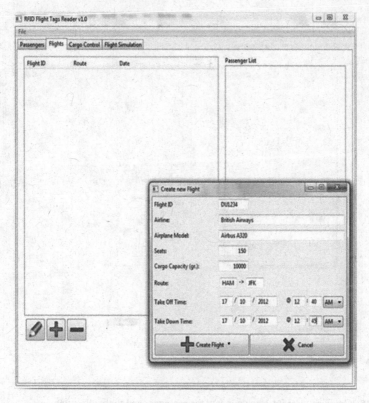

Fig. 4.17 Graphical user interface window for flight-specific data

So far, the tag has been written and the luggage control process started. The luggage is positioned at the start position of the conveyer belt to allow reading of the tag ID when the luggage passes the reader.

Choosing a flight all passengers on the flight and the checked luggage are controlled. Clicking the Scan Luggage button starts the luggage control process. The motor of the conveyer belt starts. When the luggage arrives at the reader, the ID of the luggage is scanned. After scanning, the software shows a different behavior:

- *Luggage is bound for the flight*: a second conveyer belt (marked green in Fig. 4.13) starts moving counterclockwise transporting the luggage to the right luggage wagon.
- *Luggage is not bound for the flight*: another conveyer belt (marked red in Fig. 4.13) starts moving to the right transporting the luggage back for inspection (Fig. 4.18).

The physical demonstrator of the RFID-based luggage transportation system is shown in Fig. 4.19.

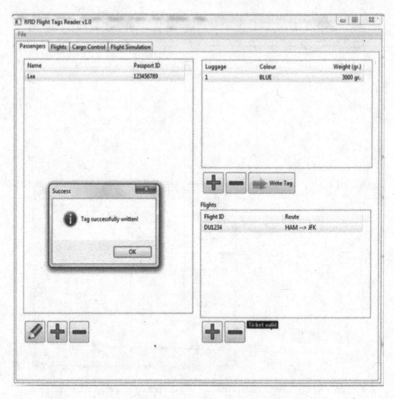

Fig. 4.18 Graphical user interface window for passenger and flight allocation

Fig. 4.19 Demonstrator of the luggage transportation system

4.7 Exercises

What is meant by the term *Internet of Things*?
Describe the architectural structure of the Internet of Things.
What is meant by *object or things in the Internet of Things*?
Give an example of objects or things in the IoT.
What is meant by *sensing and actuating technology*?
Give an example of sensing and actuating technology.
What is meant by the term *transponder*?
Describe the architectural structure of a transponder.
What is meant by the term *RFID*?
Give an example of an RFID architecture.
What is meant by the term *RFID reader*?
Give an example of an RFID reader's architecture.
What is meant by the term *capacitive coupling*?
Give an example of a capacitive coupling.
What is meant by the term *inductive coupling*?
Give an example of an inductive coupling.
What is meant by the term *smart label*?
Give an example of a smart label architecture.
What is meant by the term *energy supply of transponder*?
Give examples of the different forms.
What is meant by the term *RFID middleware*?
Describe their features.
What is meant by the term *passive transponder*?
Describe the architectural structure of a passive transponder.
What is meant by *RFID tag prices and field of operation*?
Give an example of passive tags.
What is meant by the term *RFID tag classes*?
Give an example of passive tags.
What is meant by *RFID architectural model*?
Give an example of it.
What is meant by *home automation*?
Give an example of the benefits of home automation.
What is meant by the term *sensor*?
Describe the structure and the function of a strain gage sensor.
What is meant by the term *contact sensor*?
Give an example of a contact sensor.
What is meant by the term *contactless sensor*?
Give an example of a contactless sensor.
What is meant by the term *analog-to-digital converter*?
Describe the quantization function of an *analog-to-digital converter*.
What is meant by the term *sensor network*?
Give an example of it.
What is meant by the term *sensor node*?

Give an example of a sensor node and describe the benefits of it.

What is meant by the term *wireless sensor*?

Describe the structure and the function of a wireless sensor.

What is meant by the term *wireless sensor network*?

Describe the structure and the function of a wireless sensor network.

What is meant by *a routing in a wireless sensor network*?

Give an example of a routing in a wireless sensor network.

What is meant by *fixed routing and adaptive routing*?

Give examples of fixed and adaptive routing schemes in wireless sensor networks.

What is meant by the terms *deadlock* and *livelock*?

Give examples for deadlock and livelock in wireless sensor networks.

What is meant by *clustering in wireless sensor networks*?

Give an example for it.

What is meant by the term *powerline communication*?

Describe the architectural structure of a powerline system.

What is the *purpose of powerline communication*?

Give an example of high bit-rate powerline communication PLC architecture in the home.

What is meant by the term *smart meter*?

Give an example of a smart meter.

What is meant by *home automation*?

Describe the benefits of home automation.

References

(Ahmed 2013) Ahmed, W.: Power line Communication Energy Management System, Project work at TU Clausthal, 2013

(Andrew et al. 2012) Andrew, S., Balandin, S., Koucheryavy, Y. Eds.: Internet of Things, Smart Spaces, and Next Generation Networking, 12th International Conference NEW2AN 2012 Proceedings, Springer Publ., 2012

(Auto-ID 2013) http://www.autoidlabs.org/; accessed February 12, 2013

(Berlin and Gabriel 1997) Berlin, A. A., Gabriel, K. J.: Distributed MEMS: New Challenges for Computation. IEEE Computational Science and Engineering, Vol. 4, pp. 12–16, 1997

(Biddlecombe 2005) Biddlecombe, E.: BBC News, 17.11.2005

(Botero and Chaouchi 2010) Botero, O., Chaouchi, H.: RFID Applications and Related Research Issues. pp. 129–156, In: The Internet of Things,. Ed. H. Chaouchi, Wiley Publ., 2010

(Bumiller et al. 2010) Bumiller, G., Lample, L., Hrasnica, H.: Power Line Communication Network for Large Scale Control and Automation Systems, IEEE Communications Magazine, Vol. 48, pp. 106–113, 2010

(Carcelle 2006) Carcelle, X.: Power Line Communications in Practice. Archtech House Publ. 2006

(Chaouchi 2010) Chaouchi, H.: The Internet of Things – Connecting Objects to the Web. J. Wiley Publ., 2010

(Cisco 2008) Wi-Fi Location-Based Services 4.1 Design Guide, Cisco Systems Inc. 2008, Text Part Number: OL-11612-01

(Dargie and Poellabauer 2010) Dargie, W., Poellabauer, C.: Fundamentals of Wireless Sensor Networks: Theory and Practice, John Wiley Publ., 2010

(Echelon 2015) Echelon 2015: http://www.echelon.com/

(Efthymiou and Kalogrids 2012) Efthymiou, C., Kalogrids, G.: Smart Grid Privacy via Anonymization of Smart Metering Data. In: Proceed. 1st IEEE International Conference on Smart Grid Communication, pp. 238–243, IEEE Publ. 10.1109/SMARTGRID.2010.5622050, 2010

(Floerkemeier et al. 2008) Floerkemeier, C., Langheinrich, M., Fleisch, E., Friedemann, M., Sarma, S. E. Eds.: The Internet of Things, 1st International Conference IOT Proceedings, Springer Publ. 2008

(Giusto et al. 2010) Giusto, D., Iera, A., Morabito, G., Atzori, L., Eds.: The Internet of Things, 20th Tyrrhenian Workshop on Digital Communications, Springer Publ. 2010

(Goldstein et al. 2005) Goldstein, S. C., Campbell, J. D., Mowry, T. C.: Programmable Matter. Computer, Vol. 38, pp. 99–101, 2005

(Gubbi 2013) Gubbi, J., Buyya, R., Marusic, S., Palaniswamia, M.: Internet of Things (IoT): A Vision, Architectural Elements, and Future Directions, In: Future Generation Computer Systems, Vol. 29, pp. 1645–1660. Elsevier, 2013

(Harish 2010) Harish, A., R.: Radio Frequency Identification Technology. In: The Internet of Things – Connecting Objects to the Web. Chapter 2, pp.35–52, J. Wiley Publ., 2010

(IATA 2013) Medical Manual 2013, 6th Edition, ISBN 978-92-9252-195-0

(IOT 2013) http://postscapes.com/internet-of-things-history; accessed February 12, 2013

(Kahn et al. 2000) Kahn, J. M., Katz, R. H., Pister, K. S. J.: Emerging Challenges: Mobile Networking for Smart Dust. J. Communication and Networks, Vo. 2, pp.188–196, 2000

(Kaplantzis and Sekercioglu 2012) Kaplantzis, S., Sekercioglu, Y. A.: Security and Smart Metering. European Wireless 2012 (http://titania.ctie.monash.edu.au/papers/06216815.pdf)

(Kärkkäinen 2003) Kärkkäinen, M.: Increasing efficiency in the supply chain for short shelf life goods, International Journal of Retail & Distribution Management, Vol. 31, Number 10, pp.529–536, 2003

(Lewis 2004) Lewis, F. E.: Wireless Sensor Networks. In: Smart Environments: Technologies, Protocols, and Applications, pp. 1–18. Eds.: Clark, D. J., Das, S. K., John Wiley Publ. 2004

(Mehrotra et al. 2004) Mehrotra, S., Butts, C., Kalashnikov, D., Venkatasubramanian, N., Rao, R., Chockalingam, G., Eguchi, R., Adams, B., Huyck, C.: Project RESCUE: Challenges in Responding to the Unexpected. J. Electronic Imaging, Displays, and Medical Imaging, pp. 179–192, 2004

(Miles et al 2008) Miles, S. S., Sarma, E., Williams, J.: RFID Technology and Applications, MIT University Press, 2008

(Moeller and Vakilzadian 2014a) Moeller, D. P. F., Vakilzadian, H.: Ubiquitous Networks: Power Line Communication and Internet of Things in Smart Home Environments, In: Proceed. IEEE International Conference on Electro Information Technology, pp. 596–601, DOI:10.1109/EIT.2014.6871832, IEEE Conference Publications, 2014

(Moeller and Vakilzadian 2014b) Moeller, D. P. F., Vakilzadian, H.: Wireless Communication in Aviation through the Internet of Things and RFID, In: Proceed. IEEE International Conference on Electro Information Technology, pp. 602-607, DOI:10/1109/EIT.2014.6871833, IEEE Conference Publications, 2014

(Niyato et al. 2011) Niyato, D., Lu, X., Ping, W.: Machine-to-Machine Communications for Home Energy Management System in Smart Grid, IEEE Communication Magazine, Vol. 49, pp. 53–59, 2011

(Poslad 2009) Poslad, S.: Ubiquitous Computing, John Wiley Publ., 2009

(Roberti 2005) Roberti, M.: The History of RFID Technology, RFID Journal, 2005; http://www.rfidjournal.com/articles/view?1338/2

(Schönefeld and Moeller 2012) Schönefeld, J., Moeller, D. P. F.: Runway Incursion Prevention Systems: A Review of Runway Incursion Avoidance and Alerting System Approaches. In: Progress in Aerospace Sciences: Vol. 51, pp. 31–.49, 2012

(Sohraby et al. 2007) Sohraby, K., Minoli, D., Tnati, T.: Wireless Sensor Networks: Technology, Protocols, and Application, John Wiley Publ., 2007

(Stockmann 1948) Stockmann, H.: Communication by Means of Reflected Power,

(Torriti 2012) Torriti, J.: Demand Side Management for the European Supergrid, Energy Policy, Vol. 44, pp. 199–206, 2012

(TUC   2013)   https://video.tu-clausthal.de/videos/iasor/vorlesung/iot-ss2013/   20130621/iot-20130621.html

(Waldmann 2013) Waldmann, L. RFIF for procise identification of air cargo: Optimization and risk analysis and follow up development of a prototype RFID workbench. Bacher Thesis (in German), Hamburg, 2013

(Wang and Zhang 2012) Wang, Y., Zhang, X. Eds.: Internet of Things, International Workshop Proceedings, Springer Publ., 2012,

(Weber and Weber 2010) Weber R. H., Weber, R.: Internet of Things – Legal Perspectives, Springer Publ. 2010

(Weiser 1991) Weiser, M.; The Computer for the 21st Century. In: Scientific American, pp. 94–100. 1991

(Yick et al. 2008) Yick, J., Mukherjee, B., Ghosal, D.: Wireless sensor network survey, Computer Networks, Vol. 52, pp. 2292–2330, 2008

(Zaheeruddin and Mandviwalla 2005) Zaheeruddin, A., Mandviwalla, M.: Integrating the Supply Chain with RFID: A Technical and Business Analysis, In: Communications of the Association for Information Systems, Vol. 15, pp. 393–427, 2005

(Zhao and Guibas 2004) Zhao, F., Guibas, Wireless Sensor Networks: An Information Processing Approach. Morgan Kaufmann Publ. 2004

Links

(http1 2015) http://en.wikipedia.org/wiki/Sensor_node; accessed January 21, 2015

(http2   2014)   http://www.sita.aero/knowledge-innovation/industry-surveys-reports/baggage-report-2010

(http3 2014) http://www.sita.aero/content/baggage-report-2008

(http4 2015) http://www.ups-scs.com/solutions/white_papers/wp_RFID_in_healthcare.pdf

Ubiquitous Computing

5

This chapter begins with a brief overview of ubiquitous computing. Section 5.1 defines ubiquitous computing and illustrates some of its applications. Section 5.2 introduces ubiquitous communication fundamentals and key research areas that are shaping the field. Thereafter, Sect. 5.3 describes smart devices and services in ubiquitous computing. Section 5.4 covers the important topics of tagging, sensing, and controlling in ubiquitous computing and possible applications. Section 5.5 focuses on autonomous systems, analyzing their behavior and composite structure and describing fault-tolerant behavior. A case study of the use of ubiquitous computing with robotic manipulators is described in detail in Sect. 5.6. Section 5.7 contains comprehensive questions from the ubiquitous computing area, followed by references and suggestions for further reading.

5.1   Ubiquitous Computing History to Date

The original term *ubiquitous computing* was coined by Mark Weiser in 1988 at Xerox's Palo Alto Research Center (PARC) while he served as the director of the Computer Science Lab. He envisioned a future in which computing technologies were embedded in everyday artifacts, supporting daily activities and equally applicable to our workplaces, homes, and much more (Krumm 2009).

The foundational articles of ubiquitous computing written by Mark Weiser are not many in number but are broad in the scale of their implications. A concise summary of ubiquitous computing, as it was originally referred to by researchers at PARC, is Weiser's most relevant article published in 1991 in *Scientific American*, pp. 94–100, entitled "The Computer for the 21st Century" (Weiser 1991). Other write-ups by Weiser on different aspects of ubiquitous computing can be found in the following references: Weiser (1993a, b, 1994).

Therefore, ubiquitous computing (also referred to as pervasive computing) describes how current technologies (smart things or objects) with some kind of attachment, embedment, blending of computers, sensors, tags, networks, and others

© Springer International Publishing Switzerland 2016
D.P.F. Möller, *Guide to Computing Fundamentals in Cyber-Physical Systems*,
Computer Communications and Networks, DOI 10.1007/978-3-319-25178-3_5

like smart devices (mobile, wearable, wireless), smart environments (embedded computing systems) (see Chap. 2), sensor-actor networks, and smart interaction, (tight integration of and coordination between devices and environments, anything with everything) relate to and support a computing vision for a greater availability and range of computer devices. These devices are used in a greater scope of cyber-physical systems (computing and communication systems, human, physical, and virtual) and more environments and activities (Poslad 2009). This means that ubiquitous computing can be introduced as "being everywhere at once."

The aforementioned term, *smart devices*, dominates discussions about digital devices. They are generally connected to other devices or ubiquitous networks, such as object-2-object (O2O) and thing-2-thing (T2T), via protocols such as Bluetooth-specific protocols, with the controller stack protocol Link Management Protocol (LMP) and the host stack protocol logical link control and adaptation protocol (L^2CAP), and non-Bluetooth-specific protocols, such as Object Exchange (OBEX) or the Wireless Fidelity (WiFi) protocol. WiFi is a marketing term applied to the IEEE 802.11b standard, 3G telecommunication network support services, Internet Protocol Version 6 (IPv6) unicast addresses, anycast addresses, and multicast addresses. These protocols can operate interactively and autonomously to some extent. Furthermore, smart devices can emerge as smartphones, tablets, smart watches, and others. Within a very short period of time, smart devices have exceeded any other form of smart computing and communication and are useful enablers of the Internet of Things (IoT) (see Chap. 4). Smart devices can support a variety of form factors, possess a range of properties pertaining to ubiquitous computing, and are used in smart environments, such as cyber-physical system environments, human-centered environments, and distributed computing environments (http1 2015).

In general, cyber-physical systems (CPS) (see Chap. 3) feature a tight integration of and coordination between computation, networking, and physical objects or things, in which various devices are networked to sense, monitor, and control the physical world. Furthermore, there is no universal definition of a cyber-physical system. Some of the most common ones possess characteristics such as cyber capability in physical objects or things, networking at multiple and extreme scales, complexity at temporal and spatial scales, high degree of automation, and more dependable even certifiable operations (Xia and Ma 2013). Many researchers and practitioners have pointed out that cyber-physical systems transform how we interact with the physical world (CPSRR 2008; Xia et al. 2008).

The term *human-centered environment* refers to the physical and psychological needs of humans enabling them to work at the highest level possible. Environmental solutions include communication links and interfaces and the essential aspects of the physical environment for meeting the needs and abilities of users. Furthermore, a human-centered environment is also grounded in information about people which will be used to create environmental solutions. Research findings and data on cognitive abilities, physical abilities and limitations, social needs, and task requirements will be used to create environmental solutions that enable users to live and work at their highest capacity, regardless of age or ability.

The term distributed computing environment is currently evolving into a new frontier, where computation is no longer decoupled from its environment. This stems from the need for integrating external physical data and processes with computations for the sake of ubiquitous control of the surrounding environment (Kobåa and Andersson 2008).

Finally, the term smart environment refers to a greater variety of device environments which can be differentiated as described by Poslad (2009) into:

- *Virtual computing environments*: enable smart devices to access pertinent services anywhere, anytime.
- *Physical environments*: may be embedded with a variety of smart devices of different types, including tags, sensors, and controllers. These can have different form factors ranging from nano- to micro- to macro-sized.
- *Human environments*: humans, individually or collectively, inherently form a smart environment for devices. Humans may be accompanied by smart devices, such as smartphones, smart watches, surface-mounted devices (wearable computing), and embedded devices that can sense and control. Wearable computers are especially useful for applications that require more complex computational support than just hardware-coded logic. Today, wearable computing is a research topic that includes user interface design, augmented reality, pattern recognition, and more.

With regard to this variety of environments, a core research topic in ubiquitous computing systems today refers to the so-called volatile execution environment which seeks technologies and standards that enable devices to discover each other, set up communication links, and start using each other's services (Bardram and Friday 2010). When portable devices enter a smart or ubiquitous space, an essential need for a volatile execution environment is created because the devices want to discover and use nearby resources. This can be easily achieved through service discovery technologies that are available for on-demand usage, including Jini, Universal Plug and Play (UPnP), and Bluetooth discovery protocol. It is particularly important that ubiquitous computing systems and applications are primarily distributed as they entail interaction between different devices, such as:

- Smart mobile devices
- Smart embedded devices
- Smart server-based devices
- Smart environments

These devices use different networking capabilities creating a fundamental challenge due to ubiquitous computing's volatile nature and the essential need for volatile execution environmental services (Couloris et al. 2005).

The three forms of devices for ubiquitous systems, as proposed by Marc Weiser (1991), are macro-sized tabs, pads, and boards. With the advent of technological innovation, the range of forms introduced by Weiser has been expanded into much more diverse, potentially more useful, high-performance ubiquitous devices, introduced by Poslad (2009) as follows:

- *Smart dust*: A system of many tiny microelectromechanical systems (MEMS), ranging from millimeters to micrometers to nanometers. They include sensors that detect, for example, physical or chemical quantities or can be embedded into smart clothes, the integration of sensors, actuators, computing power, power sources, etc., into the cloth—the whole of which serves as an interactive communication network. However, smart dusts usually operate on a wireless computer network and are distributed over an area to perform tasks, such as sensing via radio frequency identification (RFID) (see Chap. 4). Without an antenna much larger than itself, the range of tiny smart dust communication devices can be a few millimeters to centimeters. These devices may be vulnerable to electromagnetic disablement and destruction by microwave exposure (http2 2015).
- *Skin*: fabrics based on light-emitting, conductive polymers and organic computing devices, which can be formed into more flexible nonplanar display surfaces, such as organic light-emitting diode (OLED) displays and products that can be used for intelligent clothes and curtains. Another approach being investigated is to put ubiquitous computing capabilities onto clothes. Traditional parts of a computer, such as keyboards, liquid crystal displays (LCDs), batteries, hard drives, and mice, have evolved into new forms suitable for wearing on clothes. Most wearables are connected through a wireless infrastructure or fabric area network (FAN). Using an antenna with different functionalities, a FAN connects with sensors and/or memory and a FAN base station with out-of-body connectivity (Hum 2001). This technology can be used by rescue personnel for communication, navigation, or search and rescue as well as in disease management and prevention, rehabilitation, and overall lifestyle management (i.e., sports, fitness, weight control, stress management, therapy, etc.). This has become possible due to the OLED display, which can work without a backlight. Thus, it can display deep black levels and can be thinner and lighter than an LCD. In low ambient light conditions, an OLED screen can achieve a higher contrast than an LCD. In general, OLED devices can generate colors across the visible spectrum. Together with simple monolithic fabrication on a range of different substrates, these diverse material properties give OLEDs key advantages over existing display and lighting technology (Buckley 2013). MEMS devices can be sputtered or painted onto various surfaces so that a variety of physical surface structures can act as networked MEMS.
- *Clay*: ensembles of MEMS can be formed into arbitrary 3D shapes, as artifacts resembling different kinds of physical objects. Such an ensemble is called a tangible user interface (TUI) through which a person interacts with digital information in the physical environment. Such an interface can be a part of e-textiles, a new form of fiber materials where sensing and communication are integrated into a woven structure to monitor the signals and variables in the area of interest.

There is quite a broad vision for ubiquitous computing, but there is a danger that it may become too encompassing. Ubiquitous computing sometimes is seen as the opposite of virtual reality. In virtual reality, people or objects are inside a computer-generated world. In contrast, ubiquitous computing forces the computer to be outside in the world with people or objects. A definition of virtual reality can be derived from the meanings of the two terms: virtual and reality. As used here, virtual means "near" and reality is what people experience. Hence, the basic meaning of virtual reality is "near reality," referring to a specific type of reality emulation. In contrast, ubiquitous computing is a difficult integration of human factors, computer science, systems engineering, software engineering, and social sciences.

In summary, the aforementioned term "ubiquitous computing" refers to the so-called third wave of computing. The first wave consisted of mainframe computers followed by personal computers, with a user and machine staring uneasily at each other across the desktop. In the area of ubiquitous computing, or the age of calm technology, technology recedes into the background of peoples' lives. Against this backdrop, ubiquitous computing was the impetus for the recent boom in mobile and wearable computing research, although it is not the same thing as mobile computing or a superset of it.

Based on Mark Weiser's seminal vision of future technological ubiquity, the increasing availability of processing power will be accompanied by its decreasing visibility. As he observed, *the most profound technologies are those that disappear ... they weave themselves into the fabric of everyday life until they are indistinguishable from it.*

5.2 Ubiquitous Computing Fundamentals

Ubiquitous computing is an emerging computing discipline which exists at the intersection of computing, networking, and embedded computing systems (see Chap. 2). Today, almost all computers are connected to the Internet. Thus, ubiquitous computing can be described as a method that connects the remaining things to the Internet in order to provide information on anything at anytime and from anywhere. Since things are getting smaller and smaller, ubiquitous computing is characterized as the omnipresence of tiny, wirelessly interconnected computers embedded almost invisibly into any kind of everyday thing or object. Besides tiny computers, sensors are also getting smaller and smaller, resulting in smart embedded computing systems capable of detecting their surroundings and enabling them with information processing and communication capabilities. This results in completely new and innovative applications for smart systems (e.g., finding out where the smart system is located, what other objects are in close vicinity to the smart system, or what had happened to the smart system in the past). They can also communicate and cooperate with other smart things and, theoretically, access all sorts of Internet resources. Things and appliances could react and operate in a context-sensitive manner and appear to be smart, without actually being intelligent (Mattern 2001).

Hence, ubiquitous computing enables completely new applications (e.g., cooperating devices creating new emergent functionalities which can have enormous economic and social implications). It also enables issues relating to technology acceptance and the creation of a technology-steeped world in which reality gets closely coupled to, or even merges with, the information-based cyberspace. This allows smart devices to close the gap between embedded computing systems and cyber-physical systems.

Recent developments in the field of science will give the computer of the future a completely different shape, meaning the computer will no longer be recognizable as such because it will blend into its surroundings. This is expected to be an outcome of nanotechnology and material sciences research. As shown in Mattern (2001), one important example is light-emitting polymers, which enable displays consisting of highly flexible, thin, and bendable plastic foils to be created. Research is also taking place into electronic ink and smart paper, which will enable the new pen and the new paper to become truly mobile input/output media. However, this technology is years away from becoming usable in practice, for example, in the form of a computer as a foldable road map. Another significant option currently under development is laser projection from eyeglasses directly onto the retina, as a replacement for traditional output media (Mattern 2001).

In contrast, *The American Heritage Dictionary* (http3 2015) introduces ubiquitous computing as *being or seeming to be everywhere at the same time* which means omnipresent. Taking this into account, it can be stated that the idea of personal computers is misplaced; and laptop machines, dynabooks, and knowledge navigators are only a transitional step. The reason is that these devices cannot make computing an embedded, invisible part of daily life. Hence, creating tiny computing devices which vanish into the environment is essential to introducing ubiquitous computing. Moreover, all computers must be able to communicate with each other from anywhere to everywhere 24/7 as well as know or remember their positions. They must also know the preferences and positions of the users, must be intuitively usable, and may use, in the near future, the natural human form of communication.

Thus, ubiquitous computing is creating a completely new paradigm of a computing environment for heterogeneous sets of devices, including invisible computers embedded in everyday things or objects, such as automation devices, cans, clothes, cups, home devices, mobile devices, personal devices, security devices, vehicles, wall-size devices, wearable devices, etc., situated in the environments, inhabited buildings, secured areas, production facilities, and more. These devices may have different operating systems (OS), networking interfaces with their required protocols, and input capabilities, such as sensing, tracking, controlling, output, and more.

The interaction of ubiquitous devices can be implicit, invisible, or by sensing natural interactions which require a wide range of sensors for speech, gesture, and more. Thus, interaction in ubiquitous computing goes beyond the O2O model prevalent for traditional PCs to the many-2-many (M2M) model in which the same person, thing, or object uses multiple devices to interact with multiple devices or only the same device.

With regard to the availability of smart devices and technologies, ubiquitous environments can be created for a large number of application domains. A few examples have been described by Bardram and Friday (2010), such as classrooms, homes, hospitals, meeting rooms, and vehicles.

5.2.1 Learning in the Ubiquitous Space

The most relevant impact on classroom education results from the availability of mobile technologies. Mobile technologies enable the workforce of tomorrow to be trained in new learning concepts, increasing their:

- Creativity
- Initiative
- Responsiveness

This will facilitate adaptability by enhancing their skills to:

- Manage uncertainty
- Communicate across and within cultures, subcultures, and communities
- Negotiate conflicts

The emphasis on learning is the ability to keep learning durable for a lifetime. Thus, durable or lifelong learning uses formal and informal learning opportunities throughout people's lifetimes to foster continuous development and improve the knowledge and skills needed for employment and personal fulfillment through:

- *Learning to know*: mastering learning rather than the acquisition of structured knowledge
- *Learning to do*: equipping people for the types of work needed now and in the future, including innovation and adaptation of learning about future work environments
- *Learning to be*: education contributing to a person's complete development: mind and body, intelligence, sensitivity, aesthetic appreciation, and spirituality

In Table 5.1, the convergence between several learning approaches and technology used is shown (Sharples et al. 2005).

Using mobile technologies, learning can be regarded as situated, collaborative, ubiquitous, pervasive, or lifelong, complementing the actual know-how to become updated with the latest knowledge in the area of concentration. Moreover, mobile technologies allow sharing of knowledge with others, independent of their location. Thus, learning becomes ubiquitous with regard to mobile technologies embedded in

Table 5.1 Convergence between the learning approach and the technology used

Learning approach	Technology used
Learning centered	User centered
Situated	Mobile
Collaborative	Networked
Ubiquitous	Ubiquitous
Lifelong	Durable

digital devices or units that perform human-oriented functions. These devices or units are also becoming more durable when it comes to storing content in whatever format or version making it possible to build up backward compatibility. This allows the digital records of human learning over a lifetime to be organized and preserved.

The continuing growth of mobile technologies, the rise of the availability of the Internet elsewhere, and the constant transformation in software and telecommunication services have led to the opportunity for connecting everything with anything. One of the first opportunities to arise was the concept of mobile and ubiquitous computing. Thus, ubiquitous computing allows computers to be embedded everywhere. Early forms of ubiquitous computing networks are evident in the widespread use of mobile devices. Therefore, for the first time ever, there are more gadgets in the world than there are people, including a growing number that only communicate with other machines. According to data from digital analysts at GSMA Intelligence, the number of active mobile devices and human beings crossed over somewhere around the 7.2 billion mark. As of today, GSMA's real-time tracker puts the number of mobile devices at 7.22 billion, while the US Census Bureau says the number of people is still somewhere between 7.19 and 7.2 billion (http4 2015).

As computers became ubiquitous, they ceased to be the focus of activity, which allowed them to vanish into the background. Introducing ubiquitous computing to ubiquitous learning (u-Learning) results in students using their mobile technology to connect with the manifold digital embedded devices and services available. Therefore, in a ubiquitous learning classroom, students browse around the ubiquitous space and connect by mobile technology to interact with the various embedded digital devices and services. Thus, u-Learning has the potential to enhance education in a sustainable manner and remove many of the constraints of traditional education, e.g., allowing customization in relation to student needs and building the basis of a mobile-technology-based ubiquitous community where everything is:

• Traceable
• Identifiable
• Connected together

Smartphones are used as digital low-cost computing and communication devices. Thus, the platform capacity created the concept of sharing information between things, objects, and devices connected to the ubiquitous space. As ubiquitous computing became a reality, a real constraint of u-Learning was facilitated through the Internet of Things (IoT) paradigm, which evolved at the point in time when more things were connected to the Internet than people (see Chap. 4). Thus, the IoT refers to uniquely identifiable things or objects and their virtual representations in an Internet-like structure. In this regard, the IoT can be viewed as a smart world with ubiquitous computing and networking, making different activities easier to attain through sensors and actuators embedded in real-world things and objects and linked through wired and wireless networks to the Internet. When things and objects in the IoT can sense the environment, interpret data, and communicate with each other, they become devices for understanding complexity and responding to events and irregularities swiftly. Hence, ubiquitous computing allows computing to be embedded everywhere in a u-Learning space or a u-Learning environment. Ubiquitous computing represents a setting for pervasive education where data are present in the form of embedded, digital mobile-technology-based things and/or objects. Things and objects are normally introduced as natural systems, physical systems, humans, sensors, actors, and computers. They just have to be there and be connected with the u-Learning space.

5.2.1.1 Ubiquitous Learning Versus Mobile E-Learning

Comparing u-Learning to mobile e-learning, the level of embedding in mobile e-learning can be low, while the level of mobility has to be high because it is implemented in lightweight devices such as smartphones and PDAs, which are flimsy, handy, and usable anywhere. Internet access with wireless communication technology, however, is necessary to enable u-Learning anytime and anywhere.

Therefore, ubiquitous learning requires the embedding of digital mobile technologies, as previously mentioned, because learners and their mobile devices are moving around, which requires an active u-Learning space supported by sensors distributed across the entire u-Learning location. Depending on the location of the sensors, the signal levels detected provide a distinctive signature of the location of the object which accessed the u-Learning space location. This is a result of the sensor network at the given location and the distance from the accessed device or person. Powerline communication positioning is also capable of providing sub-room-level positioning for multiple locations of u-Learning spaces. Current powerline communication positioning systems have a median error of 0.75 m and a 90 % accuracy of 1 m. Therefore, localization is an important issue in ubiquitous computing in general. Several solutions, such as the underlying signaling technology like infrared (IR), radio frequency (RF), load sensing, computer vision, and audition, are used today, which have different characteristics related to requirements, accuracy, and cost of scaling the solution over space and over the number of objects. It may be important to note that there is no one fit for all location systems available. Each system must be evaluated based on the application domain across the variety of dimensions discussed above. In Varshavsky and Patel (2010), an interesting table of location tracking technologies across a collection of factors used to evaluate a particular location system is given. Beyond that, dynamic learning support through communication with embedded computers in the ubiquitous learning environment is necessary (Ogata and Yano 2012).

Developing u-Learning has to take into account the outcome of the existing learning theories in terms of best practices, such as a structured relationship between information and learners' understanding in educational settings. This helps to ensure that learners are not just learning facts outside of a meaningful context. For example, if a student understands why and how something happens rather than just being told that it is true, then the information has more relevance and, therefore, more meaning for the student.

From a more technical perspective, the main components needed to implement a u-Learning space are:

- *Tiny microcontroller*: embedded in things or objects, it allows information about things or objects to be stored.
- *Server*: provides client stations with access to files and/or units and a database that stores all of the data about objects/units, users, and interactions, as shared resources to a computer network.
- *Wireless communication technology*: mostly in the form of Bluetooth and WiFi.
- *Sensors*: used to detect changes in the u-Learning space, placed adjacent to objects/units, and will be used to recognize the presence of students in the u-Learning space.

Thus, u-Learning can be defined as learning supported by embedded computer networks in everyday life based on specific types of learning environments (Lyytinen and Yoo 2002). This development has given learners access to global communications and the huge number of resources available to today's students at all educational levels. After the initial impact of tiny microcontrollers and their applications in education, the introduction of e-learning and mobile e-learning epitomized the transformations occurring in education.

The mission of ubiquitous computing in education represents another step forward with u-Learning emerging from the concept of ubiquitous computing. U-Learning is pervasive and persistent, allowing students to access educational materials flexibly, calmly, and seamlessly. In this sense, u-Learning has the potential to make education easier to achieve by removing many physical constraints present with other forms of learning. Furthermore, the integration of adaptive learning with u-Learning may lead to even more innovation in the delivery of education through customization based on individual student needs. Adaptive learning itself is an educational method which uses computers as interactive teaching devices. In this method, computers adapt the educational material based on students' learning needs as indicated by their responses to questions and tasks.

The seamless interaction between students and devices in u-Learning can be introduced as follows:

- A student arrives and observes the object in the systems engineering area of concentration. Adjacent sensors detect the student's presence and send data about the object to the student's mobile smart devices.

- Objects will access the u-Learning environment server module and request information about the student.

However, being capable of both networked and independent operation, the object can operate alone and transmit data on a student, such as whether the student has accessed data previously, the most suitable format for this particular student to be used by the u-Learning environment server systems engineering module, and more. Let's assume a student has responded correctly to information in the past. This information will be transmitted (Jones and Jo 2004) by sending the required content to the student's smart device and transmitting the student's response to the u-Learning environment server component. The communication workflow for passing the required message is as follows (Moeller et al. 2013):

Workflow step	Object accessed	Information pathway
1	Systems engineering object no. 1 is accessed by Student No. 1	
2		Content of systems engineering object no. 1 is sent to Student No. 1
3	Student No. 1 responds to the content received	
4		The student's response to the content of systems engineering object no. 1 is analyzed to identify the percentage of the student's understanding of the systems engineering content. Assume Student No. 1 understands 55 % of the content. This result is then relayed to all other objects in the u-Learning space
5	Systems engineering object no. 2 is accessed by Student No. 1	
6		The u-Learning space is aware of the student's performance and attempts to explain what has not been understood as well as some remaining content before continuing with systems engineering object no. 2

Hence, the interaction of Student No. 1 with the u-Learning space in systems engineering objects during a u-Learning access (session) can be tracked and stored in the u-Learning space on the systems engineering module server. If Student No. 1 connects to the u-Learning space again, the system is aware of the Student No. 1's knowledge and can constructively assist in Student No. 1's learning progress. This results in an enhanced learning experience and a deeper understanding of the content in the systems engineering area of concentration with regard to the social technology environment in the u-Learning space making learning local and personal.

The availability of the Internet everywhere and rapid advances in information and communication technologies has led to the opportunity to connect everything with anything, including systems engineering learning objects with students. Thus, the systems engineering learning things and objects in the IoT-based u-Learning space can sense the progress in individual learning, interpret data, and communicate with the individual student. Therefore, u-Learning has become a method for better understanding today's cyber-physical systems' complexity and for responding swiftly to events and irregularities.

The prime objective is to create a customized u-Learning process to make development, editing, and implementation sustainable. The central focus of the u-Learning approach lies in the development and testing of various integrated processes, including:

- A process for the qualification of instructors (a top-down model) which enables them to gain experience in efficiently developing and implementing u-Learning materials within a short period of time at minimal cost
- Development of instruments to simplify the qualification, production, and implementation of u-Learning processes that focuses on didactics, appropriate technology, low cost, and sustainability
- Production of high-quality, u-Learning content (learning objects) to provide a venue for more effective learning by qualified teaching personnel, efficient utilization of the ubiquitous learning course modules, and other instruments of ubiquitous learners, and replication of the system and u-Learning materials to different learners in various learning scenarios.

5.2.2 Smart Home and Powerline Communication

In ubiquitous computing environments, such as smart homes, the devices surrounding us are based on interconnected technologies that are responsive to our presence and actions. The following elements are essential to create a smart home:

- Communication and integration elements
- Control and telemetry devices
- Power consumption and security and alarm services
- Smart home control center
- Telephony and data services
- Video and audio devices

The smart home control center is connected to the Internet, and all smart home devices are wirelessly controlled via tiny microcontrollers, tablets, or smartphones (Snowdon 2009). The time control of a smart home system ensures, for example, that warming up the bathroom in the morning is achieved on demand and on time. Timing is wirelessly connected with the heating system's radiator thermostat and is automatically controlled by the smart home control center. Opening a window after the morning shower to reduce the moisture in the bathroom will immediately be registered by a window sensor and forwarded to the smart home control center which lowers the radiator temperature to save energy and heating costs. When the residents leave the house for work, the smart home control center queries all sensor information to ensure that all user constraints have been enabled, for example, adjust room temperature at a lower level for the period of absence of the resident, etc. In case the resident forgot to close the open window in the bathroom when leaving for work, the smart home control center keeps the user informed via his/her smartphone, asks whether the window should remain open or be closed, and executes the resident's response.

With regard to the aforementioned smart home control strategy and control devices, compact smart metering systems are embedded in a smart home with the emphasis on measurement and intelligent regulation of energy consumption. This means that lights will be shut off when a resident leaves a room or the home for work. In addition, room temperature can be reduced when leaving for work and can be increased to a cozy level in advance of the resident's arrival back home from work. It also can be controlled from outside to identify whether or not the home's windows are closed, whether or not the coffee maker is shut off, and more. Furthermore, the energy provider can observe the energy consumption on a daily, weekly, or monthly basis. Therefore, by using smart devices in conjunction with ubiquitous computing systems, the utilization of smart homes will be more:

- Cost-effective
- Environmentally friendly
- Reliable
- Safe
- Simple

These smart devices can collect data through interconnected sensors and actuators and react in real time on demand. Hence, ubiquitous computing will allow the automation of routine physical tasks by distributing data within the smart home technological settings, at least at the level of infrastructure, basic utilities, and home appliances (Gann et al. 1999). From a user's point of view, smart home systems should simply offer convenience in everyday home activities to liberate people from their work at home and help them live more independently.

Thus, future implementations of technologies for smart home systems will depend on the extent to which they offer improvements in the quality of life and solutions to liberate people in their smart home infrastructure, as they are not system administrators with respect to the design of smart systems, inference in case of ambiguity, and reduction of cost. For more details, see also Chap. 4.

5.2.2.1 Smart Home Services

Implementing smart home services requires a reliable communications network infrastructure. Most extended technologies in smart home networks are wired technologies. With the IEEE 1901 standard for data transmission over electrical wiring (powerline communications, in short PLC), published by the Institute of Electrical and Electronics Engineers (IEEE) as the norm 1901–2010, broadcasting over powerline networks became possible. The IEEE 1901–2010 standard defines the maximum speed (500 Mbit/s on the media), frequency range (up to 100 MHz), and data transmission flow via the powerline, for inside and outside of buildings. With IEEE 1901.2, a standard extension to define data transmission via the power grid in smart grid applications exists for the remote reading of meters for electricity, gas, or water and for networking in home automation. Therefore, the powerline communication network system at home allows smart devices to easily program applications. Thus, the aforementioned factors make ubiquitous networks and powerline communication good candidates for the network infrastructure of smart home environments because of their availability and almost no or little installation cost for powerline communication (Moeller and Vakilzadian 2014).

Even within the smart home, the protection of a powerline communication network is a major concern because the use of electrical wiring implies that the network can be accessed from outside the smart home. To protect the powerline communication network, tools such as Power Packet Utility can be used to configure the network encryption key. In order to do this, one simply connects the powerline communication devices one by one to the smart home server on which the configuration tool is installed by means of a network cable (Carcelle 2006).

In addition to powerline communication used in smart homes, based on 110–220 V/50–60 Hz powerline communication technology, this technology is also used in high and low bit-rate applications and is mostly limited to one type of wire, but some can cross between the level of the distribution network and the wiring on the premises. The main technique used to transmit powerline communication signals over this electric media is to add a modulated signal of low-amplitude to the low-voltage electrical signal around a center of carrier frequencies. Therefore, the availability of powerlines enables private and public users to utilize these lines for broadband communication without any additional infrastructure. Electrical companies make use of powerline communication for low bit-rate data transfer (<50 Kbit/s) signal transmission in the 3–148 kHz frequency band to monitor home automation products, such as smart meters and electrical meters that record consumption of electric energy in intervals of an hour or less and communicate that information at least daily back to the electric power company for monitoring and billing.

Thus, the smart meter can be introduced as a gateway to the household, with the ability to constantly monitor attached devices but, more importantly, switch them on and off. Thus, security becomes a pressing issue). As described in Kaplantzis and Sekercioglu (2012), a secure smart grid has to uphold the following requirements while managing data:

- *Confidentiality*: requires that only the sender and the intended receiver understand the contents of a message. Confidentiality of information transmitted by the smart grid is paramount to customer privacy and grid success.
- *Integrity*: requires that sender and receiver ensure the message is authentic and has not been altered during transit without detection. In smart grids, integrity means preventing changes to measured data and control commands by not allowing fraudulent messages to be transmitted through the network.
- *Availability*: requires that all data is accessible and available to all legitimate users. Since the smart grid is not only communicating usage information but also control messages and pricing information, the availability of this information is crucial for successful operation and maintenance of the grid.
- *Nonrepudiation*: requires that sender and receiver not deny they were the parties involved in the transmission and reception of a message. Accountability of the members of a data transaction is critical when it comes to financial interactions. However, data generated by the network can be owned by different entities (customer, data management services, billing systems, utilities) at different times of the data life cycle.

The increased demand for energy has resulted in the need for an efficient system to manage energy consumption at the household level. There are different load control programs for individual appliances, such as water heating systems, house cooling units, etc. The model proposed for smart home environment energy management systems (SHE^{2}MS) monitors and controls appliances according to a well-defined set of energy usage limits and defined comfort levels as determined by the homeowners. Different home appliances run at different times and may not be needed all of the time. They may not need to run at the same time, or some of them could be turned off at peak times of energy consumption so their operation does not interfere with the comfort level of the home. To balance the load and/or reduce expenses, the load may need to be shifted to other time slots. Definition of these different time spans can use different factors, such as appliance usage pattern, comfort level of the home's residents (in case an appliance can be turned off affecting the home's operations/needs), energy prices at different times, or available energy load at specific time spans.

The priority for an appliance's operation must be defined. As SHE^{2}MS will turn some devices off to balance the load or to reduce expenses at peak times, devices with the lowest priority will be turned off first, such as an e-car which needs to be recharged by 8 or 9 a.m. the next day. Charging the e-car by the end of the night has the highest priority if it has not been done before then. A heating or cooling appliance can have the lowest priority during the day when no one is at home. Usage of other appliances, such as a refrigerator or dishwasher, can be prioritized based on the needs of the user.

Comfort level preference versus cost savings must also be defined because there will be particular times when the residents do not want a specific appliance to be turned off, even if its use is causing an excess in load. Therefore, SHE^2MS work is based on the availability of a potential energy limit. Thus, for all running appliances, the condition demand limit needs have to be met. Exploiting the availability limits results in distraction in appliance operation. As the demand for energy from operating appliances exceeds the available set limits of the SHE^2MS, the devices are turned off, starting with the lowest-priority devices. Hence, the workflow of SHE^2MS can be summarized by the following steps, as described in a project as part of the IoT class at TUC (http5 2015):

- Check the available energy limits.
- Check the status of all appliances.
- Calculate the energy consumption.
- Check the priority of operating devices.
- Analyze the appliances' operating conditions.
- If appliances are operating below the defined comfort level, turn those devices on.
- If the load is exceeding the limits, turn the low-priority devices off.
- Resume operation of low-priority devices when the minimum required time by the high-priority devices has been completed or appliance operation is no longer required.

Therefore, an energy-efficient building solution requires real-time monitoring, measurement, and management of building systems through implementation of different energy management control techniques and algorithms so that energy consumption in the building can be reduced (Snowdon 2009).

5.2.3   Core Properties of Ubiquitous Computing

The main feature that distinguishes ubiquitous computing from Distributed Information and Communication Technology (DICT) systems is that ubiquitous computing is part of human-centered, personalized environments interacting without attracting the attention of users. Therefore, ubiquitous computing refers to a wide range of research topics, such as distributed computing, mobile computing, location computing, mobile networking, context-aware computing, sensor networks, human-computer interaction, and artificial intelligence. Ubiquitous computing, at its core, is envisioned as small, inexpensive, robust networked processing devices, distributed at all levels throughout everyday life and generally turned to distinctly common-place ends. Against this background, a smart home ubiquitous computing environment might interconnect lighting and environmental controls based on personal biometric sensors woven into clothing so that illumination and heating conditions in a smart home environment are modulated, imperceptibly and continuously.

Given that ubiquitous computing systems are part of, and used in, physical environments, sensing and monitoring the physical environment in more detail is easier to achieve than with traditional information and communication systems. Furthermore, ubiquitous computing systems can adapt to the environment by acting on it and controlling it with regard to the following core properties (Poslad 2009):

- Tiny computers need to be networked, distributed, and transparently accessible.
- Human-tiny-computer interaction needs to be hidden.
- Tiny computers need to be context aware to optimize their operation in the respective environment.
- Tiny computers need to operate autonomously, without human intervention, to distinguish ubiquitous computing from human-computer interaction.
- Tiny computers need to handle a multiplicity of dynamic activities and interactions self-controlled by intelligent decision-making and intelligent organizational interaction operating with:
  - Incomplete and nondeterministic interactions
  - Cooperation and competition between members of organizations
  - Richer interaction through sharing of context, semantics, and goals

Thus, ubiquitous computing represents challenges across computer engineering and computer science in systems and software engineering, software engineering, systems modeling, and user interface design. Therefore, contemporary human-computer interaction models, whether command line, menu driven, or graphical user interface (GUI) based, are inappropriate techniques and inadequate for use in the ubiquitous computing paradigm. This suggests that the logical progression is a natural interaction because it is much more appropriate for ubiquitous computing which has yet to emerge because ubiquitous computing is envisaged as a system where billions of mini- and/or micro-sized ubiquitous intercommunication devices will spread out worldwide.

Ubiquitous computing is an approach consisting of many layers, each with its own rules and roles which together make up the ubiquitous system. The functionality of these layers can be introduced as follows:

- Task Management Layer (TML)
  - Monitoring user tasks, sensing, and indexing
  - Mapping users' tasks needed for services in the environment
  - Managing complex, user-dependent actions
- Environment Management Layer (EML)
  - Monitoring resources and capabilities
  - Mapping service needs and user level states of specific capabilities
- Environment Layer (EL)

- Monitoring relevant resources
- Managing reliability of resources

With regard to decentralized planning of ubiquitous computing power, it is essential that devices can spontaneously self-configure networks as well as change them. This usually requires using mobile devices instead of immobile devices, which increases the possibility of using the equipment and optimizing networks and adapting them to environmental needs. When using ubiquitous computing for communication, human behavior has to be taken into account, which calls for simplifying the use of devices, an essential aspect of ubiquitous computing as users do not like to adapt to the use of a computer. This constraint seems to be technically difficult to realize because it requires not only understanding the language but also interpreting the meaning. The same validity is required if using human gestures and facial expressions for ubiquitous computing communication. Therefore, the current state of the art beyond the classical keyboard, screen, and mouse does not consist of the core components that fit the needs of the ubiquitous computing user interface in human-computer interactions (HCIs), whose input and output technologies have not yet been envisaged.

Ubiquitous computing user interfaces will be built around a next-generation technological paradigm that reshapes a user's relationship with his/her personal information, environment, artifacts, and even friends, family, and colleagues. The challenge is not in providing the next-generation mouse and keyboard but in making the collection of inputs and outputs operate in a fluid and seamless manner. The ubiquitous computing user interface will draw on the GUI tradition but will move beyond the special-purpose device, such as a PDA, smartphone, or laptop, into supporting the activities in users' lives, some of which are not currently supported by computation (Quigley 2010).

In general, a ubiquitous computing user interface must consider a broader range of inputs compared with current mobile or gaming devices and applications. Examples of data or knowledge that a ubiquitous computing user interface may rely on include (Quigley 2010):

- Activities like supervision, conducting interviews, creating schedules, and compiling agendas
- Computing resources
  - Network bandwidth
  - Memory
- Data that can be mined and inferred
- Environmental information
  - Noise
  - Light
- Identity of users and others in the vicinity
- Intentions
- Physiological information
  - Hearing
  - Heart rate

- Preferences
- Resource availability
  - Printers
  - Fax
  - Wireless access
- Social information
  - Meeting
  - Party
- Spatial information
  - Location
  - Speed of movement
- Temporal data
  - Time of day or year
- User identification profile

Furthermore, privacy and safety in ubiquitous computing communication has too also be guaranteed because the data stored on the devices of the ubiquitous computing communication network can be accessed by authorized as well as unauthorized people. This calls for data privacy and security in wireless networks for tiny devices with low computing power. To come up with an idea for privacy solution, how ubiquitous computing affects privacy and what technical methods can be used to counter or mitigate this influence need to be considered. However, this requires a clear understanding of what exactly should be protected. Only then can the particular effects on privacy of ubiquitous computing and the required technological countermeasures be addressed. As described in Langheinrich (2010), privacy issues of traditional computer systems, such as databases, are often used in the same way in ubiquitous systems because most ubiquitous computing applications are built on standard components. Technical solutions for such ubiquitous computing can be limited to examples on how particular threats induced by ubiquitous computing technology can be addressed, as introduced in Langheinrich (2010).

5.2.4 Ubiquitous Computing Formalisms for Use Cases

Ubiquitous computing weaves computing into our everyday life, devices, and environments. Hence, people and their use of smart devices are the focus of ubiquitous technologies research, specifically understanding people's needs and reactions to new ubiquitous computing applications and services. This research is in close connection to human-computer interaction which includes psychology and anthropology research to understand how people interact with technology, based on field studies, focus groups, ethnology, and heuristic evaluations. So far, case studies are conducted outside of a research laboratory or controlled environment (i.e., in the field—field studies) (Brush 2010) and with regard to specific usability proofs integrated into business processes—use cases.

In Brush (2010), a very good description of ubiquitous computing field studies is given which is based on the three types of ubiquitous computing field studies:

- *Studies of current behavior*: What are people doing with new options?
- *Proof-of-concept studies*: Does the new technology function in the real world?
- *Experience using a prototype*: How does the use of a prototype change people's behavior or allow them to do new things?

As mentioned in Brush (2010), other types of field studies exist, including those exploring playful interaction, or ludic engagement, with ubiquitous computing technologies, as described by Gaver et al. (2006, 2007). The field studies described by Brush (2010) focus on current behavior with the use cases in home technology sharing and use of home technology and the proximity of users to their mobile phones. In her proof-of-concept field studies, the use cases are context-aware power management and team aware. So far, the type of field study has been determined. The next step is to design the field study with regard to the most important questions to be answered by the study:

- How long is the field study planned to run?
- What data have to be collected?
- What do participants of the field study have to do?

These questions require identifying the participants' profiles, planning the study, establishing control conditions, logging data, interviewing participants, and compiling the necessary statistics.

In contrast, a use-case representation is based on the concept of scenarios of collections of use cases that represent the prominent characteristics of the event (Lee et al. 1998). Therefore, a use case contains a description which is typically unstructured or semistructured text and a sequence of actions performed by the system (Camp 2000). Furthermore, the use-case concept should be valid and actions should lead to an observable result, whereupon incomplete or multiple sequences are excluded. Finally, the observable result of the use case should be of value to an actor. Use cases can be represented using a graphical notation. They serve as projections of future ubiquitous computing system usage and as projected visions of interaction with a designed ubiquitous computing system. They are considered by some authors to be scalable to large complex systems, as they can be improved and expanded incrementally with little or no loss of prior information (Gunstone 2011). Use cases can also be used in such a way as to facilitate requirement traceability throughout design and implementation (Lee et al. 1998).

5.3   Smart Devices: Components and Services

Any type of equipment, instrument, or machine that has its own computing capability is considered to be a smart device. They are digital devices which are connected to other devices or networks via different wireless protocols, such as Bluetooth, NFC, WiFi, and others, which can operate to some extent interactively and autonomously. Hence, smart devices can perform intelligent operations on their own behalf with respect to their functionality and relevant surrounding environments. They are part of a wide range of products and need only a minimum set of physical components to be categorized as smart devices. Those components are:

- *Power*: any source of power being provided to a device.
- *Memory*: to store operations.
- *Processing*: adequate processing performance with regard to the operation requirement to be executed quickly and more efficiently.
- *Communications interface*: to communicate with other devices and services within a smart space. This is an important component because if a device is able to interact with other devices within a smart space and let other devices and services interact with it, it must provide a means of communication to these other devices (Davy 2015).

Smart devices are usually connected to other devices or networks via the protocols mentioned above and can operate to some extent interactively and autonomously. Some of the popular smart products are the smart watch, smart TV, smart camera, and smartphones, as shown in Fig. 5.1 (http6 2015).

If a smart device participates in invoking a smart device service or subscribes to state change notifications, it must contain a control component which effectively acts as a client of the services smart devices offer to the smart ubiquitous space. The control component contains various internal components which allow it to interact with smart devices.

The service component of a smart device is made up of various internal components. Together these components allow a smart device to publish its services to a smart ubiquitous space on entering and allow other smart devices and services to understand and invoke these services in an abstract way. The service

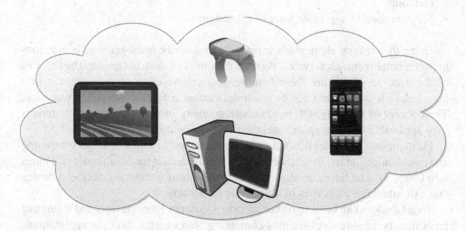

Fig. 5.1 Selection of smart devices

component allows other smart devices and services within the smart ubiquitous space to access the internal operations of a device without requiring the manufacture of specific controls or drivers. This is done by using an abstract service definition to describe to other smart devices and services within the smart ubiquitous space on how to invoke the smart device's operations. Through this abstract interface, a smart device's internal operations can be invoked (Davy 2015).

The Web Services Description Language (WSDL) is a rapidly emerging technology for describing services. WSDL provides a way of offering a service, regardless of its implementation, to other heterogeneous systems over the Internet. Web Services Description Language lets a client talk to a service through a common protocol known as Simple Object Access Protocol (SOAP), a protocol specification for exchanging structured information in the implementation of Web services in computer networks. WSDL and SOAP use the XML Information Set, a W3C specification describing an abstract data model of an XML document in terms of a set of information items, for messages. As long as the client and server can parse and understand the XML documents, the implementation of the client and server can be completely independent.

A WSDL document is made up of the following major elements (Davy 2015):

- *Type*: provides data type definitions used to describe the message exchanged between endpoints.
- *Message*: represents an abstract definition of the data being transmitted. A message consists of logical parts, each of which is associated with a definition within some type of system.
- *Port type*: set of abstract operations. Each operation refers to an input message and output messages.
- *Binding*: specifies concrete protocol and data format specifications for the operations and messages defined by a particular port type.
- *Port*: specifies an address for a binding, thus defining a single communication endpoint.
- *Service*: used to aggregate a set of related ports.

Using these major elements, it is possible to describe homogeneous interactions between heterogeneous services. As manufacturers of devices develop the services of these devices, offering these heterogeneous services to an environment in a homogeneous way would not be possible without a level of service abstraction. The concept of using WSDL to hide heterogeneity of service implementations is very applicable to smart spaces and the interoperability of smart devices.

Furthermore, smart devices will outnumber any other forms of smart computing and communication by, in part, acting as a useful enabler for the Internet of Things (see Chap. 4). The term can also refer to a ubiquitous computing device, a device that exhibits some properties of ubiquitous computing.

Smart devices can be designed to support a variety of form factors and a range of properties pertaining to ubiquitous computing, can execute multiple applications, supporting different degrees of mobility and customization, and can be used in different system environments such as:

- Physical world
- Human-centered environments
- Distributed computing environments

Smart devices can be characterized as follows:

- Consolidations of system hardware and software resources which fulfill specified smart device constraints and are mostly tiny
- Remote access to external services and execution
- Smart environment access, such as human and physical and cyber-physical world interactions
- Ubiquitous computing properties which mean devices need to be networked, distributed, and usually hidden but accessible
- Enabler for context awareness of an environment in order to optimize their operation in that environment

Thus, the sensor-based information processing and communication of smart devices can help to operate, to some extent, autonomously, i.e., without human intervention, which means self-controlled. In smart devices, smart sensors collect and consolidate information and deliver the results (e.g., to a superior planning and decision system). Therefore, the main advantages are as follows: (1) a simple collection, interpretation, and dissemination to the superior system status and environmental information from anywhere to the superior planning and control system and (2) direct access to the actual data and decisions.

However, it is not easy to introduce a closed set of properties that define all ubiquitous computing devices because of the sheer range and variety of ubiquitous computing research and applications. With the development of the Internet into a service network, the paradigm of the IoT was created. The idea is to connect all of the devices to the Internet and then to create networked Internet-based services and smartness. The idea in ubiquitous computing was to shrink the size of a computer so much that a computing system can be embedded into anything. Smart devices and services in ubiquitous computing make intensive use of communication. Therefore, device interoperability is important for smart ubiquitous spaces which can be introduced in a top-down approach of the three levels required:

- *Smart world*: represents smart ubiquitous spaces, and thus the information world, where the term interoperability means that the information has the same meaning in different devices. In general, a smart ubiquitous space is a physical space rich in devices and services that is capable of interacting with users. The physical environment and services originated outside the smart ubiquitous space. The aim of the smart ubiquitous space is to orchestrate the use of integrated physical and computing environment to bring tangible benefits to users in supporting their tasks.

- *Service world*: represents service domains, where the applications are able to use the services across device boundaries.
- *Device world*: represents device networks with the physical level interoperability and device networks.

From this it is evident that interoperability is an important feature for smart devices and services in ubiquitous computing.

5.4   Tagging, Sensing, and Controlling

Digital components continue to become smaller, faster, and cheaper to manufacture; and more low-power components can easily be deployed on a massive and pervasive scale (Poslad 2009). Ongoing work on MEMS enables sensing and actuating at the scale of a nanometer. The possibility for miniaturization migrates into all facets of daily work and life. The embedding of computing and communication technology into everything and everywhere is becoming a reality. Hence, MEMS becomes a powerful enabler for the vision of smart ubiquitous computing devices, services, and environments making greater use of the expanded Internet Protocol version 6 (IPv6) address space (see Chap. 4).

Therefore, the primary goal of tagging, sensing, and controlling can be introduced as supporting a variety of cyber-physical awareness and analysis services in the smart ubiquitous space. This requires collecting and integrating a variety of sensing information from diverse sources. The massive natural and social sensing data intensively gathered is analyzed and transformed as useful, actionable information to give appropriate feedback to things and/or objects and people in the physical world.

Hence, the major technical challenge is to create user-defined smart sensors by combining sensors and analysis services. The goal is to create simple nodes of a variety of sensors that can be configured for collecting a spectrum of integrated sensor data. This requires the design of smart sensors encapsulating the very specific and intricate details of the sources. The goal of this design is simply to compose smart sensors that access heterogeneous sensing sources. Another important requirement is to design a composite scalable smart sensor that can combine heterogeneous sensors on demand. This requirement is needed to ensure comprehensive monitoring of the environment allowing responders to assess the overall impact from a variety of information sources.

In order to rely efficiently on decoupled service components, the interfaces to these components should be based on well-defined, interpretable, and , unambiguous standards. Further, standardization of interfaces will allow for easy provisioning of various services by a number of cyber-physical systems envisioned today and in the future (Simmon et al. 2013).

5.4.1 Tagging

Physical tags are digital tags, which are networked devices with an identity, such as radio frequency identification (RFID) tags. RFID systems are used for identification and tracking purposes and include the tag, a read/write device, and a host system for data collection, processing, and transmission. An RFID tag consists of a chip, some memory, and an antenna. When these tags are attached to or linked to physical objects, they provide a way to audit physical spaces and processes. Thus, RFID is an emerging technology for identifying objects or personnel, often recognized as one of the technologies capable of realizing a ubiquitous computing network due to its strong benefits and advantages over traditional means of identification, such as barcode systems.

Compared to the barcode system, RFID tagging has some advantages including rapid identification, flexibility with regard to the manifold applications, and a highly intelligent degree (Wang et al. 2007; Xia et al. 2008). RFID-enabled systems interpret the data and make some decisions. Furthermore, they can work under a variety of environmental conditions. It has recently found a tremendous demand due to emerging, as well as already existing, applications requiring more and more automatic identification techniques that facilitate management, increase security levels, enhance access control and tracking, and reduce the labor required. A brief listing of possible RFID applications that can be used on a daily basis is (Mobashsher et al. 2011):

- Automatic payment transactions
- Automotive industry
- Health care , and hospitals
- High-value asset tracking and management
- Manufacturing
- Marine terminal operation
- Military
- Pharmaceutical management systems
- Public transportation
- Retail inventory management
- Toll roads
- Warehouse management systems

Hence, RFID tags identify and track physical objects. Applied to real-world items of the application areas mentioned above makes the physical object an energy consumer because RFID technology uses energy alimented active tags, or some energy that is applied to passive tags in order to activate those for sending data. The quantity of energy scavenged is very small, but any physical object becomes an energy consumer. Physical objects are being introduced to the digital world and also to the digital energy world. RFID , technology becomes the transforming function able to map physical objects to the digital world by the simple injection. Several implications are generated by this simple assertion: the digital representation of a real-world object, the digital clone of it, can be introduced as the object's second life.

RFID tags come in a large variety of forms and functional characteristics. One useful way of classifying tags is, as introduced in Chap. 4, to divide them into active and passive tags:

- *Active RFID tags whose read/write range is longer*: are available in various shapes and sizes and offer a number of optional functions, such as motion sensing, call buttons, and temperature sensing. Active means the tag has an internal source that powers the tag in order to transmit a frequent RF signal. A network of readers placed strategically throughout the area to be covered receives the tag's RF transmission. Active tags can be read up to several hundred meters depending on the environment and on the asset characteristics. Some tags use wireless access points as readers, while others use proprietary readers.
- *Passive RFID tags with a shorter read/write range*: are available in various formats depending on the application. They are used in many industries for auto ID and security purposes. One such example is for inventory systems in supermarkets and libraries to automatically sense when an item is being taken or returned and/or track whether or not an item will be taken out of the supermarket without paying. Passive tags contain at least two parts: an integrated circuit for storing and processing information, modulating and demodulating a radio frequency (RF) signal and other specialized functions, and an antenna for receiving and transmitting the signal, as shown in Fig. 5.2. To provoke signal transmission, an external source is required.

However, passive tags are much cheaper than active tags and are, therefore, more widely used. Tags represent a big portion of the cost in any RFID implementation. One tag may cost anywhere between 0.1 and 10 US$ depending on factors like form, operating frequency, data capacity, range, presence or absence of a microchip, and read/write memory.

Fig. 5.2 Passive tag from
Harland Simon (http7 2015)

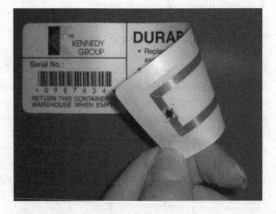

With regard to the location of , mobile resources, WiFi real-time location (RTL) can provide benefits for this area of concentration which may result in huge cost savings by allowing expensive equipment to be located for use or servicing. Furthermore, real-time information on the location of mobile resources can eliminate calls, waiting, and the frustration normally associated with equipment shortages.

5.4.2 Sensing

Sensors detect variables of certain physical properties, such as temperature, humidity, pressure, brightness, acceleration, and thermal radiation; or chemical properties, such as pH, ionic strength, and electrochemical potential; or physical conditions or changes of its environment. These qualities are detected by physical or chemical effects and transformed into a processable electrical signal for measurement or control.

Sensors are used in innumerable everyday applications of which most users are unaware. With the availability of sensors fabricated in between 100 nm and 100 µm in size, sensors, like accelerometers, are available that detect when a car has hit an object and trigger an airbag or a smartphone that is turned from its horizontal to its vertical position and the screen moves in the same way.

Easy-to-use microcontroller platforms enable sensors to be used that have expanded beyond the more traditional fields of temperature, pressure, or flow measurement. Sensors can also be grouped based on the domains to which they belong, such as , angular rate, chemical, electrical, gravity, magnetic, mechanical, radiant, and thermal and are arranged into sensor arrays, such as magnetic, angular rate, and gravity, the so-called MARG sensors (http8 2015).

The direct output of a pure sensor is generally not available in a form that computers can process. Thus, signal conditioning is required to convert sensor output to an appropriate form. The most important signal conditioning functions besides amplification and filtering is either raw or preprocessed signal conversion. Amplification means that the small raw signal detected is boosted by an amplifier. Sensor signals also may encounter interference from noise or undesirable inputs. Therefore, either active filters, which consist of resistors, capacitors, and amplifiers, or passive filters, which consist of resistors, inductors, and capacitors, are used in their different arrangements. Applications include airplanes and aerospace, cars, manufacturing systems, medical systems, and robotics.

Sensor sensitivity is an important feature because it indicates how much the sensor's output changes when the input quantity measured changes, but some sensors also have an impact on what they measure. Hence, sensors need to be designed which finally have no or less effect on what is measured.

As mentioned in the introductory part of Sect. 5.4, technological progress allows fabricating sensors on a microscopic scale as microsensors, the so-called MEMS sensing and actuating technology. In most cases, a MEMS-based microsensor reaches a significantly higher speed and sensitivity compared with macroscopic sensors, as well as a better detectability when detecting single small entities, such as bacteria, viruses, nanoparticles, or individual molecules. Sensors that work on larger scales typically measure changes in physical or chemical substances uniformly distributed over the sensor. As a result, sensitivities for quartz crystal microbalance and surface plasmon resonance techniques are often discussed in terms of mass per unit area. While this value can be translated into a value for total mass measured, the size of these devices restricts absolute mass sensitivity to the range of nanograms to pictograms (Waggoner and Craighead 2007). This results in exciting new developments of sensors which go beyond the realm of simple sensing of , movement or capture of images to delivering information, such as location in a built environment, the sense of touch, and the presence of particles or chemicals. These sensors unlock the potential for smarter systems, allowing objects to interact with the world around them in more intelligent and sophisticated ways.

Collections of millions of cooperating sensing, actuating, and locomotive mechanisms can be introduced as a new form of programmable matter because these can self-assemble in arbitrary three-dimensional shapes (Goldstein et al. 2005) and form the basis for much more fluid and flexible computers and human-computer interfaces, flexible devices, and tangible computer interfaces, for example (Poslad 2009). Therefore, technologies for smart sensors and sensor fusion are important advancements in sensors in many automotive, chemical, and environmental monitoring and industrial and medical applications. They allow to describe the increasingly varied number of sensors that can be integrated into arrays. It examines the growing availability and computational power of communication devices supporting the algorithms needed to reduce the raw sensor data from multiple sensors and convert it into the information needed by the sensor array to enable rapid transmission of the results to the required point (Yallup and Iniewski 2014).

A smart sensor can be viewed as a system which:

- Provides a digital output signal, often through a standardized interface, in the case of autonomous systems through a wireless connection
- Is accessible by an address and a bidirectional digital interface
- Performs commands and logical functions
- Have comprehensive , adjustment and diagnosis capabilities
- Mostly has data storage facilities and is a self-sufficient system in the case of an autonomous device

Smart sensors are increasingly being employed, not only in industrial settings but in every aspect of human life. They are used to monitor health, detect viral infections, select the appropriate cycle for washing machines, monitor traffic, manage intelligent buildings, and control processes in a manufacturing plant. Smart sensors are used to gather a wealth of information from the process that can improve operational efficiency and product quality. Furthermore smart sensors include features such as communication capability and onboard diagnostics.

Sensor fusion is the combination of sensor data or data derived from sensor data from disparate sources such that the resulting information is in some sense more efficient than would be possible when these sources are used individually. The term "more efficient" in this case can mean more accurate, more complete, or more dependable or refer to the result of an emerging view of deep information by combining two-dimensional images from two cameras at slightly different viewpoints.

In recent years, a new discipline has evolved to solve various problems with common characteristics: multisensor data fusion or distributed sensing. Multisensor data fusion is a technology which deals with the problem of combining data from multiple and different types of sensors to draw conclusions about physical events, activities, or situations. Multisensor data fusion attempts to combine data from multiple sensors to enable conclusions which would not be , possible from the results of only a single source.

The primary benefit of multisensor data fusion is the improvements in the quality of information in a synergetic process where the resulting whole is more than the sum of its parts because a fusion of data does not intrinsically mean a synergy. According to Mitchell (2007), multisensor data fusion enhances the information available to a ubiquitous system by using sensors in different ways (Schönefeld 2014):

- *Representation*: Information obtained during, or at the end of, the fusion process has an abstract level, or a granularity, higher than each input data set. The new abstract level/granularity provides richer data semantics than each initial source of information.
- *Certainty*: If S is the sensor data before fusion and $p(S)$ is the a priori probability of the data before fusion, then the gain in certainty is the growth in $p(S_F)$ after fusion. If S_F denotes data after fusion, then it can be expected that $p(S_F) > p(S)$.
- *Accuracy*: Standard deviation on data after the fusion process is smaller than the standard deviation provided directly by the sources. In case the data are corrupted by noise or error, the fusion process tries to minimize climate noise and errors. In general, the gain in accuracy and the gain in certainty correlate.
- *Completeness*: Bringing new information to the current knowledge base on a ubiquitous computing environment provides a more complete view. In general, if the information is redundant and constant, there could also be a gain in accuracy.

Therefore, the process of multisensor data fusion can be described by a functional model defined as a set of functions that can compromise any , data fusion system and the relation between them; but a functional model is not a process model that describes the flow of a process in detail. The general functional model of multisensor fusion, suggested in Meier (1998), is shown in Fig. 5.3. This model contains the basic functions necessary to perform the data fusion process: data association, data fusion, data abstraction, and knowledge representation of the current ubiquitous system state.

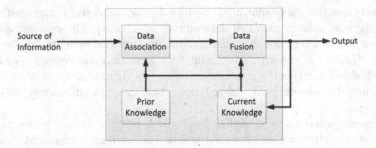

Fig. 5.3 Functional model of multisensor data fusion

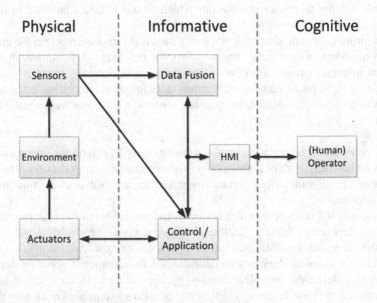

Fig. 5.4 Model of the three multisensor data fusion domains

Several authors have introduced a more detailed view on multisensor data fusion systems by dividing them into three major domains, physical, information, and cognitive, which results in the data flow between these domains, as shown in Fig. 5.4.

So far, the formal processes and the resulting data flows have been designed; the next challenge is to implement the multisensor data fusion system according to the constraints of the scenario of the area of concentration.

The applications range from industrial applications, such as controlling complex equipment, to automated military manufacturing issues, such as automatic target identification and battlefield exercises and threat analysis. Data from different sensors and sensor types are interdisciplinary and linked by certain techniques, from areas such as:

- Artificial intelligence
- Cognitive psychology
- Information theory
- Pattern recognition
- Signal processing
- Statistics

The sensor data to be processed can , include parametric location data, such as angle information (azimuth, height, and image coordinates).

Data fusion can be compared with the cognitive process used by humans to draw conclusions about their environment from the data flowing steadily to their senses.

Sensor data is analyzed based on what the sensors recorded to determine the environment. The recognition of a friend after a long time, for example, may require an analysis of factors such as the overall face shape, the identification of certain visual characteristics, and classification and detection of voice patterns or even of certain types of actions or gestures.

Thus, data fusion applications cover a wide range, including military applications, such as surveillance, tactical situations, and threat analysis, , and also nonmilitary applications, such as automation, remote sensing, and others. Furthermore, one can distinguish between direct fusion, indirect fusion, and fusion of the outputs of the former two (http9 2015):

- *Direct fusion*: the fusion of sensor data from a set of heterogeneous or homogeneous sensors, soft sensors, or virtual sensors; a common name for software where several measurements are processed together
- *Indirect fusion*: uses information sources such as a priori knowledge about the environment and human input

Let s_1 and s_2 denote two sensor measurements with noise variances σ_1^2 and σ_2^2, respectively. One way of obtaining a combined measurement s_3 is to apply the so-called central limit theorem, which is also employed within the Fraser-Potter fixed-interval smoother, as follows:

$$s_3 = \sigma_3^2\left(\sigma_1^{-2}s_1 + \sigma_2^{-2}s_2\right)$$

where $s_3 = \sigma_3^2\left(\sigma_1^{-2}+\sigma_2^{-2}\right)^{-1}$ is the variance of the combined estimate. It can be seen that the fused result is simply a linear combination of the two measurements weighted by their respective noise variances (http10 2015).

In more quantitative terms, the goal of data fusion is to improve the accuracy of conclusions, such as the estimation of states (e.g., the position or the declaration of identities). Typically, pulse radar has the ability to determine the exact radial distance from the radar antenna to the observed object; but the accuracy in determining the angular position (i.e., the , direction of a target) is limited. In comparison, a so-called forward-looking infrared radar determines the angular position of an object observed relatively accurately (i.e., its position within a two-dimensional image), while the uncertainty regarding the range of distance is large. When the data from both sensors are fused, the uncertainty of the estimated position of the target is less than the uncertainty of a measurement alone. Hence, sensor fusion combines the advantages of (both) sensors and improves the estimation of a position and reduces the uncertainty with regard to the position.

To summarize: sensors are omnipresent and ubiquitous. Leading edge research in sensors has been propelled by the advancements made in MEMS technology in the last decade. The scientific world is now on the verge of delivering sensors with radically new capabilities. In many scenarios, multitudes of smart sensors, with limited power and processing capabilities, are being organized in networks to perform more challenging tasks.

5.4.3 Controlling

A control system for physical world tasks must perform many complex information processing tasks in real time. It often operates in an environment where boundary conditions may change rapidly. The usual approach to designing control systems for physical world tasks is to decompose the underlying real-world problem in a series of functional units such as:

- Task planning
- Task execution
- Task control

They are embedded between sensor outputs and actuator inputs, the two main components of a physical world control system. After identifying the computational requirements for the physical world task, the primary decomposition of the real-world problem has to be decided, followed by the secondary decomposition, and so on.

With regard to the requirements of the physical world control system, a number of constraints can be identified, such as:

- *Multiple sensors*: this can result in inconsistent readings in case sensors overlap in the physical quantities they measure.
- *Robustness*: in case sensors fail, there must be a guarantee that they will adapt by relying on those still working to achieve robust behavior for the physical world control system.

These constraints may have a specific impact on the complex physical world control system design but need not necessarily result in a complex control system. The complex behavior may be a reflection of a complex environment (Simon 1969). For robustness, the physical world control system must perform when one or more of its sensors start to fail or give erroneous information, meaning recovery must be quick. This implies that built-in self-calibration must occur at all times to avoid external calibration steps (Brooks 1985).

To attain more configurability and flexibility in physical world control systems, programmable controllers have been developed. The hardware architecture is usually based on microcontroller devices (MC) or field-programmable gate arrays (FPGA). A microcontroller is a semiconductor chip containing the processor kernel and peripherals. In many cases, even the main memory and program memory are partially or completely on the same chip. Hence, a microcontroller is a one-chip computer system or a system on a chip (SoC). They often contain complex peripheral functions, such as controller area network (CAN), universal serial bus (USB), local interconnect network (LIN), serial peripheral interface (SPI), pulse width modulation (PWM) outputs, liquid crystal display (LCD) controller and driver, analog-to-digital converter and vice versa, and more. Some microcontrollers also have programmable digital and/or analog function blocks. Field-programmable gate arrays (see Sect. 2.2.2) are integrated circuits designed for customer configuration after manufacturing by using a so-called hardware description language (HDL) to specify the FPGA-specific configuration. As described in Sect. 2.2.2, FPGA contains programmable logic components, so-called logic blocks, and a hierarchy of reconfigurable interconnects that allow the blocks to be connected together. Logic blocks can be configured to perform application-specific combinatorial functions. In most FPGAs, the logic blocks also include memory elements, which are simple flip-flops or more complete blocks of memory. In Fig. 5.5, the Xilinx FPGA Spartan XC3S400 is shown as an example of an FPGA chip.

As described in Sect. 2.4, the concept of feedback describes a control system in which the system's output, whose dynamic depends on the chosen system's physical world model, is forced to follow a reference input while remaining relatively insensitive to the effects of disturbances. In the case of a difference between both signals, the summing point of the feedback loop generates an error signal which is transferred to the controller input. The controller acts on the error with regard to a control strategy and manipulates the physical world model to make it track the reference input. Moreover, this closed-loop feedback forces the system's output to follow the reference input with regard to present disturbance inputs. The simple controller used is the proportional or P-type controller, the simple integral or I-type

Fig. 5.5 Xilinx FPGA
Spartan XC3S400 (http11
2015)

controller, and the simple derivative or D-type controller. A more complex controller is the proportional-integral-derivative (PID) controller which can be used for ubiquitous control systems. The controller design is based on the assumption that some basic knowledge exists of the physical world problem to be controlled. That becomes the starting point for the controller design and can be tuned using feedback to address the uncertainties of the physical world parameters. There are several sources of uncertainty that can occur in controllers (Poslad 2009):

- *Approximate computation* (e.g., a linearized solution of a nonlinear problem in a part of its range)
- *Environmental dynamics* (e.g., stochastic effects)
- *Inaccurate models* (e.g., nondeterministic impacts)
- *Random action effects* (e.g., individual specific effects are not correlated with the independent variables)
- *Sensor limitations* (e.g., improper sensor placements or poor signal reception)

5.4.3.1 Parameter Identification in Adaptive Control

The prospects for controlling uncertain physical world tasks are adaptive and robust control. The adaptive control method uses a controller that adapts to a controlled system with parameters which vary or are initially uncertain. Thus, a control law is needed that adapts itself to the intrinsic conditions of the physical world task. The foundation of adaptive control is parameter identification. The identification of the parameters of a p-dimensional vector $\underline{\Theta}_{RWPS}$ of a real-world physical system can be characterized by an error criterion, defining the way in which the components of the parameter vector $\underline{\Theta}_{\text{IM}}$ of the identification model can be adjusted to coincide with

$$\underline{\Theta}_{\text{RWPS}} = \underline{\Theta}_{\text{IM}}. \tag{5.1}$$

Due to the implementation of the identification method on a computer, a time-discrete description of a time-continuous model is subsequently used. For linear dynamic systems and piecewise-constant system inputs, a description can easily be deduced from the set of n first-order differential equations as an equivalent set of n first-order difference equations or, alternatively, as one difference equation of the nth order, which is a simplification in the computation of the model output for identification purposes. Note that this is not the case for nonlinear time-continuous models.

Common methods for parameter identification include least squares and gradient descent methods. For the output-error least square method, the error criterion chosen is

$$J_N\left(\underline{\Theta}_{\text{IM}}\right) = \sum_{k=n}^{N} \left(\hat{v}_k\left(\hat{\underline{\Theta}}\right) - \xi(v_k)\right)^2 \rightarrow \text{Min}, \tag{5.2}$$

where N is the number of measurements. Defining the vectors

$$\hat{\underline{v}}^N\left(\hat{\underline{\Theta}}\right) := \left[\hat{v}_n\left(\hat{\underline{\Theta}}\right), \hat{v}_{n+1}\left(\hat{\underline{\Theta}}\right), \ldots, \hat{v}_N\left(\hat{\underline{\Theta}}\right)\right]^T \tag{5.3}$$

and

$$\underline{v}^N := [v_n, v_{n+1}, \ldots, v_N]^T, \tag{5.4}$$

of the estimated and the real-world output errors, respectively, (5.2) can be written as

$$J_N\left(\hat{\underline{\Theta}}\right) = \left(\hat{\underline{v}}^N\left(\hat{\underline{\Theta}}\right) - \xi\{\underline{v}^N\}\right)^T \left(\hat{\underline{v}}^N\left(\hat{\underline{\Theta}}\right) - \xi\{\underline{v}^N\}\right) \rightarrow \text{Min}. \tag{5.5}$$

Consider a stationary stochastic process $\{v_k\}$ with $\xi(v_k) = 0$, (5.2) and (5.3) can be simplified as

$$J_N\left(\hat{\underline{\Theta}}\right) = \sum_{k=n}^{N} \left(\hat{v}_k^2\left(\hat{\underline{\Theta}}\right)\right) \rightarrow \text{Min}, \tag{5.6}$$

and

$$J_N\left(\hat{\underline{\Theta}}\right) = \left(\hat{\underline{v}}^N\left(\hat{\underline{\Theta}}\right)\right)^T \left(\hat{\underline{v}}^N\left(\hat{\underline{\Theta}}\right)\right) \rightarrow \text{Min}. \tag{5.7}$$

Substituting $\underline{\hat{v}}^N(\hat{\Theta})$ in (5.6), we obtain the least squares output-error criterion in its well-known notation

$$J_N(\hat{\Theta}) = \sum_{k=n}^{N} \left(Y_{\text{Meas},k} - \hat{Y}_k(\hat{\Theta})\right)^2 \rightarrow \text{Min.} \qquad (5.8)$$

Defining $\hat{\underline{Y}}^N(\hat{\Theta})$, and $\underline{Y}_{\text{Meas}}{}^N$ in an analogous way, as $\underline{\hat{v}}^N(\hat{\Theta})$, and ν^N, respectively, the corresponding formulation of the criterion in (5.8) in matrix form becomes

$$J_N(\hat{\Theta}) = \left(\underline{Y}^N_{\text{Meas}} - \hat{\underline{Y}}^N(\hat{\Theta})\right)^T \left(\underline{Y}^N_{\text{Meas}} - \hat{\underline{Y}}^N(\hat{\Theta})\right) \rightarrow \text{Min.} \qquad (5.9)$$

$\hat{\underline{Y}}^N(\hat{\Theta})$ is a nonlinear function of $(\hat{\Theta})$ the parameter vector $(\hat{\Theta})^N_{\text{Min}}$, which minimizes $J_N(\hat{\Theta})$, which can be determined by numerical optimization methods.

A reasonable interpretation of the error criterion in (5.7) can be given if the output-error $\{v_k\}$ is assumed to be Gaussian. Hence, the probability of ν^N has the form

$$p(\underline{v}^N) \approx \exp\left[-\frac{1}{2}(\underline{v}^N - \xi(\underline{v}^N))^T \sum_{\nu^N}^{-1} (\underline{v}^N - \xi\{\underline{v}^N\})\right]. \qquad (5.10)$$

Consider the estimated output-error sequence $\{\hat{v}_k(\hat{\Theta})\}$ as a realization of the output-error criterion and determine $(\hat{\Theta})$ such as

$$p(\underline{\hat{v}}^N(\hat{\Theta})) \rightarrow \text{Max,} \qquad (5.11)$$

where $(\hat{\Theta})$ is the so-called maximum likelihood estimate, since we find from (5.10) that (5.11) is equivalent to

$$(\underline{\hat{v}}^N(\hat{\Theta}) - \xi(\underline{v}^N))^T \sum_{\nu^N}^{-1} (\underline{\hat{v}}^N(\hat{\Theta}) - \xi\{\underline{\hat{v}}^N\}) \rightarrow \text{Min.} \qquad (5.12)$$

Assuming a white-noise process $\{v_k\}$ stationary with

$$\text{var}(v_k) = \sigma_v^2, \qquad (5.13)$$

yields

$$\sum_{v} N = \sigma_v^2 \cdot \underline{I}; \qquad (5.14)$$

hence, (5.12) is equivalent to (5.13), and the parameter estimate $\left(\hat{\underline{\Theta}}\right)^{N}_{Min}$, which minimizes the sum of the squared distances of $\left\{\hat{v}_k\left(\hat{\underline{\Theta}}\right)\right\}$ from the expected value of v_k, becomes the so-called maximum likelihood estimate.

The output-error least squares criterion can be given as follows:

$$\underline{e}(t_j) := \underline{Y}_{MM}(t_j) - \underline{Y}_{RS}(t_j); \quad j = 1, \ldots, k, \quad (5.15)$$

with $\underline{e}(t_j)$ as the error function to be minimized. The performance criterion can be

$$J_N = \frac{1}{k} \cdot \sum_{i=1}^{m} d_i \sum_{j=1}^{k} \left| e_i(t_j) \right|^q, \quad d_i > 0. \quad (5.16)$$

Usually one selects $q = 2$, which results in the output-error least square estimation. For $q > 2$ the maximum error is minimized. By means of the weighting coefficients d_i, $i = 1, \ldots, m$, different error variances of each component of $\underline{Y}_{MM}$ may be taken into account. The model output Y is also a function of the model parameters $\underline{p}$, and the performance criterion in (5.16) can be rewritten in the form

$$J_N\left(\underline{p}\right) = \frac{1}{k} \cdot \sum_{i=1}^{m} d_i \sum_{j=1}^{k} \left| Y_{MMi}\left(t_j\right) - Y_{RSi}\left(t_j, \underline{p}\right) \right|^q. \quad (5.17)$$

Hence, the parameter identification problem requires the solution of the mathematical problem

$$J_N\left(\underline{p}\right) \overset{!}{=} Min. \quad (5.18)$$

5.4.3.2 Robust Control

Robust control is a branch of control theory where the approach to controller design explicitly deals with uncertainty as a consequence of disturbances or unexpected set-point sequences. Therefore, the idea, beyond a robust controller design, is that the controller is insensitive to all uncertainties of the physical world control task; and the final design of the controller has a fixed structure. Then a robust controller is suitable for dealing with small uncertainties (Yu and Lloyd 1997).

Robust control methods aim to achieve robust performance or stability in the presence of bounded errors. In contrast to an adaptive control policy, a robust control policy is static. Rather than adapting to measurements of variations, the controller is designed to work assuming that certain variables will be unknown but, for example, bounded (http10 2015).

With regard to the demands for set-point sequences and disturbance compensation, the feedback control loop is able to hold stability and robustness. The stability of the feedback control loop and its integral behavior is preserved even if a change in the parameters of the controlled system is in the large range. The feedback control loop is, therefore, robust with respect to the properties' disturbance compensation and set-point sequences.

5.5   Autonomous Systems in Ubiquitous Computing

Autonomy or autonomous behavior is a term used to refer to unmanned vehicles that act with no outside intervention. They accomplish this by executing tasks based on their own ability to make decisions or through a method of decision-making preprogrammed into them. Thus, an unmanned vehicle is a vehicle without a human being on board who is responsible for guiding the vehicle. Unmanned vehicles can either be remote-controlled or remote-guided vehicles, or they can be autonomous vehicles which are capable of sensing their environment and navigating on their own. They operate independently, without external intervention. Furthermore, unmanned vehicles represent the increasing challenge of autonomy, as shown in Fig. 5.6 (http12 2015).

In general, an autonomous (unmanned) vehicle performs behaviors or tasks with a high degree of autonomy, which is particularly desirable in field vehicles such as those shown in Fig. 5.6. Applications for autonomous robotic systems include space exploration, specifically the discovery and exploration of celestial structures in outer space by unmanned robotic systems such as the Mars Pathfinder rover, Sojourner. A NASA Discovery mission, Mars Pathfinder, was designed to be a demonstration of the technology necessary to deliver a lander and a free-ranging

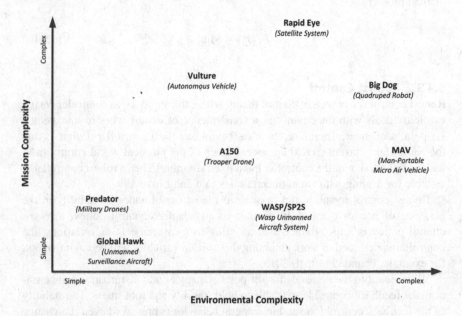

Fig. 5.6 Unmanned vehicles—the challenge of autonomy

Fig. 5.7 Mars Pathfinder rover Sojourner and rock "Yogi" (Image from NASA/Jet Propulsion Laboratory http13 2015)

robotic rover to the surface of Mars in a cost-effective, efficient manner. Pathfinder not only accomplished this goal but also returned an unprecedented amount of data and outlived its primary design life. From landing until the final data transmission on September 27, 1997, the Mars Pathfinder returned 2.3 billion bits of information, including more than 16,500 images from the lander and 550 images from the rover, as well as more than 15 chemical analyses of rocks and soil and extensive data on winds and other weather factors. Findings from the investigations carried out by scientific instruments on both the lander and the rover suggest that Mars was, at one time in its past, warm and wet, with water existing in its liquid state and a thicker atmosphere (http13 2015). In Fig. 5.7, the Mars Pathfinder rover Sojourner can be seen in front of the rock "Yogi." This image from the NASA/Jet Propulsion Laboratory is available on the Web (http13 2015).

Thus, highly autonomous vehicles must perform well under significant uncertainties in an environment for extended periods of time; and they must be able to compensate for system failures without external intervention. Hence, highly autonomous control systems provide high-level adaptive behavior to changes in the environment and control objectives. To achieve autonomy, the methods required for control system design should utilize algorithmic methods with regard to control, identification, estimation, and communication tasks, developed for continuous or discrete-state systems and decision-making methods, whereby artificial intelligence is preferred. In regard to supervising and tuning, the autonomous control system must provide a high degree of tolerance against failures, which means the fault-tolerant systems design approach is important in autonomous systems design.

Furthermore, the autonomous controller must be capable of planning the necessary sequence of control actions to accomplish a complicated task.

The need for methods to model and analyze the dynamic behavior of autonomous vehicles presents significant challenges well beyond current knowledge (Antsaklis and Passino 1993).

Besides the modern space robots, traditional industrial robots, which are automatically controlled, reprogrammable, multipurpose manipulators, and programmable in three or more axes of freedom, modern industrial robots are autonomous systems within the strict confines of their direct environment. The workspace of these robots is challenging and can often contain unpredictable variables.

More in general, the autonomous and fault-tolerant controls required for these advanced robots, as well as unmanned vehicles, such as those shown in Fig. 5.6, require the development and implementation of advanced algorithms. The algorithms allow the completeness and correctness of the autonomous system behavior to be observed, which is part of a supervised control approach where each functionality can be designed, implemented, and tested independently of the remaining system. The algorithms for such supervisory functionality constitute an increased risk for failures within the software implemented. Thus, the overall reliability can only be improved if the supervisory level is absolutely trustworthy. Hence, a seven-step design procedure was introduced in Blanke et al. (2001). The experience from this was that the design of an autonomous supervisor relies on an appropriate architecture that supports clear allocation of methods to different software tasks. This is a crucial task for both design and verification. The latter is vital since testing the supervisory functions in an autonomous control system is an important task. Therefore, architectural design of supervisory functions of a control system and their implementation and verification is not a trivial task because the architecture has to accommodate the implementation of several functions, such as (Blanke et al. 2001):

- Support of the overall coordinated unmanned vehicle control in different phases of the process
- Support of all control modes for regular operation and modes for operation with foreseeable faults of the unmanned vehicle
- Autonomous monitoring of the operational status, control errors, process status, and conditions of the unmanned vehicle
- Fault diagnosis, accommodation, and reconfiguration, as needed, which is done autonomously with status information to vehicle-wide coordinated control

The aforementioned functions can adequately be implemented in a supervisory structure with the autonomous controller and communication to the unmanned vehicle-wide control. Thus, the autonomous supervision is composed of two levels, written in Italian characters, which takes care of fault diagnosis, logic, and state control and effectors for activation or calculation of the appropriate remedial actions:

- Autonomous controller lower level with control loop inputs and outputs

- Autonomous supervision level with algorithms for fault diagnosis and effectors to fault accommodation
- Autonomous supervision level for supervisory logic
- Autonomous controller upper level with unmanned vehicle-wide control and coordination

To help promote these challenges, the Defense Advanced Research Projects Agency (DARPA) has launched the Cyber Grand Challenge, a competition seeking to create automatic defense systems capable of reasoning about flaws, formulating patches, and deploying them on a network in real time. By acting at machine speed and scale, these technologies may someday overturn today's attacker-dominated status quo. Just as the first autonomous ground vehicles fielded during DARPA's 2004 Grand Challenge weren't initially ready to take to the highways, the first generation of automated network defense systems won't be able to meaningfully compete against expert analysts or defend production networks (see Chap. 3). The Cyber Grand Challenge aims to give these groundbreaking prototypes a league of their own, allowing them to compete head to head to defend a network of customized software. DARPA plans to model the contest on today's elite cybersecurity tournaments. The program envisions numerous future benefits, including:

- Expert-level software security analysis and remediation, at machine speeds on enterprise scales
- Establishment of a lasting competition community for automated cyber defense
- Identification of thought leaders for the future of cybersecurity
- Creation of a public, high-fidelity recording of real-time competition between automatic systems

Competitors would navigate a series of challenges starting with a qualifying event in which a collection of software is automatically analyzed. Competitors would qualify by identifying, proving, and repairing software flaws. A select group of competitors who display top performance during the qualifying event would be invited to the Cyber Grand Challenge final event, slated for early to mid-2016. Each team's system would automatically identify software flaws, scanning the network to identify affected hosts. Teams would be scored against each other based on how capably their systems can protect hosts, scan the network for vulnerabilities, and maintain the correct function of software. Realization of this vision will require breakthrough approaches in a variety of disciplines, including applied computer security, program analysis, and data visualization.

5.6 Case Study: Robot Manipulator

A robot, as a system, consists of the following elements, which are integrated together to form a whole:

- *Manipulator*: the main body of the robot consisting of links, joints, and other structural elements of the robot. With other elements, the manipulator alone is not a robot.
- *End effector*: connected to the last joint of a manipulator which generally handles objects, makes connection to other machines, or performs the required tasks.
- *Actuators*: are servomotors, stepper motors, pneumatic cylinders, and hydraulic cylinders. Other actuator types are used for more specific applications.
- *Sensors*: collect information about the internal state of the robot or communicate with the outside environment.
- *Controller*: controls the robot manipulator motions. It receives data from the embedded computing system, controls the motions of the actuators, and coordinates the motions with the sensory feedback information.

Therefore, a robotic manipulator system is a device that can identify its own state and take action based on it. It consists of a linked chain of rigid bodies that are linked in an open kinematic chain at joints. The robotic manipulator shown in Fig. 5.8 can be decomposed such that the mechanical part of the robot consists of revolute joints, bodies, and the load. The rigid body manipulator system can have up to 6° of freedom of movements. This comprises three translational movements, such as moving up and down, moving left and right, and moving forward and backward. It also comprises three rotational degrees of freedom: tilting up and down, turning left and right, and tilting from side to side.

A body component describes the mass and inertia effects of the body. The joints of the robot are given by the axis. An axis is a key component that describes the motor and gear box that drives the joint, the control system, and the reference generation. A possible representation of the robotic manipulator is shown at an abstract level in Fig. 5.9, which contains the reference acceleration of the axis qddRef as an input value of the connector and a mechanical flange to drive a shaft on the output-side connector.

The decomposition of the axis shows that the reference acceleration (qddRef) will be integrated twice in order to derive a reference velocity (qdRef) and a

Fig. 5.8 Schematic diagram of the robotic manipulator

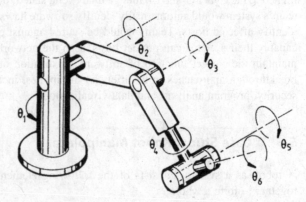

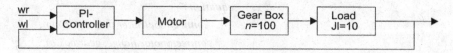

Fig. 5.9 Composition diagram of one axis of the robot manipulator in Fig. 5.8 (For details see text)

Fig. 5.10 Schematic diagram of the motor drive of the robotic manipulator

reference position (qRef). The reference values are fed into a controller (irControl), while the controller output is the reference current of the motor (irMotor) driving the gear box (irGear). The driving part of the gear box is a mechanical flange to which the axis of a shaft or a robot joint can be connected.

Typical for the axis controllers (irControl) are the velocity and the position controller, the output of which is the desired reference current of the motor. The current of the motor is approximately proportional to the motor torque produced and the quantity to be controlled. The irMotor model of the electric motor consists of an analog current controller, which can be realized using operational amplifiers, and the DC motor with the components Ra, La, and EMF. The output current of the current controller represents the input signal of the motor. The DC motor produces a torque that drives a mechanical flange.

The composition diagram of the gearbox irGear of the driving system is modeled by the motor inertia, a rotational spring to model the gear elasticity, an ideal gear box representing the gear ration, and load inertia to model the rotational inertia of all parts at the driven side of the gear. A friction component connected between the motor shaft and the shaft bearings models the Coulomb friction of the bearings.

Describing how to model the details of a component, we can consider a simple motor drive system, as shown in Fig. 5.10. The system can be built as a set of connected components: controller, motor, gear box, and load.

The model shown in Fig. 5.10 is a composite model that specifies the topology of the robotic manipulator system to be modeled in terms of components and connections between the components. For example, the statement gear box ($n = 100$) declares a component gearbox of class gear box and sets the default ratio, n, to 100. The complete model of the robotic manipulator system shown in Fig. 5.8 can be described in Modelica, as shown in Fig. 5.11.

The connections in Fig. 5.10 specify the interactions between several components, as shown in Fig. 5.9. A connector contains all of the quantities needed to describe the interaction.

The Modelica model of the motor drive, shown in Fig. 5.11, represents a typical feedback loop for which the continuous-time transfer function can be calculated

Fig. 5.11 Modelica model of the motor drive system in Fig. 5.10

model *MotorDrive of Robotic Manipulator*
  PI   *controller;*
  Motor   *motor;*
  Gearbox   *gearbox (n=100);*
  Shaft   *J1 (J=10)*
  Tachometer   *wl;*
equation
   connect(*controller.out, motor.inp*);
   connect(*motor.flange, gear box.a*);
   connect(*gearbox .b, J1.a*);
   connect(*J1.b, wl.a*);
   connect(*wl.w, controller.inp*);
  end *MotorDrive;*

using computer algebra as an efficient simulation code. The Modelica model of a continuous-time transfer function is as follows:

```
partial block SISO
  input Real u;
  output Real y;
end SISO;
block TransferFunction
  extends SISO;
  parameter Real a[:]={1,1} "Denominator";
  parameter Real b[:]={1} "Numerator";
protected
  constant Integer na=size (a,1);
  constant Integer nb(max=na)=size (b,1);
  constant Integer na=na-1 "System order";
  Realb0[na]=cat (1, b, zeros (na -nb)) "Zero expanded vec-
  tor";
  Real x[n] "State vector";
Equation
  //Controllable canonical form
  der (x[2:n])=x[1:n-1];
  a[na]·der (x[1]+a[1:n]·x=u;
  y=(b=[1:n]-b0[na]/a[na]·a[1:n])·x+b0[na]/a[na]·u
end TransferFunction;
```

Besides the aforementioned Modelica model of the robotic manipulator motor drive system for simulation purposes, the robotic manipulator transfer function can be derived as required for the mathematically based proportional-integral

(PI) controller design. The PI algorithm is described for an ideal, continuous PI-controller programmable interface controller (PIC) as follows:

$$\mathrm{PIC} = K_P \cdot e(t) + \frac{K_I}{T_I} \int e(t)\mathrm{d}t \qquad (5.19)$$

where PIC is the PI-controller output signal; $e(t)$ current controller error, defined as $w - w_d$; w input variable (set point); w_d measured control variable; K_P P-controller gain, a tuning parameter; K_I I-controller gain, a tuning parameter; and T_I reset time, a tuning parameter.

The first two terms to the right of the equal sign in Eq. (5.19) are referred to the P-controller part; the last term in Eq. (5.19) is the integral part of the controller. Its function is to integrate the controller error, $e(t)$, over time. The reset time tuning parameter, T_I:

- Provides a separate weight to the integral term so the influence of integral action can be independently adjusted
- Is in the denominator so smaller values provide a larger weight of the integral term

For only a P controller, Eq. (5.19) will hold:

$$\mathrm{PC} = K_P \cdot e(t)$$

The only way to add or subtract in the P-controller equation above is if $e(t)$ is not zero. If $e(t)$ is not steady at zero, then w_d does not equal w and an offset result. To avoid an offset in the controller of the robotic manipulator system, the PI controller is used, the equation of which is shown in Eq. (5.19). From Eq. (5.19), it can be seen that the integral sum of the error can have a final or residual value after the response time is complete. This is important because it means that $e(t)$ can be zero. Nevertheless, something can still be added or subtracted to receive the final controller output, PIC. As long as there is any error, meaning that $e(t)$ is not zero, the integral term will grow or shrink in size which has the reactive impact on PIC. The changes in PIC will only cease when w_d equals w, and then $e(t) = 0$, for a sustained period of time. At that point, the integral term can have a residual value. This residual value from integration creates a new overall bias value that corresponds to the new level of operation. Thus, integral action continually resets the bias value to eliminate offset as operating level changes. Hence, the challenges using the proportional-integral controller algorithm are:

- The tuning parameters interact with each other and their influence must be balanced by the designer.
- The integral term tends to increase the oscillatory or rolling behavior of the process response.

Although the tuning parameters interact with each other, it can be challenging to achieve the best tuning values. Implementing the robotic manipulator proportional-integral controller can be done following the best practice four-step design procedure (Cooper 2008).

1. Determine design level of operation (DLO) for w and w_d.
2. Collect process data around the DLO for PIC (w and w_d are steady near the design level of operation), and apply a test signal to force PIC for a distinct response.
3. Describe the process dynamic by fitting an approximate first-order time (FOT) model to test the data obtained from Step 2. The model parameters are process gain K_P, K_I, and reset time T_I.
4. Use the FOT parameter to complete the robotic manipulator proportional-integral controller design, based on the ideal form of the PI-controller algorithm in Eq. (5.19). For the reset time, different responding controller behavior can be investigated by identifying the best fit of the PI controller for the robotic manipulator. In case of a fast-responding controller, some overshoot and oscillation can be tolerated as w_d settles out, which means that T_I should be small and result in a so-called aggressive tuning or aggressive response. In the case of a slow controller, that will move the error in the proper direction; but quite slowly, a so-called conservative tuning or conservative response will be chosen for the robotic manipulator controller. In contrast with the two aforementioned tuning types, a third type, the moderate tuning of a controller, will move w_d reasonably fast while generating little or no overshoot, which is called moderate tuning or moderate response for the robotic manipulator controller.

A process that is naturally direct acting requires a controller that is reverse acting to remain stable. In spite of the direct acting process and reverse acting control, both gain factors Kp and K_I should have positive values. In most commercial controllers, only positive K_I values are used. The sign of the controller is then assigned by specifying that the controller is either reverse acting or direct acting to indicate a positive or negative K_I, respectively.

The ability of the PI controller to react to changes in the robot manipulator acting sequences results in the aggressive controller type in a more energetic PIC action and thus a more active w_d response. This results in some overshoot and oscillation in the robot sequencing process response, while the moderate response may better fit the robot sequencing process response.

5.7   Exercises

What is meant by the term *ubiquitous computing*?
List and define the main characteristics for ubiquitous computing.
What is meant by the term *smart devices*?
Give examples of smart devices.

What is meant by the term *ubiquitous network*?

List and define the main characteristics for ubiquitous networks.

What is meant by the term *smart environment*?

Give an example of a smart environment.

What is meant by the term *human-centered environment*?

Give an example of a human-centered environment.

What is meant by the term *distributed computing environment*?

Give an example of a distributed computing environment.

What is meant by the term *smart dust*?

Give an example of smart dust.

What is meant by the term *MEMS*?

Give an example of MEMS.

What is meant by the term *ubiquitous learning space*?

Give an example of a ubiquitous learning space.

What is meant by the term *smart home*?

Give an example of powerline communication in a smart home application.

What is meant by the term *Web Services Description Language*?

List and define the main characteristics for the Web Services Description Language.

What is meant by the term *tagging*?

Give an example of tagging.

What is meant by the term *sensing*?

Give an example of sensing.

What is meant by the term *MARG sensor*?

List and define the main characteristics for the MARG sensor.

What is meant by the term *sensor fusion*?

Give an example of sensor fusion.

What is meant by the term *multisensor data fusion*?

Give an example of a multisensor data fusion.

What is meant by the term *artificial intelligence*?

Give an example of artificial intelligence.

What is meant by the term *controlling*?

List and define the main characteristics of controlling.

What is meant by the term *parameter identification*?

Give an example of parameter identification.

What is meant by the term *robust control*?

Give an example for robust control.

What is meant by the term *autonomous system*?

Give an example of an unmanned autonomous system.

What is meant by the term *fault tolerance*?

List and define the main characteristics for fault tolerance.

What is meant by the term *robot manipulator*?

List and define the main characteristics of a robot manipulator.

What is meant by the term *robot degrees of freedom*?

List and define the main characteristics of robot degrees of freedom.

What is meant by the term *robot workspace*?

List and define the main characteristics of a robot workspace.

References

(Antsaklis and Passino 1993) Antsaklis, P. J., Passino, K. M. (Ed.): An Introduction to intelligent and autonomous control, Kluwer Academic Publ. 1993

(Bardram and Friday 2010) Bardram, J., Friday, A.: Chapter 2: Ubiquitous Computing Systems, pp. 37–94, In: Krumm, J.: Ubiquitous Computing Fundamentals, CRC Press, 2010

(Blanke et al. 2001) Blanke, M., Frei, C., Kraus, F., Patton, R. J., Staroswiecki, M.: Fault-tolerant Control Systems. Chapter 8, In: Control of Complex Systems. Åström, K., Albertos, P., Blanke, M., Isidori, A., Schaufelberger, W., Sanz, R. (Eds.), Springer Publ. 2001

(Brooks 1985) Brooks, R. A.: A Robust Layered Control System for a Mobile Robot. MIT A. I. Memo 864, 1985

(Brush 2010) Brush, A. J. B.: Ubiquitous Computing Field Studies. In: Ubiquitous Computing Fundamentals, pp.161–202, CRC Press, 2006

(Buckley 2013) Buckley A.: Organic Light Emitting Diodes (OLED): Materials, Devices, and Applications, Elsevier B.V.

(Bumiller et al. 2010) Bumiller, G., Lample, L., Hrasnica, H.: Power line communication network for large scale control and automation systems, IEEE Communications Magazine, Vol. 48(4), pp. 106.113, 2010

(Camp 2000) Camp, P. J.: Supporting Communication and Collaboration Practices in Safety-Critical Situations. Proceed. Human Factors in Computing Systems

(Carcelle 2006) Carcelle, X.: Power Line Communications in Practice. Artech House Publ. 2006

(Cooper 2008) Cooper, D. J.: PI Disturbance Rejection Of The Gravity Drained Tanks. Source: http://www.controlguru.com/2006/p111806.html

(Coulouris et al. 2005) Coulouris, G., Dollimore, J., Kindberg, J.: Distributed Systems: Concepts and Design. Addision-Wesley, Reading, MA, 2005

(CPSSR 2008) Cyber-Physical Systems Summit Report, http://varma.ece.cmu-edu/summit/, 2008

(Davy 2015) Davy, A.: Components of a smart device and smart device interactions. http://www.m-zones.org/deliverables/d234_1/papers/davy-components-of-a-smart-device.pdf

(Gann et al. 1999) Gann, D., Barlow, J., Venables, T.: Digital Futures Making Homes Smarter. Joseph Rowntree Foundation. Published by Chartered Institute of Housing, 1999

(Gaver et al. 2006) Gaver, W. W., Bowers, J., Boucher, A., Law, A., Pennington, S., Villar, N.: The History Tablecloth: Illuminating domestic activity. In. Proceed. Designing Interactive Systems: Processes, Practices, Methods, and Techniques, pp.199–208, DIS Press, 2006

(Gaver et al. 2007) Gaver, W.W., Sengers, P. Kerridge, T., Kaye, J. J., Bowers, J.: Enhancing ubiquitous computing with user interpretation: Field testing the home health horoscope. In: Proceed. ACM CHI Conference on Human Factors in Computing Systems, pp. 537–546, ACMPress, 2007

(Goldstein et al, 2005) Goldstein, S. C., Campbell, J. D., Mowry, T. C.: Programmable matter, Computer, Vol. 38, No. 6, pp. 99–101, 2005

(Gunstone 2011) Gunstone, R. E.: Formalisms for Use Cases in Ubiquitous Computing. In. Proceed. Internat. Conf. on Mobile Services, Resources, and Users, pp. 103–107, IARIA Press, 2011

(Hum 2001) Hum, A. P. J.: Fabric area network - A new wireless communications infrastructure to enable ubiquitous networking and sensing on intelligent clothes. Computer Networks, Vol. 35, pp.391–399, 2001

(Jones and Jo 2004) Jones, V., Jo, J. H.: Ubiquitous learning environment: A adaptive teaching system using ubiquitous technology. In: Proceedings of the 21st ASCILITE Conference, pp. 468-474, 2004

(Kaplantzis and Sekercioglu 2012) Kaplantzis, S., Sekercioglu, Y. A.: Security and Smart Metering, In: Proceed. European Wireless, Poland, 2012

(Kobåa and Andersson 2008) Kobåa, A., and Andersson, B.: A Vision of Cyber-Physical Internet. In: 2008 I.E. International Conference on Sensor Networks, Ubiquitous, and Trustworthy Computing

(Krumm 2009) Krumm, J.: Ubiquitous Computing Fundamentals, CRC Press, 2009

(Langheinrich, 2010) Langheinrich, M.: Privacy in Ubiquitous Computing, Chapter 3, In: In: Ubiquitous Computing Fundamentals, Krumm, J. (Ed.), CRC Press, 2010

(Lee et al. 1998) Lee, W. J., Cha, S. D., Kwon, Y. R. : Integration and Analysis of Use Cases Using Modular Petri Nets in Requirements Engineering. IEEE Transaction on Software Engineering, Vol. 24, pp. 1115–1130, 1998

(Lyytinen and Yoo 2002) Lyytinen K., Yoo, Y.: Issues and Challenges in Ubiquitous Computing, 2002, http://info.cwru.edu/ubicom/document/acm.pdf

(Mattern 2001) Mattern, F.: The Vision and Technical Foundations of Ubiquitous Computing. Upgrade, Vol. 2, No. 5, pp. 1–4, 2001

(Meier 1998) Meier, C.: Datenfusionsverfahren für die automatische Erfassung des Rollverkehrs auf Flughäfen. PhD Thesis TU Braunschweig, 1998

(Mitchell 2007) Mitchell, H. B.: Multi Sensor Data Fusion – An Introduction, Springer Publ., 2007

(Mobashsher et al. 2011) Mobashsher, A. T., Islam, M. T.,, Misran, N.: RFID Technology: Perspectives and Technical Considerations of Microstrip Antennas for Multi-Band RFID Reader Operation, In: Current Trends and Challenges in RFID, Turcu, C. (Ed.), ISBN: 978-953-307-356-9, InTech, Available from: http://www.intechopen.com/books/current-trends-and-challenges-in-rfid/rfid-technologyperspectives-and-technical-considerations-of-microstrip-antennas-for-multi-band-rfid

(Moeller 2004) Moeller, D. P. F.: Mathematical and Computational Modeling and Simulation – Fundamentals and Case Studies. Springer Publ. 2004

(Moeller et al. 2013) Moeller, D. P. F., Haas, R., Vakilzadian, H. : Ubiquitous Learning: Teaching Modeling and Simulation (M&S) with Technology, In: Proceedings Summer Simulation Multiconference GCMS 2013, pp. 125–132, Eds. H. Vakilzadain, R. Crosbie, R. Huntsinger, K. Cooper, Curran Publ. Red Hook, NY, 2013 Best Paper Award

(Moeller and Vakilzadian 2014) Moeller, D. P. F., Vakilzadian, H.: Ubiquitous Networks: Power Line Communication and Internet of Things in Smart Home Environments, In: Proceed. IEEE International Conference on Electro Information Technology, pp. 596–601, DOI:10.1109/EIT. 2014.6871832, IEEE Conference Publications, 2014

(Ogata and Yano 2012) Ogata, H., Yano, Y.: Context-aware Support for Computer-Supported Ubiquitous Learning, 2012 http://140.115.126.240/mediawiki/images/e/e9/Context_Awareness.pdf; Accessed: 07042013

(Poslad 2009) Poslad, S.: Ubiquitous Computing: Smart Devices, Environments and Interactions, John Wiley & Sons, 2009

(Quigley 2010) Quigley, A.: From GUI to UUI: Interfaces for Ubiquitous Computing. Chapter 6, In: Ubiquitous Computing Fundamentals, Krumm, J. (Ed.), CRC Press, 2010

(Schönefeld 2014) Schönefeld J.: Architectures for Embedded Multimodal Sensor Data Fusion Systems in the Robotics- and Airport Traffic Surveillance. PhD Thesis, TU Clausthal, 2014

(Sharples et al. 2005) Sharples, M., Taylor, J., Vavoula, G.: Towards a Theory of Mobile Learning. 2005. http://www.mlearn.org/mlearn2005/CD/papers/Sharples %20Theory%20of%20Mobile. pdf; Accessed: 07.04.2013

(Simmon et al. 2013) Simmon, E., Kim, K-S., Subrahmanian, E., Lee, R., de Vaulx, F., Murakami, Y., Zettsu, K.: A Vision of Cyber-Physical Cloud Computing for Smart Networked Systems, NISTIR 7951 Report, 2013

(Simon 1969) Simon. H. A.: Sciences of the Artificial. MIT Press 1969

(Snowdon 2009) Snowdon, J. L.: IBM Smarter Energy Management Systems for Intelligent Building, Report IBM T.J, Watson Research Center, 2009

(Varshavsky and Patel 2010) Varshavsky, A., Patel, S.: Location in Ubiquitous Computing, Chapter 7, In: Ubiquitous Computing Fundamentals, Krumm, J. (Ed.), CRC Press, 2010

(Waggoner and Craighead 2007) Waggoner, P. S., Craighead, H. G.: Micro- and nanomechanical sensors for environmental, chemical, and biological detection. In: Lab Chip Vol. 7, pp. 1238–1255, 2007

(Wang et al. 2007) Wang, H. H., Wang, G., Shu, Y.: Design of RFID reader using multi-antenna with difference spatial location. In: Proceedings if the WiCom International Conference, pp. 2070-2073, 2007

(Weiser 1991) Weiser, M.: The Computer for the 21st Century, In: Scientific American, pp. 94–100, 1991

(Weiser 1993a) Weiser, M.: Hot Topics: Ubiquitous Computing, In: IEEE Computer, 1993

(Weiser 1993b) Weiser, M.: Some Computer Science Problems in Ubiquitous Computing, In: Communications of the ACM, 1993, Reprinted as Ubiquitous Computing, Nikkei Electronics, pp. 137–143, 1993

(Weiser 1994) Weiser, M.: The world is not a desktop, In: Interactions, pp.7–8, 1994

(Xia and Ma 2013) Xia, F., and Ma, J.: Building Smart Communities with Cyber-Physical Systems. In: 13th ACM International Conference on Ubiquitous Computing Symposium on Social and Community Intelligence, Beijing, China, pp: 1–6, September 2011

(Xia et al. 2008) Xia, F., Ma, L., Dong, J., and Sun, Y.: Network QoS Management in Cyber-Physical Systems. In Proc. ICESS 2008, pp. 302–307, IEEE Press, 2008

(Yallup and Iniewski 2014) Yallup, K., Iniewski, K.: Technologies for Smart Sensors and Sensor Fusion, CRC Press, 2014

(Yu and Lioys 1997) Yu, H., Llooyd, S.: Variable structure adaptive control of robot manipulators. In: IEE Proceed. IEE Processes and Control Theory and Applications, Vol. 144, pp. 167–176, 1997

Links

(http1 2015) http://en.wikipedia.org/wiki/Smart_device; accessed February 16th 2015

(Http2 2015) http://en.wikipedia.org/wiki/Smartdust; accessed February 16th 2015

(http3 2015) https://ahdictionary.com/; accessed February 16th 2015

(http4 2015) http://www.independent.co.uk/life-style/gadgets-andtech/news/there-are-officially-more-mobile-devices-than-people-in-the-world-9780518.html

(httpp5 2015) https://video.tu-clausthal.de/videos/iasor/vorlesung/iot-ss2013/20130621/iot-20130621.html

(http5 2015) http://en.wikipedia.org/wiki/Smart_device; accessed February 16th 2015

(http6 2015) smarttechpkr.blogspot.com; accessed February 16th 2015

(http7 2015) http://www.harlandsimon.com; accessed February 16th 2015

(http8 2015) http://en.wikipedia.org/wiki/Sensor; accessed February 16th 2015

(http9 2015) http://en.wikipedia.org/wiki/Sensor_fusion; accessed February 16th 2015

(http10 2015) http://en.wikipedia.org/wiki/Robust_control; accessed February 16th 2015

(http11 2015) http://en.wikipedia.org/wiki/Field-programmable_gate_array; accessed February 16th 2015

(http12 2015) http://www.darpa.mil/Our_Work/I2O/Programs/Cyber_Grand_Challenge_%28CGC%29.aspx; accessed February 16th 2015

(http13 2015) http://www.nasa.gov/mission_pages/mars-pathfinder/; accessed February 16th 2015

Systems and Software Engineering

6

This chapter begins with a brief introduction to systems engineering in Sect. 6.1, which describes systems engineering as an interdisciplinary field of engineering primarily focused on how to successfully design, implement, evaluate, and manage complex engineered systems over their life cycles. It also introduces the International Organization for Standardization (ISO) and the International Electrotechnical Commission (IEC) ISO/IEC 15288:2008 standard. Section 6.2 describes the design challenges of cyber-physical systems (CPS) and their impact on systems engineering with reference to requirements definition and management using Cradle®. Cradle is a requirements management and systems engineering tool that integrates the entire project life cycle into one, massively scalable, integrated, multiuser software product. Section 6.3 introduces the principal concept of software engineering with special focus on the V-model and Agile software development methodology. Section 6.4 introduces the different requirements in software design in . CPS. It also includes the software requirements standard American National Standards Institute/Institute of Electrical and Electronics Engineers (ANSI/IEEE) 29148-2011. Section 6.5 provides a maritime area case study which focuses on tracking and monitoring containers at ports and on ships as well as tracking and monitoring containers transported from a sea gate port to a dry port. Section 6.6 contains comprehensive questions from the introduction to systems engineering topics, followed by references and suggestions for further reading.

6.1   Introduction to Systems Engineering

As introduced in Chap. 1, a system can be understood as a set of interrelated components which interact with one another in an organized form toward a common purpose. Therefore, a system is a unit that cannot be divided into independent parts without losing its essential characteristics, meaning that a system's essential characteristic properties are the product of the interactions of its parts and not the actions of the individual parts. Hence, a system provides an operational

© Springer International Publishing Switzerland 2016   235
D.P.F. Möller, *Guide to Computing Fundamentals in Cyber-Physical Systems*,
Computer Communications and Networks, DOI 10.1007/978-3-319-25178-3_6

capability to satisfy a stated need or objective. This means that a system is fully defined by the combination of resources operating in its operational environment in order to achieve the requested properties. In a physical sense, the term *system* can be synonymous with the term *product*, meaning that in the case of a design project, a system or a product will be delivered at the end. However, a system normally is comprised of a number of products. As defined in the ANSI/Electronic Industries Alliance (ANSI/EIA) 632 standard (see Sect. 6.1.2), a system is comprised of operational end and enabling products.

Engineering is the application of scientific, economic, social, and practical knowledge to invent, design, build, maintain, research, and improve structures, machines, devices, systems, materials, and processes. Hence, engineering can be extremely broad and encompasses a wide range of specialized fields, such as chemical engineering, civil engineering, control engineering, electrical and computer engineering, and mechanical engineering. Each of these fields emphasizes particular areas of applied science and technology and types of applications. Against this background, engineers use their knowledge of economics, logic, mathematics, science, and more as well as appropriate experience or tacit knowledge to find suitable solutions to problems while designing a system or a product.

Developing an appropriate mathematical model of a problem in engineering allows it to be analyzed and potential solutions to be tested by simulation methods. Therefore, models are a major step forward in the development of CPS; and there are several advantages to working with models, which can be seen in CPS design processes, including the model-based design and model-driven design approach (Derler at al. 2011). Models also allow a design to be tested in a safe environment, enabling engineers to determine if any design defects exist, which is of great benefit before a prototype of the CPS is developed. Modeling CPS requires the inclusion of models of the physical processes and software models, computational platforms, and networks (Johnson et al. 2014). For this reason, simulation is used in the conceptual analysis phase of the problem-solving cycle in engineering, an interactive procedure that strongly depends on the type of application and planning issues. Therefore, a model-based systems engineering process with functional model analysis is an approach that primarily focuses on engineered systems over their life cycles, as described in Baker (2015). This calls for requirements definition and management activities for systems and/or products, which can be done using Cradle software (CRADLE-7 2014).

Systems engineering is, as previously mentioned, an interdisciplinary approach to designing complex technical systems based on certain thought patterns and basic principles. Thus, systems engineering is an integration of disciplines because it is rarely possible for a complex technical system to be designed by only a single discipline, ensuring that all likely aspects of the CPS under design are considered and integrated into a whole. Systems engineering overlaps many disciplinary boundaries, such as:

- Control engineering
- Electrical engineering

- Software engineering
- Project management

The systems engineering approach gives insight into how engineering concepts and procedures could be applied to individual and/or special requirements of CPS. This finally results in a wide range of definitions of systems engineering, each of which is subtly different because it tends to reflect the particular focus of its source. The following are some of the more accepted and authoritative definitions of systems engineering from relevant standards and documents (Faulconbridge and Ryan 2014):

> *Systems engineering is the management function which controls the total system development effort for the purpose of achieving on optimum balance of all system elements. It is a process which transforms an operational need into a description of system parameters and integrates those parameters to optimize the overall system effectiveness.* (DSMC 1990)
> *An interdisciplinary collaborative approach to derive, evolves, and verifies a life cycle balanced solution which satisfies customer expectations and meets public acceptation.* (IEEE-STD-1220-1994 1995)
> *An interdisciplinary approach encompassing the entire technical effort to evolve and verify an integrated and life cycle balanced set of system, product, and process solutions, that satisfy customer needs. Systems engineering encompasses: the technical efforts related to the development, manufacturing, verification, deployment, operations, support, disposal, and user training for, system products and processes; the definition and management of the system configuration; the translation of the system definition into work breakdown structures, and development of information for management decision making.* (EIA/IS-632-1998 1994)
> *Systems engineering is the selective application of scientific and engineering efforts to: transform an operational need into a description of the system configurations which best satisfies the operational need according to the measures of effectiveness; integrate related technical parameters to ensure compatibility of all physical, functional, and technical program interfaces in a manner which optimizes the total system definition and design; and integrate the efforts of all engineering disciplines into the total engineering efforts.* (SECMM-95-01 1995)
> *Systems engineering is an interdisciplinary, comprehensive approach to solving complex system problems and satisfying stakeholder requirements.* (Lake 1996)
> *System engineering is an interdisciplinary approach and means to enable the realization of successful systems. It focusses on defining customer needs and required functionality early in the development cycle, documenting requirements, and then proceeding with design synthesis and system validation while considering the complete problem: operations, cost and schedule, performance, training and support, test, manufacturing, and disposal. Systems engineering considers both the business and the technical needs of all customers with the goal of providing a quality products that meets the user needs.* (Haskins 2010)

It is interesting to note that the above definitions come principally from earlier standards. Systems engineering standards, such as SITEC 15288 (see Sect. 6.1.1), ANSI/EIA-632 (see Sect. 6.1.2), and IEEE-STD-1220 (see Sect. 6.1.2), do not contain any definition of systems engineering but refer more generically to engineering of systems (Faulconbridge and Ryan 2014).

With larger and much more complex engineered systems, simulation is used in various steps as part of the regular problem-solving cycle of systems engineering based on conceptual analysis and synthesis, which refers to:

- Description of the application
  - Problem definition
  - Formulation of objectives
  - Situation analysis
  - Optional design of tests or experiments
- Modeling
  - Data collection
  - Model design
  - Verification of model implementation
  - Validation of model results
- Simulation
  - Parameter variation
  - Parameter optimization
  - Output interpretation
- Recommended solutions including documentation and implementation

The close interaction that CPS has with the physical world indicates that time and other constraints can be handled using the event-based approach in simulation. An event-based modeling approach in CPS uses events as units for computation, communication, and control in the system. CPS basically consists of a distributed set of proper system components, like sensors, which detect events, actuators, which assign the essential performance of an action, and others, which operate in their own individual reference frames. A common frame of reference does not currently exist for event-based CPS models because of their mostly heterogeneous nature (Johnson et al. 2014).

Another advantage of CPS is the real-time interaction of output and input. This allows CPS to operate in changing environmental conditions in conjunction with sensing and actuating capabilities and information processing. This assumes that a control loop indicates a notable change has occurred in the environmental conditions in the physical part of the overall system. Initially, this information has been sensed, pre- and post-processed, and, thereafter, transferred as an actuator assignment to take action within a specific time frame, influencing the changed environmental conditions in real time in the right way. To guarantee the proper influence and performance by the aforementioned the following sequence of procedures is required as part of systems engineering:

- Requirements engineering (see Sects. 3.3.1, 6.2.1, and 6.2.2)
- Reliability analysis
- Coordination of design teams from different disciplines
- Evaluation
- Maintainability
- Testing

Table 6.1 Important dates in the origins of systems engineering as a discipline

Year	Action items
1937	British multidisciplinary team analyzed the air defense system
1939–1945	Bell Labs supported NIKE development
1951–1980	MIT defined and managed the SAGE air defense system
1956	RAND Corporation invented systems analysis
1962	*A Methodology for Systems Engineering* published
1962	Jay Forrester modeled urban systems at MIT
1994	Perry memorandum urges military contractors to adopt commercial practices, such as IEEE
2002	ISO/IEC 15588 released

Source: *Systems Engineering Handbook, A Guide for System Life Cycle Processes and Activities*, edited by Cecilia Haskins, 2007, INCOSE Systems Engineering Handbook, version 3.1

Requirements definition and management is fundamental to project success in CPS design and is a primary focus of early systems engineering efforts. Once the requirements have been collected, the systems engineering process then focuses on the derivation and decomposition of these requirements from the system level right down to the lowest constituent components (sometimes referred to as requirements flow down). This process involves elicitation, analysis, definition, validation, and management of requirements. Requirements engineering ensures that a rigorous approach is taken to the collection of a complete set of unambiguous requirements from the stakeholders involved in the systems engineering design process (Faulconbridge and Ryan 2014). The requirements definition and management process for CPS design, based on the Cradle software tool, is described in more detail in Sect. 6.2.1.

The origins of systems engineering can be traced back to the 1930s followed quickly by other programs and supporters. A summary of some important highlights in the origins and history of the application of systems engineering is summarized in Table 6.1 (Haskins 2006).

6.1.1 Systems Engineering Standard ISO/IEC 15288

With the introduction of ISO/IEC 15288 in 2002, an international standard covering processes and life cycle stages in the work process of systems engineering was formally recognized as the preferred mechanism for establishing agreement on the creation of systems/products and services to be traded between enterprises. Initial planning for the ISO/IEC 15288:2002 standard started in 1994 when the need for a common systems engineering process framework was recognized. In 2004, this standard was adopted as IEEE 15288. The ISO/IEC 15288 standard was updated in 2008 and is managed by ISO/IEC JTC1/SC7, the ISO committee responsible for developing ISO standards in systems and software engineering. The standard defines the systems engineering process using the following four categories:

- Technical
- Project
- Agreement
- Enterprise

Each of these categories is defined by:

- *Activities*: Essential to fulfill the design specifications; a set of activities that transforms inputs into the desired outputs.
- *Control*: Represents the directives and constraints of the systems design.
- *Enablers*: Essential resources, such as infrastructure, workforce, tools, technologies, and more, to realize the systems design with regard to the purpose of the systems usage.
- *Inputs*: Represent the system data, information, and material.
- *Outcomes*: Represent the system functionality, e.g., processed data, products, and services.

The following (Haskins 2006) definitions of terms used frequently in systems engineering are taken from ISO/IEC 15288 (Table 6.2).

The ISO/IEC 15288:2008 is a standard for systems and software engineering life cycle processes. It establishes a common framework describing the life cycle of systems designed by engineers and defines a set of processes and associated terminology within that framework. These processes can be applied at any level in the hierarchy of a system's structure. Selected sets of these processes can be applied throughout the life cycle of a system for managing and performing the stages of its life cycle. This is accomplished through the involvement of all interacting parties, with the ultimate goal of achieving customer satisfaction.

ISO/IEC 15288:2008 also provides processes that support the definition, control, and improvement of life cycle processes used within an organization or project when acquiring and supplying systems (http1 2015). The life cycle process (LCP) is made operational through life cycle management (LCM). Life cycle management is a management approach that puts the tools and methodologies in the life cycle process into practice. Therefore, it can be understood as a product management system that helps companies to minimize the environmental and social burdens associated with their system/product or system/product portfolio during its entire life cycle. The sustainability framework, shown in Fig. 6.1, describes a scheme where sustainability is achieved through the use of the life cycle process and is supported by relevant and reliable data sets, as well as an appropriate policy framework [http2 2015].

ISO/IEC 15288:2008 concerns those systems that are man-made and may be configured with one or more of the following:

Table 6.2 Definition of frequently used terms

Term	Definition
Activity	Set of actions that consume time and resources and whose performance is necessary to achieve or contribute to the realization of one or more outcomes
Enabling system	A system that complements a system of interest during its life cycle stages but does not necessarily contribute directly to its function during operation
Enterprise	That part of an organization with responsibility to acquire and to supply products and/or services according to agreements
Organization	A group of people and facilities with an arrangement of responsibilities, authorities, and relationships
Process	Set of interrelated or interacting activities that transform inputs into outputs
Project	An endeavor with start and finish dates undertaken to create a product or service in accordance with specified resources and requirements
Stage	A period within the life cycle of a system that relates to the state-of-the-system description or the system itself
System	Combination of interacting elements organized to achieve one or more stated purposes
System element	A member of a set of elements that constitutes a system
System of interest	The system whose life cycle is under consideration
Systems engineering	Interdisciplinary approach and the means to enable the realization of successful systems. It focuses on defining customer needs and required functionality early in the development cycle, documenting requirements, and then proceeding with design synthesis and system validation while considering the complete problem. Systems engineering considers both the business and the technical needs of all customers with the goal of providing a quality product that meets the user's needs (INCOSE)

Source: *Systems Engineering Handbook, A Guide for System Life Cycle Processes and Activities*, edited by Cecilia Haskins, 2007, INCOSE Systems Engineering Handbook, version 3.1

- Data
- Facilities
- Hardware
- Humans
- Materials
- Naturally occurring entities
- Procedures (e.g., operator instructions)
- Processes (e.g., processes for providing service to users)
- Software

When a system element is a software, the software life cycle processes documented in ISO/IEC 12207:2008 may be used to implement that system element (http1 2015).

In ISO/IEC 12207:2008, the following life cycle subprocesses are defined and grouped (http3 2015):

- Agreement
  - Acquisition
  - Supply

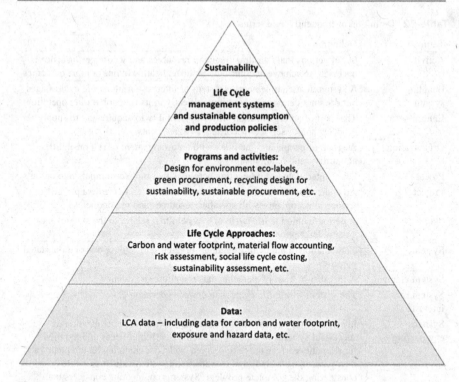

Fig. 6.1 Sustainability framework supported by the life cycle process

- Organizational project enabling
  - Life cycle model management
  - Infrastructure management
  - Project portfolio management
  - Human resource management
  - Quality management
- Project
  - Project planning
  - Project assessment and control
  - Decision management
  - Risk management
  - Configuration management
  - Information management
  - Measurement
- Technical
  - Stakeholder requirements definition
  - Requirements analysis
  - Architectural design
  - Implementation

- Integration
- Verification
- Transition
- Validation
- Operation
  - Maintenance
  - Disposal

Even after all system elements of a system of systems are deployed, product and project management must continue to account for changes in the various system element life cycles, including new technologies that impact one or more system elements and normal system replacements due to preplanned product improvement. The term system of systems refers to a collection of task-oriented or dedicated systems that pool their resources and capabilities together to create a new, more complex system which offers more functionality and performance than simply the sum of the constituent system. The standardization embodies the importance of systems engineering, which has grown with the increase in complexity of engineered systems and their highly integrated and very complex, miniaturized components. These components exponentially increase the possibility of component friction and, hence, the unreliability of the design. This requires the identification and manipulation of the properties of a system as a whole, which in complex engineering projects may differ from the sum of the parts' properties.

These factors provide motivation for applying systems engineering as an effective way to manage complexity and change. Both complexity and change have increased in today's products and services. Thus, reducing the risk associated with new systems or modification of complex systems continues as a primary goal in systems engineering. Furthermore, the complexity of the CPS being engineered can be mitigated from a technological perspective with regard to size and scale of integration, diversity of components, and diversity of disciplines involved in the design. Complexity can also be managed through the overall organization of the project (i.e., human-centered project management). The term project management refers to a set of management and organizational tasks, techniques, and means for the development of a project. It includes the planning, monitoring, and control of all aspects of a project and management of project participants to achieve safe project objectives. In this definition, functional and institutional dimensions are expressed as two viewpoints.

To summarize, systems engineering is a complex method based on its own:

- Strategies
- Procedures
- Techniques

The goal is to increase the performance of engineered systems or products. Therefore, the systems engineering methodology is applicable to the system's life cycle with regard to the conceptualization, design, production, utilization, support, and shutdown of systems or products, whether performed internally or externally to an organization. Furthermore, there is a wide variety of systems, in regard to their purpose, domain of application, complexity, size, novelty, adaptability, quantities, locations, life spans, and evolution, which require the design methods offered by the systems engineering approach to perform properly. But the evolution in design methods and the existing tools are not sufficient enough to meet the growing demands for new methods that directly address the increasing complexity of systems. Thus, the continuing evolution of systems engineering is comprised of the development of new methods and modeling which better comprehend complexity in systems engineering, such as model-based systems engineering with functional model analysis, as reported by Baker (2015) and introduced in Sect. 6.2.1. Other tools developed and often used in systems engineering are:

- Integrated computer-aided manufacturing (ICAM) DEFinition for function modeling (IDEFO)
- Unified modeling language (UML)
- Quality function deployment (QFD)

6.1.1.1 ICAM DEFinition for Function Modeling (IDEFO)

ICAM IDEF0 is a methodology for describing manufacturing functions, offering a functional modeling language for the analysis, development, reengineering, and integration of information systems, business processes, or software engineering analysis (http4 2015). The IDEF0 model consists of a hierarchical series of diagrams, text, and a glossary cross-referenced to each other. The two primary modeling components in IDEF0 are:

- Functions which are represented in a diagram by boxes
- Data and objects to interrelate the functions represented by arrows

As shown in Fig. 6.2, the position at which the arrow attaches to a box conveys the specific role of the interface. The controls enter the top of the box. The inputs, the data, or the objects acted upon by the operation enter the box from the left. The outputs of the operation leave the right-hand side of the box. Mechanism arrows

Fig. 6.2 IDEF0 building blocks

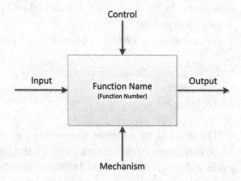

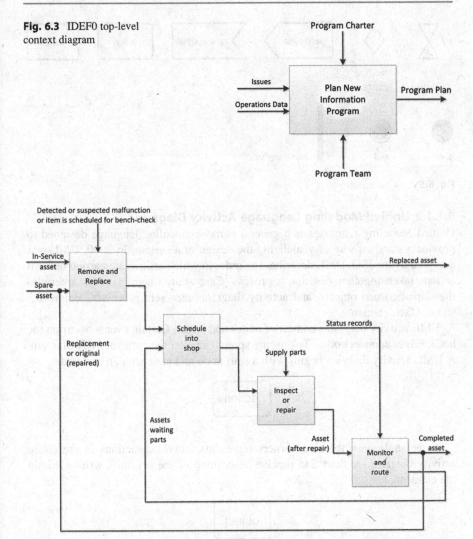

Fig. 6.3 IDEF0 top-level context diagram

Fig. 6.4 IDEF0 diagram example for maintaining reparable spares

that provide the means of support for performing the function join (point up to) the bottom of the box (http4 2015).

The IDEF0 process starts with the identification of the prime function to be decomposed. This function is identified by a top-level context diagram that defines the scope of the particular IDEF0 analysis. Figure 6.3 shows a top-level context diagram for an information system management process (http4 2015).

The purpose represented in Fig. 6.3 is assessment planning and streamlining of information management functions with regard to information integration assessment. Lower-level diagrams are generated from the diagram in Fig. 6.3. An example of an IDEF0 diagram is shown in Fig. 6.4 (http4 2015).

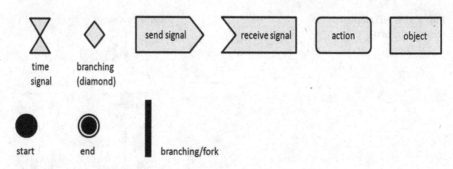

Fig. 6.5 UML elements

6.1.1.2 Unified Modeling Language Activity Diagrams

Unified Modeling Language is a general-purpose modeling language designed to provide a standard way of visualizing the design of a system. In 2000, UML was accepted by the ISO. UML describes the order in which actions are carried out. All actions taken together describe a process. Case study diagrams show interrelationships between objects, and activity diagrams represent processes. Figure 6.5 shows UML elements.

UML activity diagrams consist of nodes and edges. Certain events occur on the node. Edges connect nodes. Tokens are spread out over the entire activity diagram. A UML activity diagram begins with a start node and ends with an end node.

A rectangle with rounded corners represents individual actions in the entire activity diagram. A short and precise description of the action is written within the element.

Rectangle objects are another component of the UML activity diagram. They serve as intermediate storage units for objects. The data is moved by the preceding action to the following action, so data is passed from one action to the next.

The diamond denotes a division or branch of the path (e.g., the edges) or even a merger of two paths. The incoming token will continue but only on one of the outgoing lines. The selection criteria can be determined in advance and then recorded to the corresponding edges.

▮ branching/fork

The black bar is a branch known as a bifurcation in the road. The difference is that the token on a line not only continues on that line but on all lines by copying. Conversely, bringing an incoming line and an outgoing line together, the first incoming token must wait for the other tokens for conjunction; and then the unified token can go ahead. To allow other processes to access the elements, a signal transmitter and signal receiver are used. When a signal is sent, then all processes associated with it will not execute.

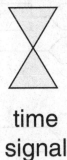

time
signal

The time event sign means that an action is time-control triggered. The time signal stops the signal flow of the diagram for a certain amount of time. Once the time has expired, the diagram or the token continues.

In Fig. 6.6, the UML activity diagram-based aircraft turnaround model is shown. The model begins with the aircraft landing and ends with the aircraft departing. After landing, the plane has to wait for the assignment of the parking position: direct to the gate, on the terminal ramp, or in a position farther away from the terminal on the ramp. After the assignment, the aircraft taxies to this position. The engines are switched off, and the brake blocks are applied. Now the (terminal) ramp-based ground-handling processes can begin.

The UML activity diagram allows the creation of models that represent an abstracted effigy of the real world to analyze the reality by reproducing important attributes that are restricted to certain periods of time. The model shown in Fig. 6.6 covers the turnaround process at Hamburg Airport, which takes place on the airport airside. From this model, the actual sequence of each process is clearly seen. The time-dependent processes and concurrencies depicted in this version of the model are close to reality. But the final model differs from this version because of the use of ground power supply and its dependence on processing time.

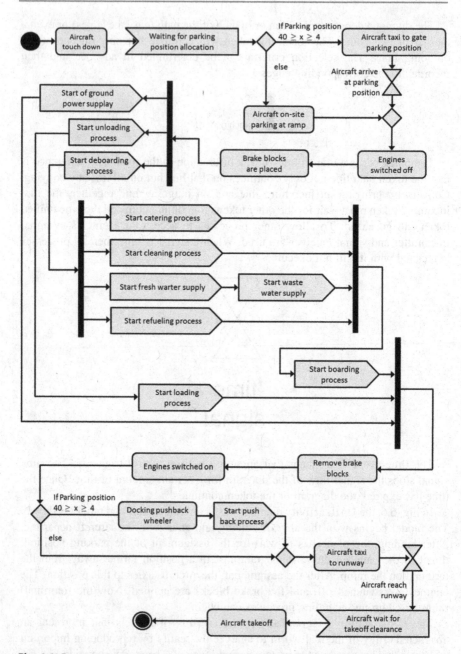

Fig. 6.6 Ramp turnaround process represented as a UML activity diagram (Moussaoui 2013)

Processing time for using the ground power supply corresponds with the sum of the processing times for all other clearance processes. Since modeling is based on the assumption of a distribution function, the sum of the processing times can only be determined at the end of the processes because the times used by MATLAB are not known in advance. This would be totally different if we assume that constant times are passed through the present model. In that case, the processing times are known in advance and can be taken out of the workspace by a MATLAB function and totaled. Thereafter, the result has to pass by one corresponding Simulink block into the model. However, for a cost analysis of ground-handling tasks, the fact that the ground power supply takes place in parallel to the remaining clearance processes has to be taken into account. Thus, for cost analysis, the ground-handling cost model has to cover all other trials because the entire clearance time must be calculated as part of the cost analysis of the ground power supply process.

6.1.1.3 Quality Function Deployment (QFD)

In Six Sigma Define, Measure, Analyze, Improve, and Control (DMAIC), quality function deployment (QFD) is a methodology and tool used in the requirements definition stage of systems engineering. It is used to (http5 2015):

- Collect customers' requirements/desires as specified by the customers in their own words
- Prioritize those desires
- Translate them into engineering/process requirements
- Establish targets to meet the requirements

Quality function deployment is a customer-driven product or service planning process. It is a methodology for translating customer requirements into company requirements at each stage from concept definition (R&D) to process engineering and production and into the marketplace. Quality function deployment is a tool to translate critical-to-customer (CTC) requirements into critical-to-quality (CTQ) requirements. Quality function deployment collects the voice of the customer (VOC) in their own words and incorporates this VOC into the cross functional team's project management of the integrated development process. The critical (C) process establishes customer objectives and measures and records them in a series of matrices, as shown in Table 6.3.

The QFD matrix translates CTCs into CTQs. For this purpose, the QFD matrix shows the relationships between requirements and options, as described in http5 (2015):

- Identify both internal and external customers.
- Create a list of customer requirements/desires (*What's*) by:
  - Asking the customer questions, such as *What are the important features of the product?*
  - Capturing the customer's own words or *voice of the customer*
  - Categorizing the *What's* into groups/buckets, if needed
- Prioritize the above collected *What's* on a scale of 1–5, with 5 being the most important. Ranking is based on VOC data. CTCs (*What's*) are listed vertically in the first column of Table 6.3, and all related CTQs (*How's*) are listed

Table 6.3 Six Sigma quality function deployment matrix (http5 2015)

Customer needs or CTCs or *What's*	Importance to customer	How's or CTQs internal to the company/process/engineering								
		Time on hold	Feedback score	How 3	How 4	How 5	How 6	How 7	How 8	How 9
Need quick response from help desk	5	5	2	3	3	1	1	3	5	2
Want satisfactory result from help desk	3	5	5			4	2		1	
What 3	2			1	2	1		4		1
What 4	1	5	1							5
What 5		2	2	4	1	3	5	2		1
What 6	3			1	2	4				4
What 7		1	1	1				2	2	5
What 8	3		3		2		1		1	2
What 9	1	2	4	2	1		1	2	5	
What 10	2	1	4	1	3	1	4			1
Score		39	47	21	38	32	34	25	36	37
Current average value of CTQ [days]		2.5	3.1	2	3	6	0	1	2	7

horizontally across the top. In the second column, assign 1–5 based on the importance of the CTCs, where 5 is the most critical to the customer.

- Score each CTQ (*How*) on how strongly it correlates to each CTC. Remember, we are looking at the absolute value of the correlation. It can be either positively correlated or negatively correlated. Use 5 for a strong correlation and 1 for a weak one. Leave it blank if there is no correlation. Some CTCs have few CTQs that relate, and the rest are unrelated.
- Compile a list of CTQs (*How's*) necessary to achieve the CTCs (*What's*).
- Translate the CTCs (*What's*) from the VOC into CTQs (*How's*).
  - Arrows show the direction for improvement (up for increasing, down for decreasing, etc.).
- Determine the correlation between each *What* and each *How*. If the correlation is strong, use 5. If it is weak, use 1. If it is in between, use 2, 3, or 4 based on the strength of the correlation.
- Next, multiply the importance rating for the CTC by the correlation score for each CTQ.
- Add up the scores vertically for each CTQ and place that value in the bottom score row.

Once the score is computed for all CTQs, the ones with the highest scores are the highest priority Six Sigma project objectives.

6.1.2 Top-Down Approach to Systems Engineering

In traditional engineering design methods, the bottom-up approach is widely used. Known components are combined into assemblies based on which of the larger subsystems can be established and can in turn be combined, sometimes at many levels, to build up the complete system prototype. This approach often resembles a seed system prototype, whereby the beginnings are small but eventually grow in complexity and completeness. Thus, all items in the design can be integrated together in accordance with the overriding system architecture to form a systems model or prototype of the final system. When the system prototype has been developed, it can be tested for the desired properties; and the design can be modified in an iterative manner until the system prototype does not fit with the desired requirements. Thus, the bottom-up approach is valid and extremely useful for relatively straightforward designs that are well defined. Hence, the bottom-up approach can be understood as a type of information processing based on incoming data to form a perception. Unfortunately, more complicated design problems cannot be solved with the bottom-up approach.

The method used today in systems engineering starts by addressing the system to be designed as a whole, which facilitates an understanding of the system, its environment, and interfaces. If the system-level requirements are clearly under-stood, the system to be designed will then be broken down into subsystems; and the subsystems will be further broken down into assemblies and then into components until a complete understanding of the system is achieved from top to bottom. Thus, this systems engineering approach is the opposite of the traditional bottom-up engineering design method because the systems engineering method uses a top-down development approach to manage the development of complex and/or complicated systems. By viewing the system as a whole initially and then progressively breaking the system into smaller elements, the interaction between the components can be understood more thoroughly. This assists in identifying and designing the necessary interface between components (internal interfaces) and between this and other systems (external interfaces) (Faulconbridge and Ryan 2014).

The first professional systems engineering society, International Council on Systems Engineering (INCOSE), was not organized until the early 1990s; and the first commercial US systems engineering standards, ANSI/EIA-632 and IEEE 1220, followed shortly thereafter. Even with the different approaches to defining systems engineering, the capability to create estimates is desperately needed by organizations. Several heuristics are available, but they do not provide the necessary level of detail that is required to understand the most influential factors and their sensitivity to cost.

The ANSI/EIA-632 standard contains five fundamental processes and 13 high-level process categories that are representative of systems engineering organizations. The process categories are further divided into 33 activities which are shown in Table 6.4 (Valerdi and Wheaton 2005).

Figure 6.7 illustrates the ANSI/EIA-632 for the top-down approach from a product perspective.

While the system development is conducted top-down, the system implementation is an integration process which is organized bottom-up, as shown in Fig. 6.8 (Faulconbridge and Ryan 2014).

The ISO and IEC form the specialized system for worldwide standardization. National bodies that are members of ISO or IEC participate in the development of international standards through technical committees established by the respective organization to deal with particular fields of technical activity. ISO and IEC technical committees collaborate in fields of mutual interest. The ISO/IEC 15288 is a systems engineering standard covering processes and life cycle stages. Initial planning for the ISO/IEC 15288:2002(E) standard started in 1994 when the need for a common systems engineering process framework was recognized. In 2004, this standard was adopted as IEEE 15288. ISO/IEC 15288 was updated February 1, 2008. ISO 15288 is managed by ISO/IEC JTC1/SC7, which is the ISO committee responsible for developing ISO standards in the area of software and systems engineering. ISO/IEC 15288 is part of the SC 7 integrated set of standards (http3 2015).

ISO/IEC 15288 also provides processes that support the definition, control, and improvement of the life cycle processes used within an organization or a project. Organizations and projects can use these life cycle processes when acquiring and supplying systems. The life cycle phases from ISO/IEC 15288, Systems Engineering—System Life Cycle Processes—vary according to the nature, purpose, use, and

Table 6.4 ANSI/EIA-632 processes and activities

Fundamental processes	Process categories	Activities
Acquisition and supply	Supply process	(1) Product supply
	Acquisition process	(2) Process acquisition
		(3) Supplier performance
Technical management	Planning process	(4) Process implementation strategy
		(5) Technical effort definition
		(6) Schedule and organization
		(7) Technical plans
		(8) Work directives
	Assessment process	(9) Product progress against plans and schedules
		(10) Progress against requirements
		(11) Technical reviews
	Control process	(12) Outcomes management
		(13) Information dissemination
Systems design	Requirements definition process	(14) Acquirer requirements
		(15) Other stakeholder requirements
		(16) System technical requirements
	Solution definition process	(17) Logical solution representations
		(18) Physical solution representations
		(19) Specified requirements
Product realization	Implementation process	(20) Implementation
	Transition to use process	(21) Transition to use
Technical evaluation	Systems analysis process	(22) Effectiveness analysis
		(23) Trade-off analysis
		(24) Risk analysis
	Requirements validation process	(25) Requirement statements validation
		(26) Acquirer requirements
		(27) Other stakeholder requirements
		(28) System technical requirements
		(29) Logical solution representations
	System verification process	(30) Design solution verification
		(31) End-product verification
		(32) Enabling product readiness
	End-product validation process	(33) End-product validation

prevailing circumstances of the system. Despite an infinite variety in system life cycle models, there is an essential set of characteristic life cycle phases that exist for use in the systems engineering domain. These phases have been slightly modified to reflect the influence of the ANSI/EIA 632 model and are shown in Fig. 6.9 (Valerdi and Wheaton 2005).

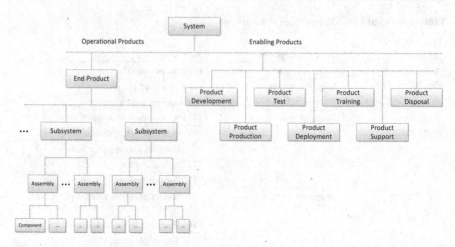

Fig. 6.7 ANSI/EIA-632 building block concept of a system comprising operational products and enabling products (ANSI/EIA-632-1998)

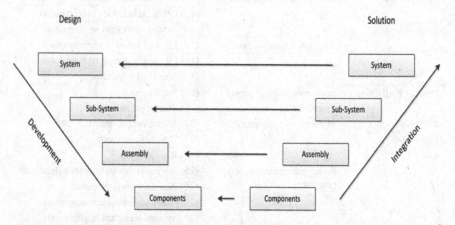

Fig. 6.8 Top-down development and bottom-up integration process

Fig. 6.9 ISO 15288 life cycle phases

Each stage shown in Fig. 6.9 has a distinct purpose and contribution to the life cycle and represents one of the major life cycle periods associated with a system. The stages also describe the major progress and achievement milestones of the system through its life cycle. These life cycle stages help answer the *when* of

Table 6.5 Systems engineering effort across ISO 15288, life cycle phases

Phase	Conceptualize	Develop	Operational test and evaluation	Transition to operation
% Effort	23	36	27	14
Standard deviation	12	16	13	9

systems engineering and the constructive systems engineering cost model. Understanding when systems engineering is performed relative to the system life cycle helps define anchor points for the model. The typical distribution of systems engineering effort across the life cycle phases for the organizations studied is shown in Table 6.5. It is important to note the standard deviation for each of the phases (Valerdi and Wheaton 2005).

6.1.3 Open Engineering Platform for Autonomous Mechatronic Automation

The complexity in construction and operation of a manufacturing facility is a central topic with which the industry must deal. The reason is that today's manufacturing lines are controlled and monitored by networked real-time computer systems. Together with the increasing complexity of manufacturing itself, the efforts associated with planning and placing into operation grow. This requires specific information and communication systems that are well adapted with regard to the needs of systems engineers in industrial automation.

This prompted creation of the so-called open engineering platform for autonomous mechatronic automation components in function-oriented architecture which is intended to set a new standard in the control architecture and engineering processes of manufacturing systems. With this new standard automation, systems engineers no longer have to think in terms of information technology when designing components but rather in terms of functions to automate the system. To achieve this goal, the standard allows the engineer to directly interact with a 3D-based engineering interface with observable physical components instead of abstract variables or input/output signals. These components include everything necessary for their operation—from mechanics, electronics, and software to standard ports and maintenance information. Together with a specific interface technology for automation components, a true "plug and produce" approach will be possible. This will allow costly installation, wiring, piping systems, configuration, and system integration to be minimized, as it was in the case of previous centralized automation systems. With regard to the plug and produce-enabled mechatronic components and the 3D model, communication across company and industrial sector boundaries is possible, contributing to the increase in the value of the whole project chain. Since all companies and industrial sectors are part of the project value chain, new business models can be developed as part of the project (http6 2015). The resulting characteristics of the new standard are:

- Open architecture
- Autonomous components
- New standards
- 3D interaction

With regard to adaptive manufacturing systems, the vision of the factory of the future requires that open platform standards be introduced. It is important to conserve resources and minimize setup times, to meet the increasing variety of products.

6.2   Design Challenges in Cyber-Physical Systems

Cyber-physical systems (CPS)   are a new category of systems engineering that combines the physical with the computational world in a holistic approach to representing and designing complex systems. However, CPSs are emerging from an intensive interaction of physical and computational components which require new systems engineering principles for this quickly emerging field. Using traditional engineering methodologies for designing complex systems may lead to various problems with regard to quality and maintainability because systems engineering methods for CPS are just beginning to evolve. Some approaches have been reported, such as:

- Utilization of an axiomatic design theory where a design is defined as mapping the process between *What* is wanted and *How* to implement the process (Suh 2001; Togay 2014).
- An automated approach to CPS design utilizing an adaptive cyber-physical system framework (ACPSF) (Tanik and Begely 2014), which provides effective guidelines for design automation which can be expanded to support CPS design.
- The CPS design challenges (Patterson et al. 2014) which introduce multidisciplinary knowledge requirements and multilayer CPS design.

With regard to the aforementioned methodological approaches, the challenges in systems engineering of CPS will be described.

When looking at CPS, two main elements must occur, the physical process and its cyber part, which processes the data from the physical process to perform the desired task. Thus, the interdisciplinarity in systems engineering in CPS design is inherently complex since the behavior of and interaction among system components is often not always immediately well defined or understood. Defining and characterizing such systems and subsystems and the interactions among them is part of systems engineering requirements. The gap that exists between requirements from users, operators, technical specifications, and more has to be successfully bridged.

The primary systems engineering requirements in CPS were introduced in Sect. 3.3.1. This section makes it clear that the complexity of CPS requires the efforts of multiple engineering disciplines to advance and solve systems engineering challenges. From a more general perspective, systems engineering requirements can be understood through a discovery process which starts with discovering the real problem that needs to be solved. The problem can be found by looking at the answers to the *How* and *What* questions. Therefore, a requirements analysis in systems engineering can begin by looking at the following statements and questions to ensure that the customer's needs are satisfied throughout a system's entire life cycle (Haskins 2006; Li 2008; http3 2015; http7 2015):

- *State the problem* which may be derived from the answers to the following questions:
  - *How* is this done the right way?
  - *What* happens when using flow diagrams for systems engineering?
  - *How* should systems engineers search the customer's environment to find out *what* customers are planning to use the system?
  - *How* should the customer's environment be searched when a priori knowledge is fuzzy in regard to interviews with customers?
  - *What* other methods can help to fulfill the requirements for stating the problem?
  - *How* can the requirements validation approach (RVA) be helpful? (http8 2015)
  - *What* are the essential issues in using the RVA?
- *Investigate alternatives* which may be derived from the answers to the following questions:
  - *How* will a decision be made after choosing a process from several alternatives?
  - *What* are the criteria to test whether a product is sufficient or not?
  - *What* criteria is the most critical?
  - *What* trade-off studies should be performed and learned thoroughly?
  - *How* can they be translated into practical problems?
- *Model the system* which may be derived from the answers to the following questions:
  - *What* is a general model for all of the systems?
  - *How is* systems engineering responsible for creating a product and a process for producing it?
  - *What* are the differences between models?
  - *What* task is most important to model a system?
- *Integrate system components* which may result from the answers to the following questions:
  - *What* is the most efficient way to integrate systems and bring subsystems together to produce the desired result?

- *How* can the interfaces between subsystems and between the main system and the external environment be designed and managed?
- *What* kind of systems engineering assumptions and comparison studies can be conducted to determine if a process is successful?
* *Launch the system* which may result from the answers to the following questions:
  - *How* should the system be launched?
  - *What* is the impact of giving stakeholders the opportunity to comment on proposed changes?
  - *How* many comments are acceptable?
  - *How* can responsibility be determined for deciding whether or not to implement the changes?
  - *What* is the impact if all systems engineering activities are documented in a common repository, the *Engineering Notebook of Knowledge*?
  - *How* is the *Notebook* really helpful for daily work?
  - *What* are people's opinions on this?
  - *What* is the impact of requiring people to always follow the *Notebook*?
* *Assess performance* which may be derived from the answers to the following questions:
  - *How* will the performance of the system be measured?
  - *How* can the system's compliance with requirements be conveniently verified?
  - *How* can it be determined if the final system satisfies each system requirement?
  - *How* can we ensure that the final build satisfies the mandatory requirements?
  - *How* well does the final build satisfy the trade-off requirements?
* *Reevaluation* which may be derived from the answers to the following questions:
  - *What* exactly does reevaluation mean in regard to verification and validation?
  - *How* should a reevaluation be conducted?
  - *What* kind of process is required and when should it be done?
  - *What* is the impact to have someone especially assigned to doing this job?

As reported in Li (2008), reevaluation is the most important task of systems engineering. For centuries, engineers have used feedback to control systems and improve performance. Reevaluation means observing outputs and using this information to modify the system inputs, the product, or the process. Reevaluation should be a continuing process with many parallel loops because systems engineering is not sequential. The aforementioned functionality can be summarized with the acronym SIMILAR which stands for state, investigate, model, integrate, launch, assess, and reevaluate (Bahill and Gissing 1998; http9 2015).

The SIMILAR tasks are performed in a parallel or iterative manner. At each step, a comprehensive set of possible engineering models are progressively combined and refined to define the target system. Several systems engineering domains use an intricate interplay of conceptual synthesis, preliminary impact analysis, and preliminary impact analysis using simulation tools and decision-making of alternative requirements and design configurations (Pollet and Chourabi 2008; Tarumi et al. 2007).

An analysis by the INCOSE Systems Engineering Center of Excellence (SECOE) indicates that the optimal effort spent on systems engineering is about 15–20 % of the total project effort (INCOSE 2007). At the same time, studies have shown that systems engineering essentially leads to a reduction in costs, among other benefits. However, no quantitative survey at a larger scale encompassing a wide variety of industries has been conducted until recently. Such studies are underway to determine the effectiveness and quantify the benefits of systems engineering (Elm 2005).

Moreover, systems engineering encourages the use of modeling and simulation to validate assumptions or theories on systems and the interactions within them. Simulation technology belongs to the tool set used by engineers of all application domains and has been included in the body of knowledge of engineering management. Modeling and simulation already help to reduce costs, increase the quality of products and systems, and document and archive lessons learned.

Furthermore, systems engineering orchestrates the development of a solution from requirements definition through operations and zero degree utilization of the system. This ensures that domain experts are properly involved, that all advantageous opportunities are pursued, and that all significant risks are identified and mitigated which require that the systems engineering process be moved from a document-centric to a data-centric approach. The document-centric approach follows this sequence (Baker 2015):

- Specifications
- Interface requirements
- Systems design
- Analysis and trade-off
- Test plans

When project information is spread across multiple documents, it is difficult to assess the completeness, consistency, relationships between requirements, design, engineering analysis, verification, and validation information. It is also difficult to establish the end-to-end traceability needed to support change impact assessments.

In order to address these document-centric limitations, more advanced systems engineering processes are transitioning to a data-centric approach, which allows all systems engineering team members to access any project or related data. Thus, the data-centric approach is a major part of the Cradle software tool, a requirements management and systems engineering tool. Cradle integrates the entire project life cycle in one, massively scalable, integrated, multiuser software product. It identifies the data to be captured, as shown by Baker (2015) in Fig. 6.10. The term *mission* used in Fig. 6.10 is synonymously used for the term *use case*.

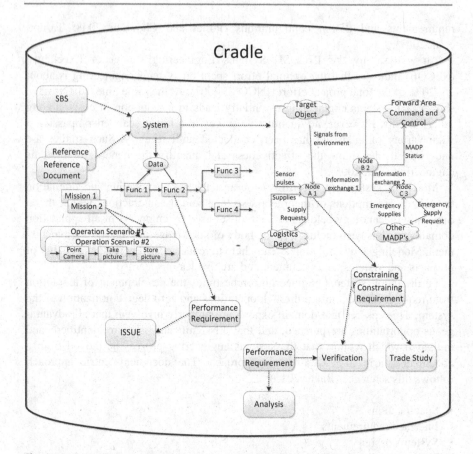

Fig. 6.10 Project data repository for the data-centric approach in Cradle-7

6.2.1 Requirements Definition and Management Using Cradle®

The Cradle software tool can be used to manage requirements definition and management activities for system development and modification. Cradle groups the requirements definition and management activities into eight stages, as shown in Fig. 6.11. The descriptions in Sects. 6.2.1, 6.2.2, and 6.2.3 are reproduced from the white paper, "Requirements Definition and Management Activities," November 2014 (3SL 2014), with permission from 3SL (3SL: Structured Software Systems Ltd.).

The *AND* nodes (circle with an embedded *A* symbol) in the diagram in Fig. 6.11 indicate that the stages between the two *AND* nodes can be performed in parallel. As illustrated, Stages 3, 4, and 5 can be worked out in parallel.

The *ITERATE* nodes (circles with embedded *I* symbol) in the diagram in Fig. 6.11 indicate that all stages between the two *ITERATE* nodes are repeated for each level in the system architecture hierarchy. Level 1 in the hierarchy identifies the system entity, as shown in Fig. 6.12. Therefore, each entity in levels

Fig. 6.11 Eight stages of requirements definition and management activities of the Cradle® software tool

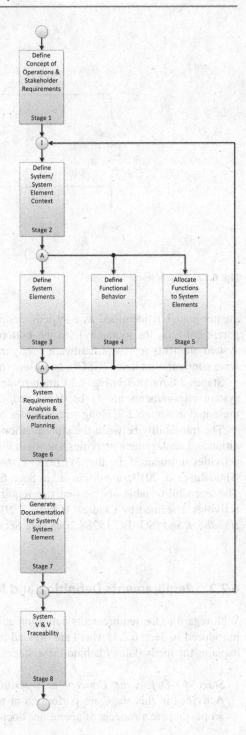

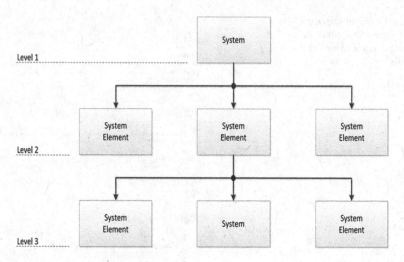

Fig. 6.12 System levels in Cradle®

greater than 1 is identified as a system element (i.e., child component parts of a parent entity in the hierarchy). At the bottom level of the system hierarchy, the system elements represent hardware configuration items (HWCI), computer software configuration items (CSCI), and users (humans).

Stages 2 through 7 in Fig. 6.11 are repeated for each system element for which system requirements are to be developed. The details of the eight stages are presented in Sect. 6.2.2, along with samples of Cradle® work products.

The traceability between the aforementioned eight stages of requirements definition and management activities, shown in Fig. 6.11, and similar technical process activities introduced in the *INCOSE Systems Engineering Handbook*, v. 3.2 (Hamelin et al. 2010), are discussed in Sect. 6.2.3 in detail, resulting in Table 6.6. The traceability table shows that the requirements definition and management activities described by Cradle® (Cradle-7 2014) are compliant with the *INCOSE Handbook* and ISO/IEC 15288:2008 (see Sect. 6.1.2).

6.2.2   Requirements Definition and Management Activities

With regard to the requirements definition and management activities in Cradle®, introduced in Sect. 6.2.1, which are divided into eight different stages, Sect. 6.2.2 explains the methodology behind these stages in detail:

- *Stage 1—Define the Concept of Operation and Stakeholder Requirements*: Activities in this stage are performed at project start-up to define the project scope, prepare a concept of operations document (ConOps), and develop a set of stakeholder requirements and disseminate them in a stakeholder requirements document (SRD).

- *Stage 2—Define the System/System Element Context*: Activities in this stage generate a system context diagram for a primary system element to identify all external entities that must interact with the system element and the required external interfaces. The external interfaces must be identified prior to beginning Stages 3 through 7 in Fig. 6.11.
- *Stage 3—Define the System Elements*: Activities in this stage define physical characteristics for the specified system element and derive the appropriate system requirements for that element. Thereafter, the breakdown of the system element into its component parts (one level down the hierarchy) is required as well as identification of a draft set of physical characteristics for each subordinate element.
- *Stage 4—Define the Functional Behavior*: System functions and their inputs and outputs are identified that satisfy the system element functional requirements identified in the previous design cycle. These functions, when integrated, describe the desired behavior the system must exhibit. In Stage 5, the functions must be allocated to specific system elements (i.e., things that must perform the functions).
- *Stage 5—Allocate the Functions to System/System Elements*: In this stage, functions and input/output (I/O) (identified in Stage 4) are assigned to different system elements and interfaces, which is known in engineering as functional allocation. This means that the behavior of the specific system element is specified.
- *Stage 6—Analyze System Requirements and Perform Verification Planning*: In this stage, the system requirements derived in Stages 3 through 5 are analyzed for clarity, completeness, consistency, and traceability to stakeholder/system requirements at the end of the previous design cycle. Also, verification planning activities should be performed for the newly created system requirements and a requirements baseline established.
- *Stage 7—Generate Documentation for the System/System Elements*: In general, this deals with the generation of deliverable documents and data packages to be used in support of each project review.
- *Stage 8—Conduct System Verification and Validation (V&V) Traceability*: Capture the status of each verification and validation activity in the database with traceability back to the requirement impacted and operational scenarios. Traceability identifies those requirements with links to design solutions that may need to be modified.

The previous explanations of requirements definition and management activity stages are described in more detail in the following subsections.

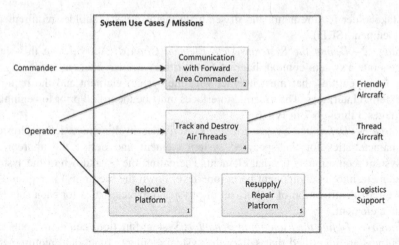

Fig. 6.13 Use-case diagram for the operational phase

6.2.2.1 Stage 1: Define the Concept of Operations and Stakeholder Requirements

The purpose of *Stage 1* is to define the requirements for a system that can provide the services deemed essential by users and other stakeholders in a defined environment. Successful projects depend on meeting these needs and requirements. The activities associated with this stage are:

- *Activity 1.1—Clearly Define and Document the Project Scope*. Create a top-level system breakdown structure (SBS) item for the project; and capture the project purpose, planned schedule, and expected stakeholders. For details see 3SL (2014).
  - Gather project-related source material and capture it in the Cradle database using a specific item type. The items created should be cross-referenced to the top-level SBS item for the system being developed.
  - Interview the stakeholders identified; and capture their needs, goals, and objectives. Link each need, goal, and objective to the top-level SBS item. For details see 3SL (2014).
  - Identify the product life cycle phases (e.g., production, operation, disposal) for which stakeholder requirements are to be defined.
- *Activity 1.2—Define Use Cases*. Develop use cases for each applicable life cycle phase. The term *use case* is used synonymously for *mission*. For example: Define a use-case diagram for the operational phase as shown in Fig. 6.13.
- *Activity 1.3—Develop the Concept of Operations*. Create the concept of operations document (ConOps) based on the use cases/missions previously defined.
  - Define one or more operational scenario(s) to accomplish each use case/ mission identified.

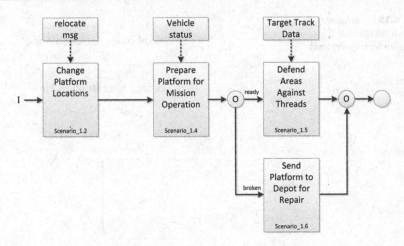

Fig. 6.14 Operational scenario no. 1

- Operational scenarios are process/activity flows that describe how a system will be used, manufactured, tested, deployed, or operated throughout the product's life cycle. A typical scenario is a stimulus-response flow of activities/operations with I/O, which is introduced as Operational Scenario No. 1 in Fig. 6.14. These scenarios are from the viewpoint of the end-use stakeholders and not the product developers. The *OR* nodes (circles with embedded *O* symbol inside) indicate that the scenarios between the two *OR* nodes can be performed in a disjunctive way.
  - Process flow diagrams (PFDs) are used to describe an operational scenario. Establish traceability between a specific use case and the set of operational scenarios modeled by PFDs. Link the operational scenario to the use case. For details see 3SL (2014).
  - Multiple scenarios are usually required to understand the operational activities and I/O associated with the accomplishment of each mission/use case.
  - Link use-case specification items to the top-level SBS for traceability purposes.
- *Activity 1.4—Create the Stakeholder Requirements.* Define requirements that identify the intended operational services to be provided by the system and the interaction the system will have with its operational environment.
  - Utilize the needs, goals, and objectives and the different operations identified during development of the operational scenarios to acquire the knowledge needed to address specific stakeholder requirements. Once a new database item is created for the proposed requirement, a "shall statement," a brief rationale, is entered and then cross-referenced to the source material (i.e., need or scenario operation). For details see 3SL (2014).
  - If a source document is provided that contains important information, such as stakeholder requirements or test cases, use the Cradle Document Loader to parse the contents of the document into individual pieces of information that

Fig. 6.15 Validation
objectives that fulfill
stakeholder requirements

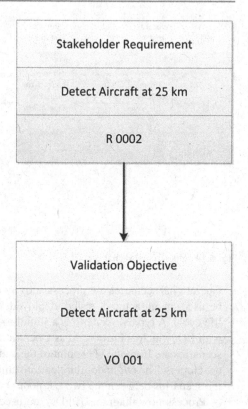

will be automatically loaded into the Cradle data repository as individual items of information.

- If one of the created items has multiple "shall" statements that were captured from the original source document, use the Cradle "split" command to automatically create new child items containing single "shall" statements.
- Use the Cradle Conformance Checker to verify the use of good and bad phrases in the requirement statements.
- Create an SRD, and review it with the customer to ensure that it is valid, meets the customer's needs, and is clearly understood by all stakeholders.
- Stakeholder requirements and associated operational scenarios are the bases for the system developer to derive system-level design requirements during Stages 2 through 6. These design requirements are known as system requirements.
- *Activity 1.5—Plan System Validation.* Plan for system validation activities to be performed during integration and test by developing validation objectives to be used to validate that the system fulfills stakeholders' needs as defined in the SRD. Each validation objective identifies the validation methods, facilities, equipment, and resources needed to accomplish the validation activity and should be linked to the stakeholder requirement being validated as shown in Fig. 6.15.

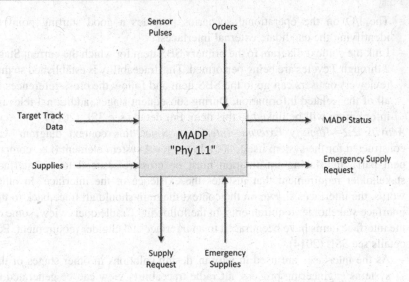

Fig. 6.16 Context diagram of Cradle MADP (MADP: mobile air defense platform)

- *Activity 1.6—Identify and Capture Design Constraints.* Identify and capture design constraints in the data repository.
- *Activity 1.7—Capture Glossary Items.* Include a common glossary item type so that project members can collect a list of terms and acronyms that can be included in the published documents.

6.2.2.2 Stage 2: Define the System/System Element Context

A system context diagram in systems engineering defines the boundary between the system, or a system element, and the external entities that interact with the system. The high-level information exchanged between the system and the external entities is also shown on the diagram. The diagram is a high-level, black-box view of the proposed system with its required external I/O.

- *Activity 2.1—Create a Context Diagram.* Using the information developed in the previous stage(s), create a system operational context diagram using the Cradle Physical Architectural Diagram (PAD). The purpose of this diagram is to define the top-level external interfaces for the system or one of the system elements.
  - Draw the proposed black-box system/system element in the center of the diagram, as shown in Fig. 6.16.
  - Draw the external environment entities as rectangles around the proposed system/system element symbol, as shown in Fig. 6.16.
  - Draw the external interfaces between the proposed system/system element and the entities outside the system boundary using annotated arrows, as shown in Fig. 6.13. These external interfaces must be identified prior to beginning Stages 3 through 7 in Fig. 6.11.

- The I/O on the operational scenarios provides a good starting point to identifying the candidate external interfaces.
- Link the context diagram to the primary SBS item for which the current Stage 2 through 7 cycles are being performed. This traceability is established so that reviewers or users can go to the SBS item and follow the cross-references to all of the related information. During subsequent stages, additional relevant information will be linked to this item. For details see 3SL (2014).

• *Activity 2.2—Identify External Interfaces.* Since this context diagram was constructed for the system itself, not a lower-level system element (i.e., component part), interfacing information must be cross-referenced to an interface stakeholder requirement that justifies the existence of the interface. In other words, the interfaces shown on the context diagram should all trace back to the interface stakeholder requirements. In the following Cradle query view, some of the interface items have been traced to an interface stakeholder requirement. For details see 3SL (2014).

- As the interfaces are used in various design solutions in other stages of the systems engineering process, a Cradle traceability view can be generated to support impact analyses as needed. For details see 3SL (2014),
- Stage 2 is repeated for each level of the system structure hierarchy. When creating a context diagram for a subordinate system element, the source material will be the allocated system requirements derived during the previous cycle.

6.2.2.3 Stage 3: Define the System Elements

A system is composed of a set of interacting system elements, each of which can be implemented to fulfill its respective system requirements. A system element has a structure (how it is built) and behavior (what it does). An example of a system element is a hardware configuration item, software configuration item, or human operator, all of which perform work.

When defining a system element, it is important to separate the structure definition from the functional behavior definition. If the desired behavior is developed independently from a predefined physical structure, then alternative component parts can be evaluated with the desired behavior mapped to each alternative, so they each exhibit the desired behavior. A trade-off analysis can then be performed to identify the best solution. The mapping of functional behavior (described in Stage 4) to a system element structure (defined in this stage) is known as functional allocation and is illustrated in Fig. 6.17.

The purpose of this stage is to define the desired physical characteristics for the specified system element and then derive the appropriate system requirements for that element. Next, the system element is broken down into its component parts (one level down the hierarchy); and a draft set of physical characteristics is identified. Going one level deeper helps to identify missing information or inconsistent system element physical characteristics as shown in Fig. 6.18.

Functional Behaviour Description

System Structure Description

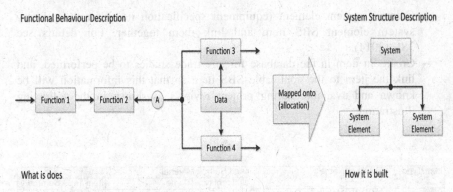

What is does

How it is built

Fig. 6.17 Functional behavior allocation

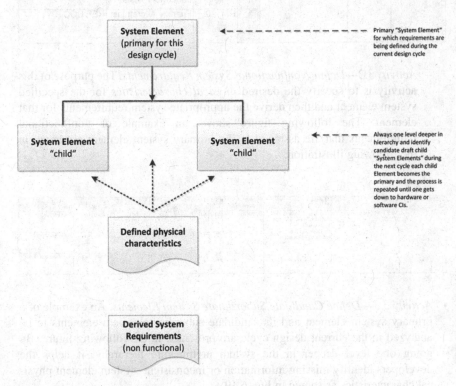

Fig. 6.18 Desired physical characteristics for system elements

The activities in this stage are as follows:

- *Activity 3.1—Analyze System Element Source Information.*
  - Using source information developed during Stage 1, or developed during Stages 3 through 7 of the previous design cycle, perform analyses and trade studies to identify any mistakes in the material. If mistakes are found, document them via an ISSUE item type and link them to the item in question.

- Create a system element (equipment specification item) for the primary system element SBS item and link them together. For details see 3SL (2014).
- Create an item in the database for any trade studies to be performed, and link the item to the applicable SBS item so that this information will be known and available during project reviews as shown in the following illustration:

Item Type	SBS ID	Name	Linked Source Material			
				Identity	Name	Item Type Summary
SBS	MADP.1	Mobile Air Defense Platform (MADP)		Ref_3	Environmental Considerations and Laboratory Tests	REF DOC
				Ref_20	Friendly Aircraft List	REF DOC

- *Activity 3.2—Define Nonfunctional System Requirements.* The purpose of this activity is to specify the desired *physical characteristics* for the specified system element and then derive the appropriate system requirements for that element. The following figure shows an example of nonfunctional requirements that are assigned to the primary system element as shown in the following illustration:

System Element ID	Name	TEXT	Type of Specification	Linked Non-Functional Reqs			
					ID	Name	Item Type	Req Category
Phy.1.1	MACP Physical Model (PAD)	The main interfaces for the Mobile Air Defense Platform are illustrated in the following diagram.	System		R.0005	Air Defense System Mounted on All Terrain	System Req	Non-Functional
					R.0011	Vehicle Safety Record	System Req	Non-Functional
					R.0012	On Road Platform Weight	System Req	Non-Functional

- *Activity 3.3—Define Candidate Subordinate System Elements.* An example of a primary system element and its candidate subordinate system elements to be analyzed in the current design cycle are presented in the following figure. By going one level deeper in the system architecture hierarchy, it helps the developers identify missing information or inconsistent system element physical characteristics as shown in Fig. 6.19.

6.2.2.4 Stage 4: Define the Functional Behavior

When developing and/or validating the primary system element's functional and performance requirements, a common technique used by systems engineers is to construct functional behavior models. These behavior models are developed to identify the functions the system element must perform. The models are developed from the viewpoint of the system developer, not the stakeholder.

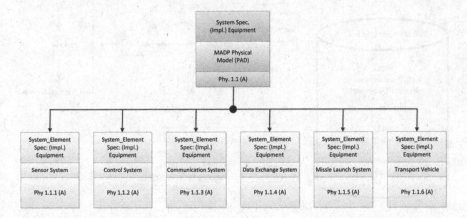

Fig. 6.19 Subordinate system elements

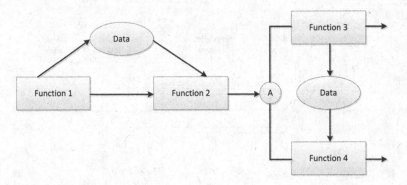

Fig. 6.20 Functional models

A behavior model consists of three things:

- *Functions*, which accept input and transform it to output. Functions are executed by the system elements.
- *Input and output* of various types.
- *Control operators*, which conditionally define the execution order of functions.

Engineers may choose to develop functional models after they have developed a textual requirements document, while many others follow the process described in the activities of Stage 4. But what is the best practice? Which comes first? As shown in Fig. 6.20.

The activities for this stage assume that models are developed first. However, if the project has a well-defined set of functions that need to be implemented in a new system, then the model development activity can be skipped, the functions list imported into Cradle, and the other activities continued.

- *Activity 4.1—Develop Functional Model for the Primary System Element.*
    - Using the source requirements from the previous design cycle, define the needed functions and I/O as illustrated in Fig. 6.21.

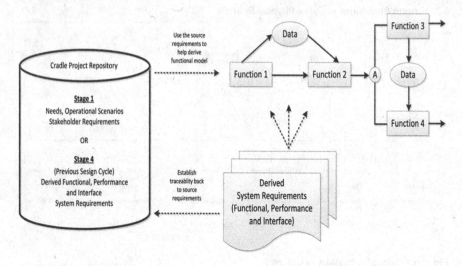

Fig. 6.21 Needed functions and I/O

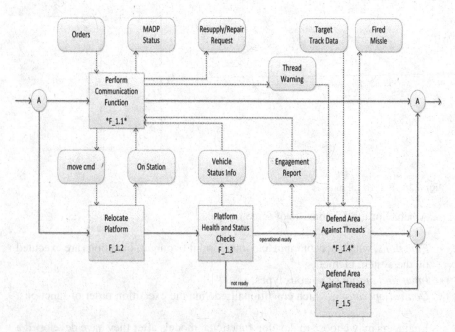

Fig. 6.22 eFFBD example

- The two most commonly used functional modeling notations are the enhanced Functional Flow Block Diagram (eFFBD) and the UML/SysML activity diagram. For this example, the eFFBD was used as shown in Fig. 6.22.
- *Activity 4.2—Create Functional Requirements and Link to Functions.*
  - Link each function to an applicable existing requirement (stakeholder or system), or create and link a new derived system requirement to the function. A traceability table (shown below) can be used to verify that each function is linked to one or more requirements. The team must review the data in the table to ensure proper traceability exists. For details see 3SL (2014).

- *Activity 4.3—Establish Traceability to Source Requirements.* All derived requirements must trace to a valid source requirement (system or stakeholder). If a source can't be identified, determine if the derived requirement should be removed. If it is determined that the derived requirement is valid, then determine if the source material needs to be modified to account for the missing source requirement. An example of a backward traceability report is shown in the following illustration:

System REQs		Identity	Name	TEXT	Req_Type	Category of Req	REQ Status	Backward Traceability of Source Reqs			
								ID	Name	Item Type	Req Category
7		R.0010	Target Identification Reliability		System Req	Non-Functional					
8		R.0011	Vehicle Safety	The transport vehicle	System Req	Non-Functional					
9		R.0012	Vehicle Weight	The Vehicle Weight shall not exceed 20000 pounds.	System Req	Non-Functional					
10		R.0013	Detect and Track	The system shall detect and track approaching aircraft.	System Req	Functional	Proposed	R.0003	Track Targets	Stakeholder Req	Functional
11		R.0014	Detect Aircraft at 25KM	The system shall detect aircraft within a 25KM radius.	System Req	Functional					
12		R.0015	100 Simultaneous Tracks	The system shall track up to 100 aircraft simultaneously.	System Req	Functional	Proposed				
13		R.0040	Target Track Data Message Content	The Target Track Data interface message shall consist of the	System Req	Interface	Proposed	R.0003	Track Targets	Stakeholder Req	Functional

- *Activity 4.4—Create and Link Interface Requirements to Data Definitions.*
  - The logical I/O in the functional models helps derive/validate interfaces between system elements in Stage 5 and helps to create interface requirements. Cradle automatically creates the following kinds of function I/O tables to provide easy visibility of interfacing information that needs to be linked to an appropriate interface requirement. The second table below in the illustration shows an example of where the users created and linked new interface requirements.

MADP Functions #1		Identity	Name	TEXT	Type of Specification	Input – Data Definition Name	Output – Data Definition Name
	Previous...						
1		F_1	Operate MADP (eFFBD)	This is the top level (root) function of the	Function		
2		F_1_1	Perform Communication Function	The first function is to communicate.	Function	Engagement Report	MADP Status
						Orders	Thread Warning
						on station	move cmd
						vehicle status info	resupply/repair request
3		F_1_1.1	Communicate with outside world		Function	Outgoing information	Incoming information
						digital cmds	
						voice cmds	

MADP Functions		Identity	Name	TEXT	Type of Specification	Inputs			Outputs		
						Data Definition ID	Linked Interface Reqs Identity	Name	Data Definition ID	Linked Interface Reqs Identity	Name
	Previous...										
1		F_1	Operate MADP (eFFBD)	This is the top level (root) function of the	Function						
2		F_1_1	Perform Communication Function	The first function is to communicate.	Function	Engagement Report			MADP Status		
						Orders	R.0052	Phone Link	Thread Warning	R.0053	Satellite Link
							R.0053	Satellite Link			
						on station			move cmd		
						vehicle status info			resupply/repair request		

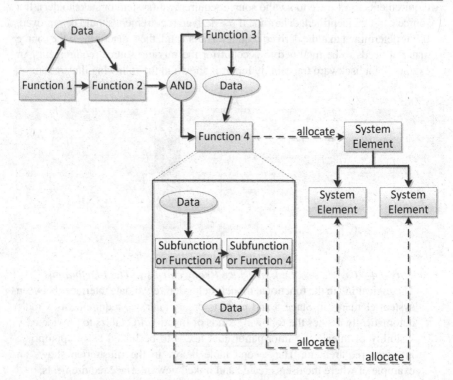

Fig. 6.23 Uniquely allocated subfunctions

- *Activity 4.5—Decompose Functions.*
  - In Stage 3, once the primary system element for the current design cycle is decomposed into its component parts (one level down the hierarchy), each top-level function defined in the first part of Stage 4 should be evaluated to see if it must be decomposed into subfunctions that can be uniquely allocated to one of the component parts. The actual allocation process takes place in Stage 5. The following Fig. 6.23 illustrates the concept.
- *Activity 4.6—Establish Flow Down of Functional Requirements and Link to Functions.*
  - Whenever a function is decomposed, the requirement linked to the parent function must be evaluated to determine whether it can flow down unchanged to the subfunction or whether it must also be decomposed as shown in Fig. 6.24.

6.2.2.5 Stage 5: Allocate the Functions to System/System Elements
In this stage, the functions and I/O (identified in Stage 4) are assigned to different system elements and physical interfaces. In Fig. 6.25, the three kinds of system requirements (functional, nonfunctional, and interface) are linked directly and/or indirectly to system element that is to satisfy the requirements.

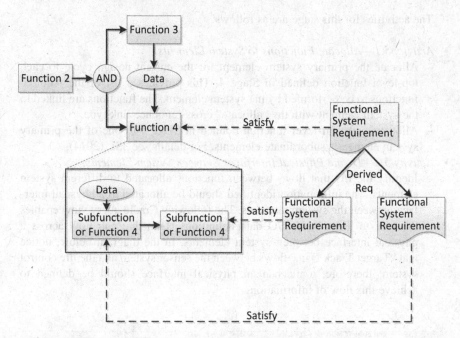

Fig. 6.24 Decomposed subfunctions

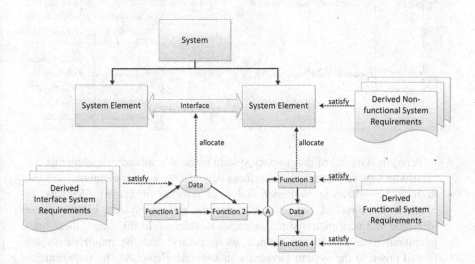

Fig. 6.25 Allocate functions to system/system elements

The activities for this stage are as follows:

- *Activity 5.1—Allocate Functions to System Elements.*
  - Allocate the primary system element for the current design cycle to each top-level function defined in Stage 4. This provides a list of the required functions to be performed by the system element. The functions are linked to the system element with the "allocate" cross-reference link type.
  - Allocate each leaf-level function defined in Stage 4 to one of the primary system element's subordinate elements. For details see 3SL (2014).
- *Activity 5.2—Define Physical Interfaces Between System Elements.*
  - Identify the I/O that flows between functions allocated to different system elements. The information identified should be allocated to a physical interface between the system elements. The following Cradle query view enables the user to identify the I/O data definitions (DD) that can flow across a physical interface between system elements. In the diagram below, notice that "Target Track Data" flows between the sensor system and the fire control system; therefore, a mechanism physical interface should be defined to achieve this flow of information.

	System Element ID	System Element Name	Type of Specification	Allocated Functions and I/O		Input DD (Interface Info) Name	Output DD (Interface Info) Name
				Function ID	Name		
2	Phy.1.1	MACP Physical Model (PAD)	System	F_1	MADP Functional Model (eFFBO)		
3	Phy.1.1.1	Sensor System	System_Element	F_1.4.1	Detect and Track Incoming Aircraft		Target Track Data
4	Phy.1.1.2	Fire Control System	System_Element	F_1.3	Perform Health and Status		vehicle status info
				F_1.4	Defend Area Against Threads	Target Track Data	Engagement Report
						Thread Warning	fire missle
				F_1.4.2	Identify Friend or Foe		
				F_1.4.4	Kill Assessment		Engagement Report
5	Phy.1.1.3	Communication System	System_Element	F_1.1	Perform Communication Function	Engagement Report	MADP Status
						Orders	Thread Warning
						on station	move cmd
						vehicle status info	resupply/repair request

  - A physical model of the primary system element's subordinate elements can also be used to identify the interfaces between the system elements.
- *Activity 5.3—Allocate Nonfunctional Requirements to System Elements.* In Stage 3, the *physical characteristics* for the primary system element were identified and nonfunctional requirements developed. In this stage, these non-functional requirements are refined, as necessary; and the requirements are flowed down to the system element's subordinate elements. The requirements R.55 and R.56 are derived from requirement R.12 and linked to subordinate system elements. For details see 3SL (2014).
- *Activity 5.4—Perform Trial Allocations.* In order to get a proper design solution, you must balance performance, complexity, and risk by performing trial allocations. This means changing the allocations and/or changing the functional model I/O flow to get a balanced design solution.

6.2.2.6 Stage 6: Analyze System Requirements and Perform Verification Planning

- *Activity 6.1—Analyze Requirements.* In this stage, the system requirements derived in Stages 3–5 should be analyzed for clarity, completeness, consistency, and traceability back to the stakeholder/system requirements established at the end of the previous design cycle.
- *Activity 6.2—Identify Verification Objectives.* Identify verification objectives to be used to verify each system requirement. Each verification objective should identify the methods, facilities, equipment, and resources needed to accomplish the verification activity. The verification objectives are linked to the system requirement being verified.
- *Activity 6.3—Establish Requirements Baseline.* Establish a requirements baseline at the end of each design cycle.

6.2.2.7 Stage 7: Generate Documentation for the System/System Elements

- *Activity 7.1—Generate Deliverable Document.*
  - Standardize the look and feel of the project's documentation using Cradle's Document Publisher.
  - All documents that can be generated from the data repository should be generated using Document Publisher, not handcrafted.
- *Activity 7.2—Generate Data Packages.* Data packages should be created from the data in the project repository in support of each project review. These packages should be generated from the data repository, not handcrafted.

6.2.2.8 Stage 8: Conduct System Verification and Validation (V&V) Traceability

Capture the status of each V&V activity in the database with traceability back to the impacted requirement and operational scenarios. The traceability identifies those requirements with links to design solutions that may need to be modified.

- *Activity 8.1—Ensure Verification Traceability.* The purpose of the verification process is to confirm that the specified design requirements (i.e., system requirements) are fulfilled by the system.
- *Activity 8.2—Ensure Validation Traceability.* The purpose of the validation process is to provide objective evidence that the services provided by a system when in use comply with stakeholders' requirements, thereby achieving the system's intended use in its intended operational environment.

6.2.3   INCOSE Systems Engineering Handbook (v.3.2.2) Traceability

This section contains a matrix that shows the traceability between the eight stages of the requirements definition and management process described in Sects. 6.2.1 and 6.2.2 and technical process activities described in the INCOSE Systems

Engineering Handbook (v.3.2.2). The handbook processes are consistent with ISO/IEC 15288:2008 life cycle processes and are identified in Fig. 6.26.

The process activities described in Sects. 6.2.1 and 6.2.2 can be traced to material in the INCOSE handbook in Sections 4.1, 4.2, 4.3, 4.6, and 4.8. The traceability is shown in the following Table 6.6 with sections 4.1–4.8.

6.3   Introduction to Software Engineering

The increasing complexity of software requires the use of different system views, to ensure efficient development of high-quality software for cyber-physical systems (CPS) in software engineering. Thus, software engineering can be introduced as the

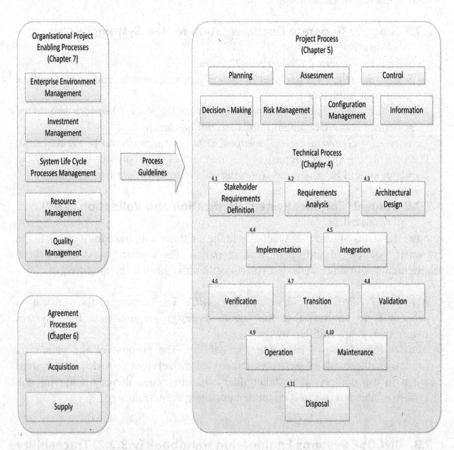

Fig. 6.26 Life cycle processes

Table 6.6 Cradle process activities traced to material in the INCOSE handbook in Sections 4.1, 4.2, 4.3, 4.6, and 4.8

INCOSE handbook		Cradle-8 stage requirements definition and management	
Section no.	Activity description	Activity no.	Activity description
4.1	Stakeholder requirements definition process		
	Purpose is to define requirements for a system that can provide the services needed by users and other stakeholders in a defined environment. It identifies stakeholders, or stakeholder classes, involved with the system throughput, the system's life cycle, and stakeholders' needs, expectations, and desires. It analyzes and transforms these into a common set of stakeholder requirements that express the intended interaction the system will have with its operational environment and that are the references against which each resulting operational service is validated		
4.1.2.1	Identify users and stakeholders	1.1	Define project scope
4.1.2.2	Define needs	1.1	Define project scope
4.1.2.3	Capture source requirements	1.4	Create stakeholder requirements
		1.6	Design constraints
4.1.2.4	Initialize requirements database	1.5	Plan system validation
		1.7	Capture glossary items
4.1.2.5	Establish concept of operations	1.2	Define use cases
		1.3	Develop concept of operations
4.1.2.6	Generate system requirements document	7.1	Generate deliverable document
4.2	Requirements analysis process		
	Purpose of the requirements analysis process is to transform the stakeholder requirements—driven view of desired services into a technical view of a required product that could deliver those services. This process builds a representation of a future system that will meet stakeholder requirements and that, as far as constraints permit,		

(continued)

Table 6.6 (continued)

INCOSE handbook		Cradle-8 stage requirements definition and management	
Section no.	Activity description	Activity no.	Activity description
	does not imply any specific implementation. It results in measurable system requirements that specify, from the supplier's perspective, what characteristics it is to possess and with what magnitude in order to satisfy stakeholder requirements. It identifies stakeholders, or stakeholder classes, involved with the system throughput, the system's life cycle, and stakeholders' needs, expectations, and desires. It analyzes and transforms these into a common set of stakeholder requirements that express the intended interaction the system will have with its operational environment and that are the references against which each resulting operational service is validated		
4.2.2.1	Identify users and stakeholders	6.1	Analyze requirements
		6.2	Verification planning
		6.3	Requirements baseline
4.2.2.2	Define needs	6.1	Analyze requirements
4.2.2.3	Capture source requirements	2.1	Create context diagram
		2.2	Identify external interfaces
4.2.2.4	Initialize requirements database	4.1	Develop functional model for the primary system element
		4.5	
		4.6	Decompose functions establish flow down of functional requirements and link to functions
4.2.2.5	Establish concept of operations	3.2	Define nonfunctional system requirements
4.2.2.6	Develop specification trees and specifications	3.1	Analyze system element source information
		3.2	
			Analyze system element source information
		3.3	Define candidate subordinate system elements

(continued)

Table 6.6 (continued)

INCOSE handbook		Cradle-8 stage requirements definition and management	
Section no.	Activity description	Activity no.	Activity description
4.2.2.7	Allocate requirements and establish traceability	4.2	Create functional requirements and link to functions
		4.3	
			Establish traceability to source requirements
4.2.2.8	Generate the system specification	7.1	Generate deliverable document
4.3	Architectural design process:		
	Purpose of the architectural design process is to synthesize a solution that satisfies system requirements. This process encapsulates and defines areas of solution expressed as a set of separate problems of manageable, conceptual, and, ultimately, realizable proportions. It identifies and explores one or more implementation strategies at a level of detail consistent with the system's technical and commercial requirements and risks. From this, an architectural design solution is defined in terms of the requirements for the set of system elements from which the system is configured. The specified design requirements resulting from this process are the bases for verifying the realized system and for devising an assembly and verification strategy		
4.3.2.1	Architectural design concepts	5.1	Allocate functions to system elements
		5.2	Define physical interfaces between system elements
		5.3	Allocate nonfunctional requirements to system elements
		4.4	Create and link interface requirements to data definitions
4.6	Verification process:	8.1	System verification and validation (V&V) traceability
	Purpose of the verification process is to confirm that the		

(continued)

Table 6.6 (continued)

INCOSE handbook		Cradle-8 stage requirements definition and management	
Section no.	Activity description	Activity no.	Activity description
	specified design requirements are fulfilled by the system		
	This process provides the information required to effect the remedial actions that correct nonconformances in the realized system or the processes that act on it		
4.8	Validation process:	8.1	System verification and validation (V&V) traceability
	Purpose of the validation process is to provide objective evidence that the services provided by a system when in use comply with stakeholders' requirements, achieving its intended use in its intended operational environment		
	This process performs a comparative assessment and confirms that the stakeholders' requirements are correctly defined. Where variances are identified, these are recorded and guide corrective actions. System validation is ratified by stakeholders		

study and application of engineering to the design, development, and maintenance of software (IEEE 1990). Formal definitions of software engineering include:

- Application of a systematic, disciplined, quantifiable approach to the development, operation, and maintenance of software, which is any set of machine-readable instructions that direct a computer processor to perform specific operations (IEEE 1990)
- Engineering discipline that is concerned with all aspects of software production (Sommerville 2007)
- Establishment and use of sound engineering principles in order to economically obtain software that is reliable and works efficiently on real machines (Hamilton 1972)

The discipline of software engineering was created to address the poor quality of software, gain control of projects exceeding time and budget, and ensure that software is built systematically, rigorously, measurably, on time, on budget, and within specifications. Engineering already addresses all of these issues; hence, the same principles used in engineering can be applied to software. The widespread lack of best practices for software at that time was perceived as a software crisis (Sommerville 2007; Naur and Randell 1968).

As programs and computers became part of the business world, their development moved out of the world of bespoke craftwork and became a commercial venture. The buyers of software increasingly demanded a high-quality product built on time and within budget. Many large systems of that time were seen as absolute failures—either they were abandoned or did not deliver any of the anticipated benefits. Hence, the software crisis occurred because a number of fundamental problems with the process of software development were identified (http10 2015):

- Software was frequently never completed, even after further significant investment had been made.
- The amount of work involved in removing flaws and bugs from completed software, to make it usable, often took a considerable amount of time—often more than had been spent in writing it in the first place.
- The functionality of the software seldom matched the requirements of the end users.
- Once created, software was almost impossible to maintain; the developer's ability to understand what they had written appeared to decrease rapidly over time.

Fortunately, there has been an awareness of the software crisis; and it has inspired a worldwide movement toward process improvement. Software industry leaders saw that consistently following a formal software process leads to better quality products, more efficient teams and individuals, reduced costs, and better morale. Meanwhile, software engineering has evolved over the past several years from an activity of computer engineering to a discipline in its own right.

With an eye toward formalizing the field, the IEEE Computer Society has engaged in several activities to advance the professionalism of software engineering, such as establishing certification requirements for software developers. To complement this work, a joint task force of the Computer Society (CS) and the Association for Computing Machines (ACM) has recently established another linchpin of professionalism for software engineering: a code of ethics. After an extensive review process, version 5.2 of the Software Engineering Code of Ethics and Professional Practice, recommended last year by the IEEE-CS/ACM Joint Task Force on Software Engineering Ethics and Professional Practices, was adopted by both the IEEE Computer Society and the ACM.

The Software Engineering Code of Ethics and Professional Practice, intended as a standard for teaching and practicing software engineering, documents the ethical and professional obligations of software engineers. The code instructs practitioners about the standards society expects them to meet, about what their peers strive for, and about what to expect of one another. In addition, the code informs the public about the responsibilities that are important to the profession. Adopted by the CS and the ACM, two leading international computing societies, the code of ethics is intended as a guide for members of the evolving software engineering profession. The code was developed by a multinational task force with additional input from other professionals from industry, government posts, military installations, and educational professions (Gotterbarn et al. 1999).

In its short version, the code summarizes aspirations at a high level of abstraction. The clauses that are included in the full version give examples and details of how these aspirations change the way we act as software engineering professionals. Without the aspirations, the details can become legalistic and tedious; without the details, the aspirations can become high sounding but empty; together, the aspirations and the details form a cohesive code.

According to the code, software engineers shall commit themselves to making the analysis, specification, design, development, testing, and maintenance of software a beneficial and respected profession. In accordance with their commitment to the health, safety, and welfare of the public, software engineers shall adhere to the following eight principles:

1. *Public*: Software engineers shall act consistently with the public interest.
2. *Client and employer*: Software engineers shall act in a manner that is in the best interests of their client and employer, consistent with the public interest.
3. *Product*: Software engineers shall ensure that their products and related modifications meet the highest professional standards possible.
4. *Judgment*: Software engineers shall maintain integrity and independence in their professional judgment.
5. *Management*: Software engineering managers and leaders shall subscribe to and promote an ethical approach to the management of software development and maintenance.
6. *Profession*: Software engineers shall advance the integrity and reputation of the profession consistent with the public interest.
7. *Colleagues*: Software engineers shall be fair to and supportive of their colleagues.
8. *Self*: Software engineers shall participate in lifelong learning (Gotterbarn et al. 1999).

From the aforementioned, a first draft can be derived for software engineering projects, which includes the following activities:

- Basic project framework, including ideas, vision, and concepts
- Project charter, including values and rules
- Business case, including costs and benefits
- Selecting the applicable method, which can be the waterfall method or Agile methods
- Scope of supply, including objects and modules
- Release and project plan with deadlines and content
- Project organization, including organization plan and responsibilities

The chosen method has a decisive influence on the software engineering project because it determines how the solution will be developed, how the project phases are planned, and how the project's progress can or should be measured. The two most frequently used methods in software engineering are the:

- *Waterfall method or V-model*: A sequential design process used in software development processes in which progress is seen as flowing steadily downward (like a waterfall).
- *Agile method*: It helps teams respond to unpredictability through incremental, iterative work cadences, known as sprints.

6.3.1 V-Model

The V-model is a linear process model for software that organizes the software development process in various solid phases. The phase results are always considered as binding targets for the next phase. Each phase must be completed so that the next can begin. For this purpose, each phase generated is a result of a detailed document or program. In Fig. 6.27, the principal structure of the V-model is shown.

The V-model, which can be considered to be an enhancement of the waterfall model with the possibility of kickbacks in a previous phase, is added. The phases of the V-model are:

- Requirements definition and/or requirements analysis
- Functional systems design also known as:
  - Functional specification
  - Software system requirements
  - Specification sheet
    These describe, from a black-box perspective, how the software system implements the user requirements. This also includes the description of interfaces such as a graphical user interface (GUI) and possibly other interfaces to the outside. The GUI description includes a description of the GUI screens, how to navigate through the screens, and how the software

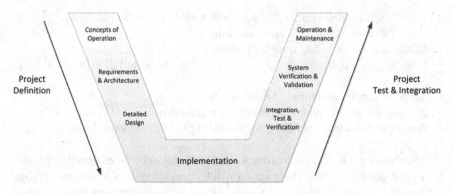

Fig. 6.27 V-model of the software engineering process

system reacts to user inputs. The functional software systems design does not describe how to technically implement the software system to achieve this black-box behavior. Other aspects within the black-box view are:
Performance, response time, and resource consumption
Robustness, behavior during overload, and incorrect entries of installation, update capability, supported platforms, etc.

- Technical systems design, also called architecture, describes how to implement the software system technically. Aspects of the architecture are, for example, the programming language and the technologies used, the component and class diagrams, and database schemas. Therefore, the architecture must be designed so that:
  - Software system requirements are implemented which require that the requirements are met for the black-box view of the software systems, including performance requirements and the portability of the software system.
  - Software system is maintainable, testable, and modifiable,
  - Components and their interfaces are recognizable.
      The architecture is usually described from multiple views, including:
  - Static view with component and class diagrams
  - Dynamic view, a description of how processes and workflows run
  - Deployment point of view which is a description of which software system and what artifacts (dll, jar, exe, servers, etc.) are installed.
- Implementation, programming, coding
- Module tests
- Integration testing
- Functional and system tests
- Acceptance testing

An important difference between the V-model and the waterfall model is that within the V-model, tests can be defined in the development process as early as possible. At this point, a distinction has to be made with regard to the terms validation and verification.

- *Validation*: Determine if the requirements have been met, that is, if the right things were done and if the user can achieve his/her goals.
- *Verification*: Determine if the software was developed as specified, e.g., if a GUI looks as specified and responds to user actions per the requirements. Verification refers to the software system and not to the usage requirements.

Since validation takes place only at the end of the development process, there is a risk that errors in the requirements remain undetected until the end. This is a major disadvantage in comparison to iterative process models. Furthermore, the V-model requires that requirements remain consistent. Once the requirements analysis phase has been completed, the V-model does not permit changes to the requirements. However, this is practical for very few projects. It is reported that 80 % of projects suffer from unstable or growing requirements. Since most projects also come under time pressure.

To summarize, the V-model represents the sequence of a real, physical project, such as building a house. At the beginning, the foundation is laid, then the walls are constructed, and finally the roof is built at the end. From this example, it can clearly be seen that the later an erroneous specification is detected, the more difficult and more expensive it will be to remedy. If false assumptions or ambiguousness are detected during the late phases of a project, termination of the project is often the only way to limit the damage. This is particularly problematic with extensive developments requiring a complete run of the model, which can take several years. By the end of the design phase, it is too late to be responsive to changing requirements.

6.3.2 Agile Software Development Methodology

The Manifesto for Agile Software Development (Beck et al. 2001), also known as the Agile Manifesto, which first laid out the underlying concepts of Agile development, was introduced in 2001. Some of the authors formed the Agile Alliance, a nonprofit organization that promotes software development according to the Manifesto's values and principles. Introducing the Manifesto on behalf of the Agile Alliance, Jim Highsmith said:

> The Agile movement is not anti-methodology; in fact many of us want to restore credibility to the word methodology. We want to restore a balance. We embrace modeling, but not in

*order to file some diagram in a dusty corporate repository. We embrace documentation, but
not hundreds of pages of never-maintained and rarely-used tomes. We plan, but recognize
the limits of planning in a turbulent environment. Those who would brand proponents of XP
(Extreme Programming) or SCRUM or any of the other Agile Methodologies as "hackers"
are ignorant of both the methodologies and the original definition of the term hacker*
(Highsmith 2001).

Thus, the meanings of the Manifesto items in regard to Agile software development
are (http11 2015):

- *Individuals and Interactions*: In Agile development, self-organization and motivation are important, as are interactions such as co-location and pair
  programming.
- *Working Software*: Working software is more useful and welcome than just
  presenting documents to clients in meetings.
- *Customer Collaboration*: Requirements cannot be fully collected at the beginning of the software development cycle; therefore, continuous customer or
  stakeholder involvement is very important.
- *Responding to Change*: Agile development is focused on quick response to
  change and continuous development.

Therefore, in this method of software engineering, planning is less detailed,
resulting in greater flexibility—also referred to as agility. Instead of a complete
requirements specification, the Agile software development method requires an
overview of the modules to be implemented. Its formulation can already be done in
the project setup. Afterward, the most important sequential iterations are specified;
and thereafter, the other parts of the software are implemented and integrated. The
sequence is not fixed at the beginning but is adapted to the current state of
knowledge. Software is tested after each iteration and is, for this reason, executable.
In some cases, the software is rolled out; in other cases, this happens only after all
iterations have been executed.

Twelve principles of Agile software development have been launched by the
Agile Alliance to supplement the Manifesto. They are as follows (http12 2015):

- The highest priority is to satisfy the customer through early and continuous
  delivery of valuable software.
- Welcome changing requirements, even late in development. Agile processes
  harness change for the customer's competitive advantage.
- Deliver working software frequently, from a couple of weeks to a couple of
  months, with a preference to the shorter timescale.
- Business people and developers must work together daily throughout the project.
- Build projects around motivated individuals. Give them the environment and
  support they need, and trust them to get the job done.

- The most efficient and effective method of conveying information to and within a development team is face-to-face conversation.
- Working software is the primary measure of progress.
- Agile processes promote sustainable development. The sponsors, developers, and users should be able to maintain a constant pace indefinitely.
- Continuous attention to technical excellence and good design enhances agility.
- Simplicity—the art of maximizing the amount of work not done—is essential.
- The best architectures, requirements, and designs emerge from self-organizing teams.
- At regular intervals, the team reflects on how to become more effective and then tunes and adjusts its behavior accordingly.

Well-known Agile software development methods include (http11 2015):

- Agile modeling
- Agile unified process (AUP)
- Dynamic system development method (DSDM)
- Essential unified process (EssUP)
- Extreme programming (XP)
- Feature-driven development (FDD)
- Open unified process (OpenUP)
- Scrum
- Velocity tracking

6.3.3 Comparison of the V-Model and the Agile Software Development Methodology

Selection of the appropriate development method is essential for the respective software. The questions to be answered include "How will the project be organized?" and "Who is involved in the project?" The software development methodology is not only determined by the expectations of the customer but also by the underlying project complexity, which can be seen from the comparison of the V-model and the Agile software development methodology in Table 6.7 (Rusche 2013).

Table 6.7 Six Sigma quality function development matrix

	V-model	Agile method
Requirements	Right choice if at least 80 % of the requirements are unique and agreed with regard to their prioritization. For smaller projects, this is often the case	Right choice when the core functions are known, but the description of the overall system or the previous prioritization, however, is difficult. For projects which are noticeably unclear, this is often the case
Project schedule	Functional and technical requirements can be implemented in accordance with the project schedule developed. Planning is controlled in each project phase and updated, but it is not fundamentally questioned or revised	Project schedule itself and all dates are fixed; however, the content of iterations is not fixed
Integration	Individual modules are integrated over several days/weeks and then tested	Integration is carried out at the end of each iteration. Singular expense; risk is lower but resources are dedicated for a longer time frame
Application	New software is not created on a green field but is, to a large extent, based on existing functions. Thus, the scope of delivery is largely known in advance. Typically, this is the case when an old solution is enhanced	Project is created from scratch. New features are developed. Scope of delivery is largely unknown. Instead of producing concept sheets, real software is developed within a few weeks with each iteration enhanced for the desired application
Functionality, budget	After the specifications are determined, the initial offer is revised and the functions are adapted to the budget; what will be delivered is determined by the agreed-upon price	Because the functionality can vary widely in the implementation, a minimum scope for the software is determined. The number of hours to produce the product is agreed upon

6.4 Requirements in Software Design in Cyber-Physical Systems

The design of cyber-physical systems (CPS) is hampered by the limited ability to deal with complex designs at a system level. There are many factors impeding system-level design, such as the lack of formalized high-fidelity models for large and complex systems, insufficient ways of measuring performance for important system components, and, finally, inadequate scientific foundations. A key factor is compositionality and modularity in the design approach. Compositionality in CPS is impacted by the strong interdependencies of software and systems engineering and often limited by an inadequate systems design.

With regard to Sect. 3.3.1, software requirements can be introduced as descriptions of services that software must provide and the constraints under which software must operate. From a more general perspective, it can be stated that software requirements range from high-level abstract statements of services or constraints to more detailed functional specifications.

To introduce the different types of requirements for software, we follow the concept described in http13 (2015):

- User requirements
  - Written for customers.
  - Statements are written in a natural language including diagrams of services the software system should provide as well as its operational constraints.
    Area from which problems can arise when using a natural language:
    - Ambiguity: writers and reader may not interpret words in the same way.
    - Over flexibility: same thing can be expressed in different ways.
    - Requirements mixture: different requirements may be mixed together.
    - Lack of modularization: natural languages are inadequate to structure software requirements.
    Alternatives to natural languages are:
    - Graphical notations: graphical language supplemented by textual annotations and often used to define functional requirements.
    - Mathematical/formal specifications: mathematical concepts such as finite state machines or sets; unambiguous specification which helps reduce arguments between customers and contractors, but customers often do not understand mathematical/formal specifications.
- Software requirements
  - Written as a contract between a client and contractor.
  - Structured document that sets out detailed descriptions of software services.
- Software specification
  - Written for developers.
  - Detailed software description which can serve as the basis for the design and/or implementation.

Besides the aforementioned requirements, functional, nonfunctional, and domain-related requirements are important in software engineering.

- Functional requirements
  - Statements of services which the software should provide, how the software should react to particular input, and how the software should behave in particular situations.
- Nonfunctional requirements
  - Constraints on the services or functions the software must assure, such as time constraints, constraints on the development process, and standards to be compiled.
- Domain requirements
  - Result from the application domain of the software reflecting the characteristics of the application domain
  - Can be functional or nonfunctional

Finally, a requirements document will be completed which contains the following elements:

- Official statement of what is required from the software developers
- Definition and specification requirements
  - Specify external system behavior
  - Specify implementation constraints
  - Be easy to change, but changes must be managed
  - Serve as a reference tool for maintenance
  - Record provident about the life cycle of the system
  - Characterize responses to unexpected events

The requirements document is not a design document; it has to state *what* the software should do rather than *how* it should do it.

The so-called software requirements specification (SRS) is an IEEE standard that was first published as ANSI/IEEE Std 830-1984. The latest version is currently IEEE Std 29148-2011. With the SRS, the IEEE has defined how the software requirements specification document should be established. The relevant section of the document thus has been determined. This document is basically divided into two sections (http14 2015):

- C-Requirement (Customer Requirement): Area comparable with the specification sheet
- D-Requirement (Development Requirement): Area comparable with the requirements specification

Below a C-Requirement, the requirements from the customer's point of view and/or the end user are considered. A D-Requirement characterizes the development needs and is the view from the perspective of a developer, in contrast to the customer's view. With regard to requirements, the qualitative and the quantitative definitions of the required software are meant from the perspective of the customer. Ideally, such a specification includes a detailed description of the purpose of the planned use, in practice as well as the required functionality of the software. Here, technical aspects, such as the following, should be taken into account.

- What is the software able to do?
- To what extent and under what conditions will the software be used?

Thus, the software requirements specification follows the IEEE standard and includes at least three main points which should be met. In practice, this is frequently modified. An exemplary structure might look like:

- Name of the software product
- Name of the manufacturer
- Version date of the document and/or software

1. Introduction
    (A) Purpose (of the document)
  ' (B) Coverage (of the software product)
    (C) Explanation (of terms and/or short cuts)
    (D) References (to other resources or sources)
    (E) Overview (how the document is built)
2. General description (the software product)
    (A) Product perspective (with regard to other software products)
    (B) Product features (a summary and overview)
    (C) User features (information on expected users, e.g., education, experience, expertise)
    (D) Limitations (for developers)
    (E) Assumptions and dependencies (factors that influence the development but do not restrict, e.g., choice of operating systems)
    (F) Distribution of requirements (not feasible and shifted to later version features)
3. Specific requirements (as opposed to 2.)
    (A) Functional requirements (depend strongly on the nature of the software product)
    (B) Nonfunctional requirements
    (C) External interfaces
    (D) Design constraints
    (E) Requirements for performance
    (F) Quality requirements
    (G) Other requirements

The difficulties that may arise in such a requirements analysis, in practice, include:

• Potential conflicts of interest, i.e., different targets by users
• Unclear or even unknown technical conditions
• Changing requirements or priorities already during the design process

To summarize, for the successful implementation of complex software projects, the following core processes are essential:

• Requirements analysis
• Prototyping
• Architecture and design
• Programming and validation
• Implementation at customer side and testing
• Maintenance schedule

As can be seen from the above diagram, which shows the traditional design approach, a two-phase design approach may be appropriate. The first phase is the

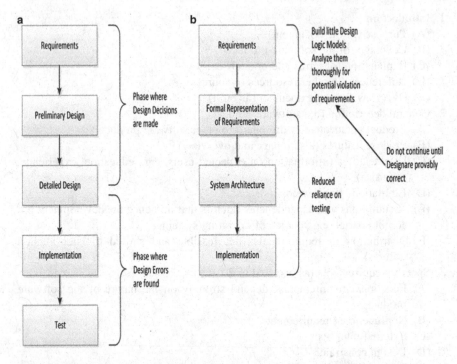

Fig. 6.28 (a) Traditional approach to design and testing, and (b) early error detection

one in which the design decisions are made. In the subsequent second phase, design errors are identified so far as is available. Since the error detection uses a relatively late stage, the cost for error detection can be high. To achieve earlier error detection for the requirements analysis, only small design logic models should be used which can easily be thoroughly analyzed for potential violation of requirements. This results in a formal representation of requirements with the constraint not allowing forward movement until the design or parts of the design have been proved as probably correct. Hence, the resulting detailed design and, thereafter, the implementation reduces reliance of testing to minor issues. This can be described in the illustration given in Fig. 6.28.

Besides the traditional design steps mentioned in Fig. 6.11, the following related processes controlling the implementation can be summarized:

- Requirements management
- Project management
- Quality management
- Configuration management
- Documentation
- Customer training

For the implementation of the software solutions, the following selected examples of developing platforms can be used:

- Agile tools for use cases and prototype implementation in the context of the IMC-AESOP approach, which envisions a service-oriented architecture (SOA)-based Supervisory Control and Data Acquisition/Distributed Control System (SCADA/DCS) infrastructure that enables cross layer, service-oriented collaboration not only at the horizontal level, e.g., among cooperating devices and systems but also at the vertical level between systems located at different levels of an enterprise software architecture (Colombo et al. 2014).
- Development with Java SE/EE/ME on Apache, Tomcat, and JBoss
- Development with C # on the .NET Framework on Microsoft platforms
- Development using C ++ for real-time and embedded systems as well as software development tools such as:
  - Eclipse IDE, MS Visual Studio
  - Polarion Application Lifecycle Management Tool
  - Polarion Requirements and Test Case Management
  - Build Management with Apache Maven and Ant
  - Test Toolsets with JUnit and SoapUI

6.5 Maritime Area Case Studies

A case study basically is an in-depth study of a particular area of concentration rather than a sweeping statistical survey. Hence, it is a method used to narrow down a very broad field of research into one easily researchable topic. A huge number of potential research topics are available in the maritime domain. Major parts in the maritime domain are ports which are the gateways to a continent or an island. In the European Union (EU), 74 % of extra-EU goods are shipped through ports. They are equally important for intra-European trade: 37 % of the intra-EU freight traffic and 385 million passengers pass through ports every year.

Over 1,200 commercial seaports operate along some 70,000 km of the Union's coasts. Europe is one of the densest port regions worldwide. In 2011, around 3.7 billion tons of cargo (more than 60,000 port calls of merchant ships) transited through European ports. Bulk traffic represented 70 % of it, containers 18 %, and roll-on/roll-off (Ro-Ro) traffic 7 %, the rest being other general cargo. The EU port industry has a significant economic impact in terms of employment and activity in the port industry itself (direct impacts), down the supply chain (indirect impacts), and in the wider EU economy (induced impacts) (http15 2015). Therefore, the case study in Chap. 6 focuses on the maritime domain and especially on the maritime domain awareness.

The maritime domain is defined as all areas and things of, on, under, relating to, adjacent to, or bordering on a sea, ocean, or other navigable waterway, including all maritime-related activities, infrastructure, people, cargo, vessels, and other conveyances. In the case of the maritime domain awareness (MDA), which is defined as the effective understanding of anything associated with the maritime domain that could impact the security, safety, economy, or environment, CPS became of interest with regard to container monitoring on container ships, at the sea gate harbor, at dry ports, and on trucks and/or trains, which refers to a very broad field of research. Hence, the case study is restricted to two applications that address the tracking and monitoring of containers. The first application deals with tracking and monitoring of containers at ports and on ships, the second with tracking and monitoring containers transported from a sea gate port to a dry port (Möller 2014).

6.5.1 Tracking and Monitoring Containers at Ports and on Ships

This case study application has a crucial background, because nations, unfortunately, face terrorism and smuggling on a daily basis. Thus, research is being conducted on tracking and monitoring containers, whether at port or on a ship, especially those shipped globally on a regular basis, to avoid terroristic attacks or smuggling. These tasks have to work remotely. Working remotely means connecting to a computer running an operating system with software from another computer running an operating system with software that is connected to the same network, or to the Internet, for which the computer has permission to connect. This allows access to all of the work computer's programs, files, and network resources from the remote computer.

Tracking and monitoring containers, whether in port or on a ship, can be accomplished via both cellular and satellite mobile networks which are connected to an embedded system containing the required measuring devices. As described in Patterson et al. (2014), two different methods are used to track and monitor the containers: cable transmission and carrier transmission. In the case of the cable transmission method, four different cables and sockets are placed strategically inside the container which measure, e.g., the compression and temperature of the container by feeding back a Boolean value. The carrier transmission method uses modems located in the container and sends back frequencies which allow monitoring of the state of the container.

As described in Patterson et al. (2014), these methods are not cost-effective and require a lot of setup time. Therefore, wireless communication is much better for tracking and monitoring containers locally and remotely. Local tracking and monitoring is based on active and passive radio frequency identification (RFID) tags (see Section 4.2), attached to the container. The RFID tags allow information to be broadcast to RFID reading equipment which instantly lists the items and specifications for the container. Local tracking uses the ZigBee application, used mostly for interconnection networks. ZigBee is a specification for a suite of high-level communication protocols used to create personal area networks built from small, low-power digital devices. ZigBee is based on the IEEE 802.154 standard. ZigBee is able to broadcast packets at a low energy level for optimal energy savings and can transmit data over long distances by passing data through a mesh of

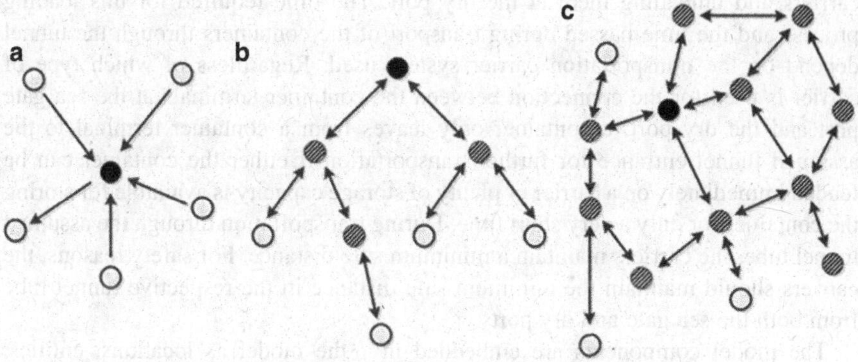

Fig. 6.29 Network configurations for ZigBee: (**a**) star, (**b**) tree, (**c**) mesh

networks of intermediate devices to reach more distant zones. ZigBee has a defined rate of 250 kbit/s, best suited for intermittent data transmissions from a sensor or input device. Hence, crewmen on the ship use ZigBee networks to monitor containers. The maritime people on land track and monitor the container by cellular gateway and satellite gateway (Bay et al. 2010). The onboard container tracking and monitoring system uses the following for remote container tracking and monitoring:

- ZigBee network
- Local monitoring center
- Cellular gateway
- Satellite gateway

The land-based container tracking and monitoring system uses:

- Cellular base station
- Remote monitoring center

A ZigBee network can be implemented as one of the three topologies shown in Fig. 6.29. A ZigBee network supports peer-to-peer communication and a star configuration. It adds a routing protocol and a hierarchical network to the 802.15.4 specification addressing a scheme that allows for cluster-tree and multihop mesh networking topologies.

6.5.2 Tracking and Monitoring Containers Transported from the Sea Gate Port to a Dry Port

Transportation of containers from the container terminal at the sea gate port and the assumed tunnel entrance at the sea gate port requires loading the containers on carriers and unloading them at the dry port. The time required for this loading process and the time passed during transport of the containers through the tunnel depend on the transportation carrier system used. Regardless of which type of carrier is used for the connection between the container terminals at the sea gate port and the dry port, a container only leaves from a container terminal to the assumed tunnel entrance for further transportation if either the container can be loaded immediately on a carrier or plenty of storage capacity is available for storing the container for only a very short time. During transportation through the assumed tunnel tube, the carriers maintain a minimum safe distance. For safety reasons, the carriers should maintain the minimum safe distance in the respective tunnel tube from both the sea gate and dry ports.

The model components are embedded in   the model as locations, entities, arrivals, and processes (LEAP).

The first step in LEAP is the definition of *locations*. *Locations* represent fixed spots in the model, such as a container yard. At these model places, operations on objects can be made. In the graphical user interface (GUI), they appear as an icon of the layout and can be fitted with a logo and an advertisement. Such a display specifies how many dynamic objects (*entities*) in the *location* are included at the current time, which makes it easier to track the system state during a simulation run. The capacity of a *location* is explicitly set in the model, which means that a *location*, at any point in time, can contain up to a certain number of *entities*.

In the second step, the definition of *entities* is carried out. *Entities* are those objects that move through the model or a part of the model during the simulation runs and may be changed. *Entities* are assigned a speed and size.

The third step is the definition of *arrivals*. *Arrivals* define when *entities* arrive in the model. Here, it is determined at which *location* this will happen and if and at what distance repetitions occur, how many repetitions of *arrivals* there are, and how many *entities* enter the simulation model per repetition set. Moreover, at *arrival*, a manipulation of the global variable can be   defined as well as the initial allocation of *attributes*.

In the final step, the processing, one or more target *locations* are issued to and from a *location* for an *entity* to reach the next *entity*. Leaving a *location* can also change the type of *entity*. Moreover, waiting periods can be specified, conditions examined, and variables and attributes manipulated. If several *locations* are specified as a target, one can be selected based on conditions. All possible paths by which an *entity* can pass through the model are defined. Should an *entity* leave the model, "EXIT" is specified as the target *location*.

The following sample shows the   source code for an attribute query of submodel input as part of processing the location in-arrival container.

```
Distribution of incoming containers based on attribute origin
 #A container arrives... #to location in_load_ship, #if he should be
    unloaded from a ship
 # Distribution of incoming containers based on attribute origin
    #A Container arrives,
```

```
          #at Location in_load_ship,
           # if he should be unloaded from a ship
IF origin = 12 THEN
{
  ROUTE 1
}
          #to Location in_to_hh,
          # if he should arrive without shipping in the port of Hamburg
  ELSE IF origin < 6 THEN
{
  ROUTE 2
}
          #at Location in_to_drport,
          #if his transportation part start in Maschen or Rade ELSE IF
(origin = 6 OR origin = 7) THEN
{
  ROUTE 3
}
          #at Location in_to_cth,
          #if his transportation path start in Magdeburg
  ELSE IF origin = 8 THEN
{
  ROUTE 4
}         #at Location in_to_dd_container,
          #if his transportation part start in Riesal
  ELSE IF origin = 9 THEN
{
  ROUTE 5
}
          #atLocation in_to_ctw,
          #if his transportation path start in Wilhelmshaven
  ELSE IF origin = 10 THEN
{
  ROUTE 6
}
```

Besides *locations*, *entities*, *arrivals*, and *processing*, the simulation software, ProModel, offers the elements of *attributes*, *networks*, *resources*, *macros*, and *global variables*.

- *Attributes*: Defined either for *locations* or *entities* and can be created as integers or real. If an *attribute* is defined for an *entity*, a value can be assigned to each individual *entity*. The value of the *attributes* can be changed during a simulation run.

- *Networks*: Used to build graphs and essentially consist of *nodes* and *paths*. *Nodes* can have either a certain limited or unlimited capacity. For a limited capacity, a value must be explicitly specified. Then the concerned *node* can contain only *entities* and *resources* up to this maximum at any time. *Nodes* can be linked to *locations*. Such connections are required if *entities* in a network pass from one *location* to another one. Paths connect two *nodes*. They have a certain length, which is explicitly specified. Both *entities* and *resources* can pass along networks.
- *Resources*: Used in the model to transport *entities* from one *location* to another. *Resources* are linked to a *network*. If an *entity* is assigned to a *resource* when changing from one *location* to another, it is set in the *processing*.
- *Macros*: Placeholders for text parts. A *macro* has a name for identification, which forms the placeholder, and a text, which is inserted and executed during execution of the simulation at the location of the placeholder. *Macros* can be defined as scenario parameters. For a scenario parameter, the text that occurs during the simulation in the place of the *macro* can be changed before beginning the simulation run. Macros are essential to define scenarios in ProModel. To create scenarios, there must be at least a *macro* that is defined as a scenario parameter. A *macro* that is defined as a scenario parameter is assigned a value for each scenario to be examined. This happens outside the scope of the model so that the model itself is not modified.
- *Global Variables*: Variables that can be accessed from any part of the model. This means that their current value can be queried and changed, allowing different model parts to communicate with each other.

Is it also possible to first develop a model in the form of more than one part or in submodels and then merge them into an overall model (Möller 2014). This process is referred to as merging. There are two kinds of merging. In a merge as a model, the pasted model is not changed. In a merge as a submodel, however, the names of all model elements, such as *locations* and *global variables*, receive a prefix or suffix that is specified during merging. *Entities* and *attributes* are excluded from these changes. Submodels can communicate only via *global variables* so that no further adjustments need to be made after the merge. If it should be possible to move *entities* between the submodels, then the submodels can be adapted for this purpose; however, the target location must be changed after the merging of the models in the processing of individual rules.

The simulation runs can be traced on the screen because movements are graphically represented by *entities* and *resources*. Also, the current values of *global variables* and the number of *entities* in the individual locations can be traced if their representation in the model is created. Through these opportunities, model building is easy to understand; and the black-box nature of the simulation is, thus, at least partly resolved. This can improve the acceptance of the results for decision-makers, who have no direct relation to the modeling and simulation methodology.

As described in Sect. 6.5.1, container tracking and monitoring is based on active and passive RFID tags (see Sect. 4.2) attached to the container. The RFID allows their information to be broadcast to RFID reading equipment, which instantly lists the items and specifications for the container. As described in Sect. 6.5.1, ZigBee is able to broadcast packets at a low energy level for optimal energy saving and can transmit data over long distances by passing data through a mesh of networks of intermediate devices to reach more distant zones. Hence, land-based haulage contractors use ZigBee networks to track and monitor the containers. The land-based haulage contractors track and monitor the container by cellular gateway and satellite gateway (Bay et al. 2010). The architecture of the remote container tracking and monitoring system contains the following for the land-based container tracking and monitoring system:

- ZigBee network
- Land-based local monitoring center
- Cellular gateway on the carrier
- Cellular base station
- Satellite gateway
- Remote monitoring center

The results obtained from the simulation give an answer to how the container will move from the sea gate port to the dry port, which gives a real-time dependent scenario of the container and its actual location, which can also be tracked and monitored by simulation. The simulation results are compared with the actual data obtained from the RFID reading equipment to identify whether or not alterations in the container sequence happened which is an indication for an unexpected event. This monitoring of an event can indicate that the container on the carrier is in the wrong order, resulting from scanning the whole sequence of containers on the carrier.

6.6   Exercises

What is meant by the term *systems engineering*?
Describe the systems engineering process in detail.
What is meant by the term *model-based design*?
Give an example of a model-based design.
What is meant by the term *model-driven design*?
Give an example of a model-driven design.
What is meant by the term *life cycle*?
Describe the life cycle process in detail.
What is meant by the term *constraint*?
Give an example of a constraint.
What is meant by the term *reliability analysis*?
Give an example of reliability analysis.

What is meant by the term *maintainability*?

Describe the maintainability process in detail.

What is meant by *ICAN definition for function modeling*?

Give an example of the ICAN definition for function modeling.

What is meant by the term *UML*?

Give an example of a UML.

What is meant by the term *quality function deployment*?

Describe the quality function deployment process in detail.

What is meant by the term *model of the system*?

Give an example of a model of the system.

What is meant by the term *reevaluation*?

Give an example of reevaluation.

What is meant by the term *software engineering*?

Describe the software engineering process in detail.

What is meant by *the term V-model*?

Give an example of the V-model.

What is meant by the term *Agile model*?

Give an example for the Agile model.

What is meant by the term *Agile unified process*?

Describe the Agile unified process in detail.

What is meant by the term *extreme programming*?

Give an example of extreme programming.

What is meant by the term *open unified process*?

Give an example of an open unified process.

What is meant by the term *vehicle tracking*?

Describe the vehicle tracking process in detail.

What is meant by the term *project schedule*?

Give an example of a project schedule for the Agile method.

What is meant by the term *integration*?

Give an example for the integration in the V-model.

What is meant by the term *ZigBee*?

Give an example for the application of ZigBee on a ship.

What is meant by the term *cellular base station*?

Give an example for the integration in the remote container monitoring system.

What is meant by the term *remote monitoring center*?

Give an example for the integration in the remote container monitoring system.

What is meant by the term *satellite gateway*?

Give an example for the integration in the remote container monitoring system.

What is meant by the term *cellular gateway*?

Give an example for the integration in the remote container monitoring system.

What is meant by the term *local monitoring center*?

Give an example for the integration in the remote container monitoring system.

References

(ANSI/EIA-632-1998) ANSI/EIA-632-1998 Processes for Engineering a System, Washington, D.C., Electronic Industries Association (EIA), 1999

(Bahill and Gissing 1998) Bahill, A. T., Gissing, B.: Re-evaluating systems engineering concepts using systems thinking. IEEE Transactions on Systems, Man and Cybernetics, Part C: Applications and Reviews, Vol. 28 No. 4, pp. 516–528, 1998

(Bai et al. 2010) Bai, Y., Zhang, Y., Shen, C.:: Remote container monitoring with wireless networking and cyber-physical system. IEEE Mobile Congress (GMC), 2010; doi 10.1109/GMC2010.5634569

(Baker 2015) Baker, L.: Model-Based Systems Engineering Process with Functional Model Analyses, Presentation AlaSim 2015

(Beck et al. 2001) Beck., K., Beedle, M., van Bennekum, A., Cockburn, A., Cunningham, W., Fowler, M., Grenning, J., Highsmith, J., Hunt, A., Jeffries, R., Kern, J., Marick, B., Martin, R. C., Melleor, S., Schwaber, S., Sutherland, J., Thomas, D.: Manifesto for Agile Software Developmen; http://www.agileAlliance.org

(Colomba et al. 2014) Colomba, A. W., Bangemann, T., Karnoukos, S., Delsing, J., Stluka, P., Harrison, R., Jammes, F., Martinez Lastra, J. L. (Ed.). Industrial Cloud-Based Cyber-Physical Systems, Springer Publ. 2014

(CRADLE-7 2014) Requirements Definition and Management Using Cradle, White Paper, 2014

(Derler et al. 2011) Derler, P., Lee, E. A., Sangiovanni-Vincentelli, A. I. : Addressing modelling challenges in Cyber-Physical Systems. Technical Report UCB/ EECS-2011-17. Berkeley, 2011

(DSMC 1990) Defense Systems Management College: Systems Engineering Management Guide, Washington, DC, U.S. Government Printing Office, 1990

(EIA&IS/632/1998) Systems Engineering, Washington, D.C., Electronic Industries Association (EIA), 1994

(Elm 2005) Elm, J. P.: Surveying Systems Engineering Effectiveness. Proceedings Systems Engineering Conference, 2005. http://web.archive.org/web/20070615160805/http://www.splc.net/programs/acquisition-support/presentations/surveying.pdf

(Faulconbridge and Ryan 2014) Faulconbridge, R. I., Ryan, M. J.: Systems Engineering Practice, Argos Press, Canberra, Australia, 2014

(Gotterbarn et al. 1999) Gotterbarn, D., Miller, K., Rogerson, S.: Computer Society and ACM Approve Software Code of Ethics. pp. 84–88, Computer, 10, 1999

(Hamelin et al. 2010) Hamelin, R. D., Walden, D. D., Krueger, M. E.: INCOSE Systems Engineering Handbook v.3.2: Improving the Process for SE Practitioners, INCOSE International Symposium 2010, Volume 20, Issue 1, pages 532–541. Published online: 4 NOV 2014: DOI: 10.1002/j.2334-5837.2010.tb01087.x

(Hamilton 1972) Hamilton, M.: Software Engineering. In: Information Processing pp. 530–538, North-Holland Publ, 1972

(Haskins 2010) Haskins, C. (Ed.): Systems Engineering Handbook: A Guide for System Life Cycle Processes and Activities, International Council oo Systems Engineering, San Diego, C. A., 2010

(Haskins 2006) Haskins, C.: Systems Engineering Handbook – Version 3. International Council of Systems Engineering, 2006

(Higsmith 2001) Higsmith, J.: History Agile Manifesto. http://agilemanifesto.org/history.html

(IEEE-STD-1220-1994) IEEE Trial-Use Standard for Application and Management of the Systems Engineering Process, NewZork, N.J., IEEE Computer Science, 1995

IEEE Standard Glossary of Software Engineering Terminology, IEEE Standard 610.12-1990, 1990

(Incose 2007) INCOSE Systems Engineering Cost Estimation by Consensus, 2007

(Johnson et al. 2014) Johnson, M., Randolph, T.: Hu, F. : On Modeling Issues in Cyber-Physical Systems. pp. 89-100, Chapter 7, In: Cyber-Physical Systems. (Ed.) Hu, F., CRC Press 2014

(Lake 1996) Lake, J.: Unraveling the Systems Engineering Lexicon. Proceedings of the INCOSE Symposion, 1996

(Li 2008) Li T.: Systems Engineering Assumptions and Comparison Tests, Graduate Report, TU-Delft, 2008

(Möller 2014) Möller, D. P. F.: Introduction to Transportation Analysis, Modeling and Simulation, Springer Publ. 2014

(Naur and Randell 1968) Naur, P., Randell, B.: Software Engineering - Report on a conference sponsored by the NATO, 1968

(Patterson et al. 2014) Patterson, C., Vasquez, R., Hu, F. : Cyber-Physical Systems: Design Challenges, pp. 15–33, Chapter 2. In: Cyber-Physical Systems. Hu, F. (Ed.), CRC Press, 2014

(Pollet and Chourabi 2008) Pollet, Y., Chourabi, O.: A formal approach for optimized concurrent System Engineering. Proceed. EngOpt – International Conference on Engineering Optimization, 2008

(SECMM-95-01) Systems Engineering Capability Maturity Model, Version 1.1, Carnegie Mellon University, Pittsburgh, P.A., Software Engineering Institute, 1995

(Summerville 2007) Summerville, I.: Software Engineering (8th ed.), Perason Education, 2005

(Rusche 2013) Rusche, C. A. (Ed.): The BSI Workbook (in German), BSI AG, 2013

(Suh 2001) Suh, N. P.: Axiomatic Design: Advances and Applications. Oxford University Press. 2001

(Tanik and Begley 2014) Tanik, U. J., Begly, A.: An Adaptive Cyber-Physical System Framework for Cyber-Physical Systems Design Automation, pp. 125–140, Chapter 11. In: Applied Cyber-Physical Systems. Suh, S. C., Tanik, U. J., Carbone, J. N., Eroglu, A. (Eds.), Springer Publ. 2014

(Tarumi et al. 2007) Tarumi, S., Kozaki K., Kitamura, Y., Tanaky, H., Mizoguchi, R.: Development of a Design Supporting System for Nano Materials based on a Framework for Integrated Knowledge on Functioning Manufacturing Process. Proceed. 10th IASTED International Conference Intelligent Systems and Control, pp. 446–454, 2007

(Togay 2014) Togay C.: Axiomatic Design Theory for Cyber-Physical System. pp. 85–100, Chapter 8. In: Applied Cyber-Physical Systems. Suh, S. C., Tanik, U. J., Carbone, J. N., Eroglu, A. (Eds.), Springer Publ. 2014

(Valerdi and Wheaton 2005) Valerdi, R., Wheaton, M.: ANSI/EIA 632 As a Standard WBS for COSYSMO, Proceedings 5th Aviation, Technology, Integration, and Operations Conference (ATIO), Arlington, Virginia, 2005

(3SL 2014) Requirements Definition and Management Using Cradle, White Paper 3SL, November 2014

Links

(http1 2015) http://www.iso.org/iso/catalogue_detail?csnumber=43564

(http2 2015) http://www.lifecycleinitiative.org/starting-life-cycle-thinking/life-cycle-approaches/

(http3 2015) http://de.wikipedia.org/wiki/ISO/IEC_15288

(http4 2015) http://en.wikipedia.org/wiki/IDEF0

(http5 2015) http://sixsigmatutorial.com/what-is-six-sigma-quality-function-deployment-qfd-download-free-excel-qfd-template/50/

(http6 2015) http://www.opak-projekt.de/index.php/projektbeschreibung

(http7 2015) http://www.incose.org/practice/fellowsconsensus.aspx

(http8 2015) http://www.cs.ndsu.nodak.edu/~hdo/pdf_files/apsec10.pdf

(http9 2015) www.incose.org

(http10 2015) http://www.chris-kimble.com/Courses/World_Med_MBA/Software_Crisis.html

(http11 2015) http://en.wikipedia.org/wiki/Agile_software_development

(http 12 2015) http://en.wikibooks.org/wiki/Introduction_to_Software_Engineering/Process/Agile_Model

(http13 2015) http://www.inf.ed.ac.uk/teaching/courses/cs2/LectureNotes/CS2Ah/SoftEng/se02.
    pdf
(http14 2015) http://de.wikipedia.org/wiki/Software_Requirements_Specification
(http15 2015) http://europa.eu/rapid/press-release_MEMO-13-448_en.htm

This chapter begins with a brief introduction to manufacturing in Sect. 7.1, identifying the enabling technologies and opportunities with regard to the sequence of industrial revolutions. It also introduces the reader to Digital Manufacturing, referring to smart and agile manufacturing and smart factories, one of the major concepts of *Digital Manufacturing/Industry 4.0*. Section 7.2 introduces the principal concept of individualized production, an important application area in smart factories. Section 7.3 describes networked manufacturing and the concept of smart supply chains that enable the sending of product data over the Internet for service purposes and more. Section 7.4 introduces the paradigm of concurrent open and closed production lines. The topic of Sect. 7.5 is cybersecurity, and Sect. 7.6 introduces several case studies on *Digital Manufacturing/Industry 4.0*. Section 7.7 contains comprehensive questions from the area of *Digital Manufacturing/Industry 4.0*, and followed by references and suggestions for further reading.

7.1 Introduction to Manufacturing

The primary objective of manufacturing is, as described in Sect. 3.4.6, the manufacture of goods that can be sold to customers. Therefore, facilities are constructed to accomplish that goal. Modern manufacturing systems include all of the intermediate processes required for the manufacturing of products and the integration of product components. It should be mentioned that some industries, such as semiconductor and steel manufacturers, use the term fabrication instead of manufacturing.

Appropriate production layouts are designed for manufacturing a specific product and are referred to as flow lines because machines are oriented in such a way that the product can flow from the first machine or workstation to the second, from the second to the third, and so on down the manufacturing line. Raw materials enter the front of the manufacturing line. Processing of the product is complete at the last machine or workstation, with the raw material finally converted into the product of the respective manufacturing line.

© Springer International Publishing Switzerland 2016
D.P.F. Möller, *Guide to Computing Fundamentals in Cyber-Physical Systems*,
Computer Communications and Networks, DOI 10.1007/978-3-319-25178-3_7

Processing at a workstation, however, may often remove part of the raw material. Hence, provision must be made to dispose of such material and any consumable tooling. Therefore, product lines are effective and efficient arrangements for manufacturing when justified by the product mix and volume. The design of the production line takes into consideration how machines are located and maintained, how parts are batched and dispatched, and how performance is measured.

Performance can be measured by using Little's Law. Little's Law is a fundamental law of system dynamics which was introduced in 1961 (Little 1961) and is the most recognized law applied to manufacturing systems. Besides setting consistent targets for staff, this law allows operators to ensure that they get consistent data on the performance of the manufacturing system. In this regard, Little's Law is simple in appearance. It states that for a given area of a manufacturing system, the following equation holds:

$$WIP = PR \cdot TT$$

where *WIP* stands for work in progress, *PR* is the production rate, and *TT* represents the throughput time. The *WIP* levels and *TT* referred to are average values. In the case of steady-state conditions, *WIP* is directly proportional to the *TT*; the proportionality constant for this case is the production rate. Suppose a manufacturing area has a 2-week *TT* and completes 30 jobs per week, then the *WIP* is:

$$WIP = PR \cdot TT = 30*2 = 60 \, \text{jobs}$$

This result can be interpreted such that at any time, the manufacturing systems operator has 2 weeks worth of jobs in the manufacturing area, or 2 * 30 jobs equals the *WIP*. Hence, this simple law allows manufacturing targets to be set for *WIP* and *TT*, establishing a required target for the manufacturing area operator of achieving a 2-week *TT*, manufacturing of 30 jobs per week, and maintaining no more than 60 *WIP* jobs. This example may seem trivial, but in the case of more complex manufacturing systems with many products and many machine work centers, it may not be apparent that set targets are physically inconsistent. In such cases, careful application of Little's Law can be a good checkpoint (Suri 1998).

Let's assume the production rate is X, and there are N jobs in the manufacturing system. This means that each of the workstations is occupied by a job. Every $1/X$ time unit, a new job arrives in the system, and each job in the system advances one place. Each time, the question will be raised, "How long will it take a job to get through the manufacturing system?" The answer can be found by applying Little's Law: spending $1/X$ time units of N workstations, the time T in the system will be

$$T = N\left(\frac{1}{X}\right).$$

Table 7.1 Cycle times of manufacturing cells

Workstation	Cycle time [sec]
1	8
2	10
3	12
4	10
5	8
6	12

Let's assume that a manufacturing area may consist of six workstations. The mean workstation cycle times are shown in Table 7.1. Unlimited work is allowed between the workstations.

Time in system (*TIS*) is defined as the available time per time period divided by customer demand per time. For example, if the available time is 8 h and customer demands is 2.880 units, then *TIS* is

$$TIS = \frac{28.800 \text{ sec}}{2.880 \text{ units}} = 10 \text{ sec}$$

Therefore, no operation can exceed 10 s in order to meet customer demand. If time in system is exceeded, then solutions must be found to improve the process, reallocate work among the operators, or improve the method. A comparison of the workstation cycle times in Table 7.1 with a time in system of 10 s, the cycle time at workstations 3 and 6 should be reduced through a continuous improvement. A Kaizen event could be used to improve the methods or reallocate some of the work content to workstations 1 and 5, to allow the process to meet the customer demand.

The daily production as a function of the number of operators is given in Table 7.2.

From Table 7.2 it can be seen that adding a sixth operator did not increase production and instead increased WIP to 1.350 parts. Therefore, the maximum production and the lowest WIP are achieved with five operators (Schroer et al. 2007).

One common approach to increasing production is to increase capacity by adding additional machines at the bottlenecks at workstations 3 and 6 in Table 7.1 because of the 12 s cycle times. If another machine is added at each of these two workstations, the total cycle time is still 60 s. Assuming six operators, the average time an operator works on a part is 60 s per six operators or 10 s. The average workstation cycle times at workstations 3 and 6 with two machines at each station were 12 s per two operators or 6 s. Production is now 28,800 s/10 s, or 2880 parts, and an average WIP of two parts.

Adding a seventh and eighth operator will not increase production since the maximum workstation cycle time, now workstations 2 and 4, is still 10 s. However, with seven or eight operators, average WIP increased to 810 parts as shown in Table 7.3 (Schroer et al. 2007).

Table 7.2 Production and WIP as function of number of operators

Daily production	WIP	Number of operators
480	0	1
960	0	2
1400	0	3
1920	0	4
2440	0	5
2440	1350	6

Table 7.3 Production and WIP as function on number of operators with two machines at workstation 3 and 6

Daily production	WIP	Number of operators
2800	2	6
2800	810	7
2800	810	8

In manufacturing, assembly lines are distinguished from transfer lines, which are themselves examples of product layouts. Product layouts have the advantage of low throughput time (*TT*) and low work in progress (*WIP*) inventories. *WIP* represents batches of parts and materials that have been released to the shop floor for manufacturing a product, but the product has not yet been completed. Therefore, product layouts are effective in avoiding costs for storage, movement, obsolescence, damage, and recordkeeping. Therefore, product layout implies dedicating the required manufacturing processes to the product (Askin and Standridge 1993).

Furthermore, manufacturing operations are of either a fabrication or assembly nature. Fabrication refers to either the removal of material from a raw stock or a change in its form for the purpose of obtaining a more useful component, etc. Plastic injection molding, aluminum extrusion, steel accessories, turning a diameter, drilling a hole, and bending a flange are examples of fabrication. Assembly refers to the combining of separate parts or raw stock to produce a more valuable unit or device. However, manufacturing systems can also fabricate parts first, and second, use these parts for assembly into products. This requires that we look at how material flows through the manufacturing system and how processes are linked to obtain the desired volume of production at the intended quality, which makes use of concurrent engineering, i.e., performing tasks concurrently. The functionalities of design engineering, manufacturing engineering, and more are integrated to reduce the time required to bring a new product to market.

Embedding digital technology is transforming traditional manufacturing into advanced manufacturing, where robots with sensors are becoming more dexterous, mobile, and aware of their surroundings on factory shop floors. However, these robots need further development so that they will be able to adequately react to information from the environment or changing situations on the manufacturing floor. If two objects on the manufacturing floor do not match, are jamming, or have play, the robot on the floor does not know what to do. Thus, an important goal is for humans and robots to work spatially closer so that the productive robot can learn from the flexible actions of the human worker. This will be accomplished by the robot watching the human worker's activities; recognizing patterns of movement, actions, and gestures; and processing this information. Thus, the robot will imitate the flexibility of the human worker by adapting his/her work sequence and combining it with its own productivity. These advanced robots will become more autonomous with cognitive abilities, enabling them to work in remote areas and then eventually to work as "mechatronic colleagues" in a common workspace with humans. The data the robots generate will be combined with the streams from countless tiny sensors embedded around and in everything in the manufacturing environment. Against this background, the elements of the digital world (hardware, software, networks, and data) are becoming pervasive in the world of manufacturing, and they are doing so quickly, broadly, and deeply, as reported in Westerman et al. (2014).

Since the sequence of industrial revolutions, often called evolutions, manufacturing has been transformative for countries and companies, as shown in Fig. 7.1. In the eighteenth century, the first industrial revolution, Industry 1.0, was characterized by mechanical production powered by water and steam. The industrial revolution in the twentieth century, Industry 2.0, introduced mass production, based on the division of labor and powered by electrical energy. In the 1970s, Industry 3.0 was set in motion by embedded electronics and information technology (IT) for further automation of production. Today, we can pursue the introduction of *Digital Manufacturing/Industry 4.0* based on cyber-physical production systems (CPPS) that combine communication, IT, data, and physical components, transforming traditional manufacturing systems into smart manufacturing systems, described in more detail in Sect. 7.1.1. To achieve this goal, standards are required for solving the problems of the heterogeneous environments of mechanical

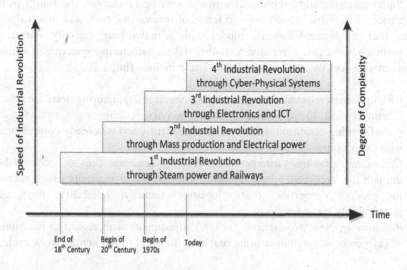

Fig. 7.1 The four stages of industrial revolution

engineering, electrical engineering, information and communication technologies, and cloud computing.

7.1.1   Smart and Agile Manufacturing

Smart manufacturing is the future of manufacturing, merging the virtual and physical worlds through cyber-physical systems (CPS) (see Chap. 3), embedding the fusion of technical and business processes. This new method of industrial manufacturing is known as *Digital Manufacturing/Industry 4.0* or *smart factory*.

Smart manufacturing provides significant real-time quality, time, resources, and cost advantages when compared to traditional manufacturing systems. Thus, smart manufacturing is designed according to sustainable, service-oriented business practices. This approach insists on adaptability, flexibility, self-adaptability and learning characteristics, fault tolerance, and risk management, which is made possible by a flexible network of CPS-based manufacturing units. These units are automatically able to oversee the whole manufacturing process as well as the product status within the smart manufacturing system. This facilitates the creation of flexible manufacturing systems which are able to respond in almost real time and can be optimized according to a global network of adaptive and self-organizing manufacturing units.

Hence, in smart manufacturing, machines are able to talk to products and other manufacturing machines to deliver decision-critical data. This enables all information to be processed and distributed in real time resulting in consolidated changes to the entire manufacturing system through the *Digital Manufacturing/Industry 4.0* paradigm.

Smart manufacturing represents a manufacturing revolution, the fourth in the sequence of industrial revolutions in terms of innovation, cost, and time savings. This bottom-up manufacturing model with a networking capacity introduces numerous advantages, compared to traditional manufacturing systems, in a much more market- or customer-driven form. These include (http1 2015):

- *CPS-optimized manufacturing processes*: Smart manufacturing units are able to determine and identify their area of activity, configuration options, and manufacturing conditions as well as independently and wirelessly communicate with other units.
- *Optimized individual customer product manufacturing*: This is accomplished through the intelligent compilation of ideal manufacturing units which account for product properties, costs, logistics, security, reliability, time, and sustainability considerations.
- *Resource-efficient production*: Tailored adjustments with regard to the human workforce enable manufacturing machines to adapt to the human work cycle.

In 2012, a research report (Lopez Research 2014) published by the World Economic Forum (WEF) and entitled "The Future of Manufacturing: Opportunities to Drive Economic Growth" stated that:

> ...manufacturing has been immensely important to the prosperity of nations, with over 70 percent of the income variations of 128 nations explained by differences in manufactured product export data alone.

The Economist claimed, in 2012, that today a new industrial revolution can be seen which is based on the digitization of manufacturing. Others refer to this as smart manufacturing.

Manufacturing in general has a huge impact on the gross domestic product (GDP), and it is expected to steadily rise from its current level. Against this background, GDP numbers are used to measure the economic performance of a whole country or region but can also measure the relative contribution of an industrial sector. It is estimated that by embedding the key technologies of CPS (see Chap. 3) and the Internet of Things (see Chap. 4) into the manufacturing process, manufacturing will grow with regard to the GDP. Success in a competitive global market like manufacturing, however, depends not only upon continuous product innovation and services but also smart manufacturing processes and, ultimately, smart factories (see Sect. 7.1.2).

As manufacturing shifts toward higher value-added activities, its major contribution is to productivity growth: the sector accounts, for example, for 65 % of Europe's business R&D and 60 % of its productivity growth. This shift toward higher value-added activities correlates with a growing servitization of manufacturing, a strategy of creating value and a greater innovative capacity by adding services to products, as reported by Lopez Research (2014). This transformation of manufacturing toward higher value-added products will result in more innovative and higher-skilled activities. The development of high-value-added activities is increasingly being done through global value chains. Participation in global value chains allows companies (and countries) to specialize in certain activities and to be more efficient, with higher productivity growth. In Europe, many companies participate also or even only in European value chains. These European value chains are integral parts of global European manufacturing chains and allow European manufacturing companies to be globally competitive (Veugelers and Sapir 2013).

In contrast to smart manufacturing, agile manufacturing represents an approach to developing a competitive advantage in today's fast-moving marketplace, where mass markets are fragmented into niche markets. This requires combining organization, workers, and technology into an integrated and coordinated whole. Hence, agile manufacturing places an extremely strong focus on fast response to customer requirements, turning speed and agility into a key competitive advantage. An agile manufacturing company is in a much better position to take advantage of short windows of opportunity and quick changes in customer demand.

With the concept of agile manufacturing, a new paradigm has been introduced in the world of manufacturing, whereby agility is characterized by:

- Cooperativeness and synergism that may result in virtual corporations
- Strategic visions that enable an organization to thrive in face of continuous and unpredictable change
- Responsive creation and delivery of customer-valued, high-quality, mass-customized products and/or services
- Resilient organization structures of a knowledgeable and empowered workforce and facilitated by a high-end information and communication technology infrastructure that links constituent partners within a unified electronic network (Sanchez and Nagi 2001)

Thus, agility in manufacturing can be seen as a strategy to be adopted by manufacturers bracing themselves for performance enhancements that will enable them to become national and international leaders in an increasingly competitive market of fast-changing customer requirements. The need to achieve synergistic competitive advantages in manufacturing without trade-offs is fundamental to the agile paradigm. The agile paradigm is an organizational form that has the processes, tools, and training to enable a manufacturer to respond quickly to customer requirements and market changes while still controlling costs and quality.

The concept of agile manufacturing focuses on rapid response to customer requirements by turning speed and agility into a key competitive advantage. This concept was first introduced in a report by the Iacocca Institute with the title "21st Century Manufacturing Enterprise Strategy" (Iacocca 1991). An enabling factor in becoming an agile manufacturer has been the development of manufacturing support technologies that allow marketing, design, and production personnel to share:

- Common databases of parts and products
- Data on production capacities and problems which are initially assumed to be negligible problems but may have larger downstream effects

Furthermore, agile manufacturing is of particular value for manufacturers in countries with large, well-developed local markets and high labor costs. It leverages proximity to the market by delivering products with an unprecedented level of speed and personalization, which cannot be matched by offshore competitors and will turn local manufacturing into a competitive advantage. There are four key elements of agile manufacturing, as shown in Fig. 7.2 (http2 2015):

- *Modular product design*: Designing products in a modular fashion that enables them to serve as platforms for fast and easy variation
- *Information technology*: Automating the rapid dissemination of digitized information throughout the company to enable lightning fast response to orders

- *Corporate partners*: Creating virtual short-term alliances with other manufacturers that improve time to market for selected product segments
- *Knowledge culture*: Investing in employee training to achieve a culture that supports rapid change and ongoing adaptation

From Fig. 7.2, it can be seen that central to agile manufacturing is the deployment of advanced information and communication technology and the development of resilient organizational structures to support highly skilled, knowledgeable, and empowered staff. Thus, agile manufacturing enterprises are expected to be capable of rapidly responding to changes in customer requirements. With agile manufacturing, customers will not only be able to gain access to products and services but will also be able to easily assess and exploit competencies, thereby enabling them to use these competencies to achieve the things they are seeking (Kidd 1994).

As described in http2 (2015), the use of agile manufacturing for any given business segment should start with answers to the following questions:

- Is there a potential market for a personalized fast-delivery version of one of the company's current products?
- Is there a new product that the company can develop that is within the company's sphere of competence (or alternately that can be codeveloped with a partner) that would strongly benefit from personalization and fast delivery?

In http2 (2015), an example is given of a "3-Day Car Project (in the UK) and a 5-Day Car Project (in the EU) which focus on the idea of transforming automotive manufacturing into a build-to-order system (i.e., each car built for a specific customer order) with delivery times measured in days instead of weeks or months. Considering that the actual manufacturing time for a car is on the order of 1.5 days, this is a realistic goal—although perhaps not yet an attainable goal. But without a

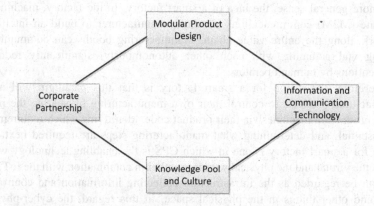

Fig. 7.2 Four key elements of agile manufacturing

doubt, the company that gets there first will have created a significant competitive advantage."

7.1.2 Smart Factory

A core element in the context of *Digital Manufacturing/Industry 4.0* is the smart manufacturing or the intelligent manufacturing system, the so-called *smart factory*. In general, *Digital Manufacturing/Industry 4.0* can be introduced as a sequence of industrial innovations beginning with the first and second industrial revolution, which was based on mechanization; the third industrial revolution, based on automation that focused on lean production; and the fourth industrial revolution, which is embedding the virtualization characterized by CPS, smart machines, smart products, and mobile devices, resulting in the smart factory paradigm.

The smart factory paradigm will allow individual manufacturing, resilient manufacturing, and augmented operators. This means that a smart factory will distinguish itself by a new intensity of socio-technical interaction of all stakeholders and resources involved in the manufacturing process. A smart factory will be centered on the interconnectedness of autonomous, situationally controlled, knowledge-configuring, sensor-based, spatially distributed manufacturing resources, such as manufacturing machines, robots, conveyor and storage systems, and utilities, including their planning and control systems. Thus, a smart factory will be characterized by a consistent engineering approach, incorporating the manufacturing process as well as the manufactured product itself, in which the cyber and the physical world seamlessly interlock. Hence, *Digital Manufacturing/ Industry 4.0* in the context of a smart factory can be interpreted as peer-to-peer information and communication between manufactured products, systems, and machines.

The basic principles behind *Digital Manufacturing/Industry 4.0* are the advent in the outcome of cyber-physical systems (see Chap. 3) and the Internet of Things (IoT) (see Chap. 4). These innovations allow smart manufacturing to be realized or, in a more general sense, the idea of a smart factory. In the factory, machine-to-machine (M2M) communications allow the manufacturer to build an intelligent network along the entire value chain. Manufacturing booths can communicate, sensing and actuating with each other autonomously, significantly reducing interventions by human operators.

Therefore, the vision for a smart factory is that this paradigm will allow manufactured products to control their own manufacturing by sharing the needs of the production machines via their product code, identifying which requirements are essential, and determining what manufacturing steps are required next. The vision for a smart factory is one in which CPS is the enabling technology which brings the virtual and the physical worlds together in combination with the IoT. The IoT can be regarded as the infrastructure collecting information and controlling itself and other things in the physical space. In this regard, the cyber-physical manufacturing system (CPMS) creates synergy among the objects of the virtual and the physical worlds by integrating middleware and cyberware on computer hardware.

Furthermore, smart manufacturing will largely control and optimize itself since the manufactured products communicate with each other and with the machines on the manufacturing shop floor to adapt or clarify the sequence of their production steps. Moreover, a smart factory will also schedule manufacturing priorities on demand, as well as urgent requirements for products in the manufacturing line. To achieve this goal, software agents, who are autonomously acting computer programs, will be able to monitor the requisite processes and ensure that production rules are respected or adapted on demand. Hence, factories will be able to economically manufacture individual products. This will result in the challenge that manufacturing machines monitor themselves; and in case of predicted failures, they will autonomously activate the respective maintenance procedure to eliminate the fault on demand.

In the forecited research record from Lopez Research (2014), it is described that companies such as Bosch, General Electric, and Johnson Controls present an Internet of Things enabled vision where machines predict failures and trigger maintenance processes autonomously rather than by relying on possibly unreliable monitoring by maintenance personnel. Another Internet of Things (IoT) example is shown for a self-organized logistic approach that reacts to unexpected changes in the manufacturing process, such as material shortages, bottlenecks, and others. Thus, manufacturers will use smart IT to realize real-time, dynamic, efficient, automated manufacturing processes.

Smart information-technology-driven processes and the IoT will deliver new added value, as described in Christensen (2015) and Lopez (2014), by connecting:

- *Managers/operators*: Sensors connected through intelligent networks will provide an unprecedented level of visibility into manufacturing operations and supply chain flows allowing deeper insight into the manufacturing line in a smart factory, more than just the high-value processes currently enabled. Therefore, the usage of the IoT paradigm in a smart factory will improve manufacturing by connecting operators with the right information on demand, using the right device at the point of demand and across enterprise boundaries, including supplier data, maintenance partner data, distribution chain data, and more.

  Hence, in *Digital Manufacturing/Industry 4.0*, plant managers/operators will have access to the smart factory cloud with the aim of pooling the information from component suppliers of the production line in order to optimize the supply chain. This will be achieved through software-based apps which run on smart devices. Such smart devices will also allow plant managers/operators to have access to essential data on manufacturing equipment and manufacturing line efficiency and to visualize data or alerts from any location on the manufacturing line at much lower cost than today's customized systems can achieve.

Moreover, in a smart factory, managers/operators will be able to control, regulate, and shape the intelligent cross-linked manufacturing resources and manufacturing steps with regard to situational and context-sensitive targets. Therefore, they will adopt the important role of quality assurance in the smart factory. Henceforth, the increasing process complexity in the smart factory will become manageable resulting in higher productivity, higher quality levels, significantly higher flexibility and resilience, and optimal attainment of resources.

- *M2M/Process communication*: Introducing the IoT into smart factories will allow them to enhance communication to enable faster information flow, faster decision-making, and greater market responsiveness by connecting devices into both operational and business software processes.

  Machine-to-machine communications in particular will enable new levels of automation. An example referred to in Christensen (2015) and Lopez (2014) is about General Motors which uses sensor data to decide if it is too humid to paint automobiles. If the system identifies unfavorable conditions, the automobile is routed to another area in the manufacturing process, reducing repainting and maximizing plant uptime. This change has saved the company much money.

  It is also conceivable that a company's internal cloud could merge with manufacturing locations all over the world, distributing machine operating data. This will allow identification and better coordination of required repairs achieving a more efficient utilization of resources. Ultimately, the individual locations will control themselves. So far, in the cloud, suitable software-based algorithms (apps) have been deposited.

- *Smart data*: The IoT will change the types of devices that are embedded in the smart factory system, connecting physical items, such as sensors, actuators, radio frequency identification (RFID) tags and readers, and others, to the Internet and to each other.

  Smart data includes processing and analytics instead of just big data—big data is a term for data sets so large, dynamic, or complex that they cannot be handled by traditional data processing applications and are characterized by the three Vs of volume, velocity, and variety. The three Vs refer to the steady increase in data volumes, the high speeds at which data is transmitted, and the formats with which data presents itself in regard to storing, analyzing, and displaying the results in an appropriate manner. These are required to analyze data from the IoT-enabled devices and services, which will be the foundation for areas such as forecasting, proactive maintenance, intelligent automation, and overall digitization, otherwise known as innovative manufacturing through data. Thus, smart data refers to the development of new applications based on intelligent or smart analysis of large amounts of data. These applications will turn smart data into executable data for the manufacturing line of the smart factory. Therefore, smart data must not only answer the question: "*What happens right now on the manufacturing line?*" but also, "*What will happen soon on the manufacturing line; and what is needed to run the product mix cost effectively?*"

In general, it is assumed that the IoT will impact every business and has the ability to radically transform industrial businesses, such as manufacturing, business-to-business marketing, and others, as the Internet is an IT that diffuses at exponential rates. In smart factories, there will be an increasing number of machines and devices in the manufacturing line that transmit small and large amounts of data. Hence, manufacturing in a smart factory will embed analytics to enable smarter decisions with regard to more efficient plant operation. The smart factory of the future will link different domains, such as:

- Material requirements planning
- Manufacturing resource planning
- Manufacturing execution systems

These domains will enable the smart factory to achieve optimal results from the manufacturing process, which is part of the common, innovative research work being conducted between industry and research facilities. Since there is no established smart factory for measuring the usage criteria as well as the success of its implementation, research facilities are investigating concepts associated with its development and implementation to provide mechanisms for measurement and control. One of the most advanced concepts is the $SmartFactory^{KL}$. Based on a feasibility study, a nonprofit registered association named "Technology Initiative $SmartFactory^{KL}$" was established in June 2005. The founding partners represented various sectors of the economy and research. Their common objectives were the development, application, and distribution of innovative, industrial plant technologies and the creation of a foundation for their widespread use in research and practice. The partnership has grown to 21 participants and includes producers and users of factory equipment as well as universities and research centers.

As a result of the "Technology Initiative $SmartFactory^{KL}$" project, a hybrid manufacturing facility has been built as a demonstration and development platform. It produces colored liquid soap. The product is manufactured and dispensers are filled, labeled, and delivered per customer order. The plant design is strictly modular, and it consists of a manufacturing process area as well as a piece goods handling area. The machinery and components are identical to those found in modern industrial plants and stem from various manufacturers. The result is a multivendor production and handling facility available for research purposes that is absolutely comparable in its complexity to real manufacturing plants. A photo of the $SmartFactory^{KL\ L}$ facility shown in Figs. 7.3 and 7.4 shows the floor plan of the $SmartFactory^{KL}$ (Zühlke 2008).

The research work with the $SmartFactory^{KL}$ facility focuses on the use of innovative information and communication technologies in automated systems and on the resulting challenges in the design of such systems. Different wireless communication systems are employed in the demonstration facility. As a consequence, a permanent wireless local area network (WLAN) connection has been implemented for the decentralized control systems of the components in the piece goods area to the higher level control center. Bluetooth, ZigBee, ultra-wideband

Fig. 7.3 Photo of the *SmartFactory*^{KL} facility—with permission of (Zühlke 2008)

(UWB), near field communication (NFC), and RFID systems (see Chap. 4) are deployed among the components, which serve as an extended link at the sensor/actuator level. The wireless communication guarantees new freedom in plant layout and reduces the planning effort since cabling is no longer required. However, the robustness of the radio communications in such a heterogeneous environment must always be proven. In this demonstration facility, several monitoring devices are installed to monitor the frequency bands and the quality of service in data transmission. Various sources of interference are used to check the electromagnetic compatibility of the systems.

Wireless communication in combination with modular construction allows the facility to operate according to the "plug'n work" principle. Every task works as a well-defined function within the process chain. Because no physical connections exist between the components other than the power supply, it is relatively simple to replace or add individual components for a modification to or extension of the manufacturing processes. The components recognize their function, position themselves within the process chain, and integrate automatically into the control systems for the manufacturing management. The configuration of the information flow becomes ever simpler because the components identify their tasks from the manufacturing situation and attune themselves to the surrounding manufacturing components.

The logical continuation of the "plug'n work" principle is the transition from traditional function-oriented to service-oriented control architectures (SOA). Using

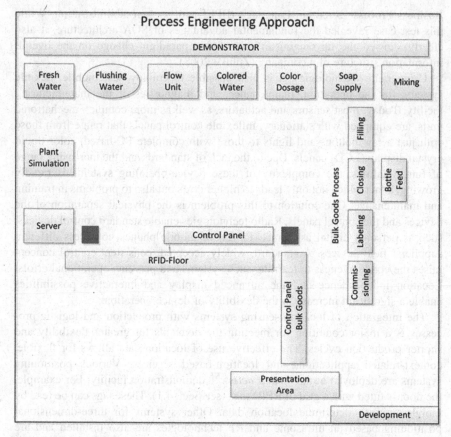

Fig. 7.4 Floor plan of the *SmartFactory^{KL}* facility modified after (Zühlke 2008)

an open SOA allows anyone to access and interact with the building blocks of such an open, service-oriented architecture platform. Thus, SOA enables the development of applications that are built by combining loosely coupled services which interoperate based on a formal definition that is independent of the underlying platform and programming language. Hence, SOA makes it easy for computers connected over a network to cooperate. Every computer can run an arbitrary number of services, and each service is built in a way that ensures that it can exchange information with any other service in the network without human interaction and without the need to make changes to the underlying program itself (Möller 2014).

The *SmartFactory^{KL}* has experimentally converted a subarea of the plant control to the SOA architecture. The purpose of this was to gain experience in the handling of this new architecture for industrial control processes. The present system is based on a Business to Manufacturing Markup Language (B2MML) model according to ISA-95, a Web Services Description Language (WSDL) model, as well as the

Business Process Execution Language (BPEL) for system administration. While this test case revealed the fundamental advantages of SOA architecture, it also clearly showed the far-ranging effects of this paradigm change on the overall information structure of a company (Zühlke 2008).

Using radio technologies, it is also possible to employ new mobile, flexible systems for the operation, maintenance, and diagnostics of the manufacturing facility. Today, most sensors and actuators, as well as more complex mechatronic units, are equipped with stationary, inflexible control panels that range from those with just a few buttons and lights to those with complete PC-based, color liquid crystal display (LCD) panels. Due to the lack of standards and the increasing range of functionalities, the complexity of these device-operating systems is rapidly growing, a fact which not only leads to higher costs but also to problems in training and maintenance. One solution to this problem is the physical separation of the devices and the control panels. Radio technologies enable standard control devices, such as personal digital assistants (PDAs) or mobile phones, to access different suppliers' field devices. A standard, widely accepted, consistent control concept raises the conduciveness to learning such systems and prevents operational errors. Location independence and the advanced display and interactive possibilities enable a significant increase in the flexibility of device operation.

The integration of location sensing systems with production and logistic processes is a major condition for meeting the demands for greater flexibility and shorter production cycles. The effective use of location data allows for flexible, context-related applications and location-based services. Various positioning systems are deployed at the *SmartFactory*^{KL} demonstration facility. For example, the floor is fitted with a grid of RFID tags (see Sect. 4.1). These tags can be read by mobile units to determine location data. Other systems for three-dimensional positioning based on ultrasonic and RF technologies are also installed and are currently being tested, especially in terms of the accuracy achievable under industrial conditions. The installed systems cover the full range of components within the automation pyramid, shown in Fig. 7.5, from field devices (sensors/actuators) and programmable logic controllers (PLC) through process management and manufacturing execution systems (MES) to enterprise resource planning (ERP) software. The entire spectrum of control technologies for industrial manufacturing is represented in the *SmartFactory*^{KL}.

The platform offered by the *SmartFactory*^{KL} served as a basis for research and development in numerous projects with various partners. For example, a demonstrator has been developed that shows the usability of commercial mobile phones for radio-based parameterization of components (Görlich et al. 2007). Using Java software, which runs on the mobile phones of several different brands, it is possible to monitor and configure a multitude of field devices in the *SmartFactory*^{KL}. The available devices and wireless links are automatically identified. Furthermore, a uniform operating philosophy facilitates handling of field devices and enables access to any device from any location on the shop floor, thereby speeding up parameterization, diagnostics, and control of field devices. Rapid switching from one device to another is possible without changing location (Zühlke 2008).

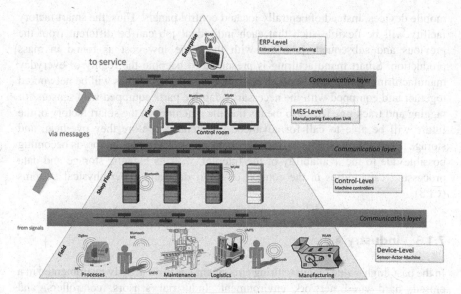

Fig. 7.5 Automation pyramid modified after (Zühlke 2008)

Once the machinery and the systems are connected in the smart factory, their information will be used to automate workflows to maintain and/or to optimize manufacturing lines without human intervention. More and more, the human task in a smart factory will be to monitor the manufacturing process with its huge amount of data based out of which the manager/operator has to create something meaningful. Thus, a characteristic of a smart factory is its interaction with mobile smart devices, such as smartphones or tablet PCs, which the manufacturing manager/operator can use at a particular sensor within the smart factory assembly line. Immediately afterward, production data will be shown on the screen, for example, how many parts were processed and when, which scrap rate was newly formed, and much more. Hence, mobile manager/operator panels and wireless technologies are one of the key factors in the future development of manufacturing automation, such as smart factories (Terwisch and Ganz 2009).

With regard to the success of today's mobile, interactive systems in the consumer sector, the use of smart mobile interaction systems are a promising approach for industrial man-machine interaction. From an economic point of view, the development of smart mobile devices and software platforms strictly for industrial use is justified due to low sales volumes and high development costs. The adaptation and integration of existing hardware from the consumer market, such as smartphones and tablet PCs, as well as software technologies, such as Android, iOS, and Windows, have turned out to be promising for use in enabling novel industrial interaction technologies.

Henceforth, in the smart factory, there will no longer be a central computer. The manufacturing machines and parts produced will control themselves. The manufacturing staff will supervise the whole manufacturing process with smart mobile devices, instead of centrally located control panels. Thus, the smart factory facility will be flexible such that each individual job can be different from the previous and subsequent jobs but with the same low cost as found in mass production. Smart manufacturing is presumed to become the reality of everyday manufacturing in the near future. For this purpose, all machines will be networked together and equipped with the necessary adaptive parts equipped with sensors for tagging and tracking. This also means that the machines in the smart factory of the future will be able to call for advanced maintenance before they overheat, and storage systems will be able to organize themselves. The smart factory is becoming possible due to the availability of the Internet, with its gigantic storage and data processing capabilities in the context of smart data and cyber-physical systems (CPS).

7.1.3 Industry 4.0

In the past, high-value manufacturing equipment has been heavily instrumented in a closed, hard-wired network environment. Industrial sensors, controllers, and networks are expensive, and upgrading projects in existing facilities is not easy. Moreover, growth in the use of the IoT (see Chap. 4) on the consumer side has driven cost reductions in sensors, controllers, and communications through high-volume semiconductor manufacturing, such as smart devices. However, standard industrial equipment is constrained by a huge installed base of legacy equipment and standards. That is why capital, energy, human resources, information, and raw material are acquired, transported, and consumed to transform the material into value-added products and components. To accomplish this goal, the industry is working worldwide to achieve the next level of innovation in industrial production. In Germany, this is promoted under the heading *Industry 4.0*. In other European countries, the USA, China, and many other countries, similar initiatives have been established to pave the way for a customized intelligent industry in the future.

When it comes to manufacturing under the label of *Industry 4.0*, *Digital Manufacturing*, or *smart factory*, the advantages are networking, flexibility, and dynamic, well-organized manufacturing lines for customizable products. This will go hand in hand with the trend of moving from stationary to mobile, universal human-machine interfaces based on modern information and communication technologies which come out of the consumer market in the manufacturing world, as mobile operator panels and wireless technologies and as one of the key factors for the future development of manufacturing automation (Terwiesch and Ganz 2009).

In addition to the new business processes and manufacturing methods based on the Internet of Things, Data, and Services (IoTDaS) are the areas of applied research and development for *Digital Manufacturing/Industry 4.0*. Thus, the progress in industrial manufacturing is ultimately based on a variety of innovative development steps. As introduced in Sect. 7.1, the history of modern industrial society is linked to an innovative development that generally is communicated as the industrial revolutions. The driving forces behind these revolutions were not charismatic political leaders but engineers, scientists, managers, and employees in the manufacturing companies responsible for the technical innovations, called revolutions, in the second half of the nineteenth century. The leading innovators represented organizations in chemistry, the optical industry, and automotive and IT.

The term *Industry 4.0* was publicly introduced for the first time in April of 2011 on the occasion of the Hannover Messe (Hannover Fair) in Germany, the world's largest industrial fair. The objective of *Industry 4.0* is the creation of the smart factory, characterized by its transformation ability, resource efficiency, and ergonomics as well as the integration of customers and business partners into the business and value-added processes. The enabling technologies are CPS (see Chap. 3), which brings the virtual and physical worlds together to build a networked world in which intelligent objects communicate and interact with each other.

As introduced in Chap. 3, CPS represents the next evolutionary step from the current embedding of computer systems and provides the basis for the creation of an IoT, later expanded to the IoTDaS, the fundamental building block of *Industry 4.0*. The *Industry 4.0* approach is based on small networks of computers equipped with sensors and actuators that are embedded (see Chaps. 2 and 5) into materials, articles, and equipment and machine parts and are connected to each other via the Internet. With *Industry 4.0*, the traditional industrial field devices will be replaced by CPS that act as intelligent agents in the IoT and represent the basic framework of a smart factory. Furthermore, in the *Industry 4.0* approach, the plant, machinery, and individual workpieces continuously exchange information through the Internet. All production and logistic processes are embedded in this communication network.

Numerous industrial applications will be derived in the future which incorporate these key technologies, allowing many processes to be coordinated and controlled in real time over long distances. However, this requires the standardization and modularization of many individual manufacturing process steps and programming modules of workable models. This represents a manufacturing revolution in terms of innovation, cost, and time as well as the creation of a bottom-up manufacturing value creation model whose networking capacity creates new and more market opportunities. However, the manufacturing advantages are not limited solely to one-off manufacturing conditions but can also be optimized according to a global network of adaptive and self-organizing manufacturing units belonging to more than one operation.

In this regard, the IoT creates the conditions for the continuous exchange of data from which situationally appropriate, automatic process adjustments are derived. Furthermore, the use of CPS allows control of the process to be decentralized by the products themselves through the processing of environmental data by means of embedded computing systems and deducing control commands. In this way, manufacturing becomes much more flexible. Flexibility is also achieved through application of open machine design that runs different functions or tools in direct chronological order. By using digital product memory (DPM) and storing order-related data directly on the item, the manufactured workpiece manages its own fabrication by choosing the manufacturing processes based on the customer's requirements. The result is a customized product.

The concept of DPM offers a more comprehensive and flexible approach to tagging physical objects with digital life cycle information. Therefore, the benefit of applying DPM to an object strongly depends on its conceptual design and implementation, which is particularly true when domain-specific requirements can be identified. As a working solution for providing value-added services, DPM benefits the product not only during manufacturing but at later stages of its life cycle as well (Stephan et al. 2013).

In addition to DPM, the *Industry 4.0* era will also enable a comprehensive semantic manufacturing memory and an integrated dynamic fine tuning of manufacturing processes along the entire value chain. Self-adjusting process networking and embedded control systems will also be incorporated, creating a model-based fault diagnosis capability. This will guarantee that the information in the company's knowledge base about devices, products, processes, services, and more will be available from any location in real time.

Henceforth, in the *Industry 4.0* era, small quantities of products will be manufactured under mass production conditions, and manufacturing processes themselves will control the product assembly. The *Industry 4.0* era differs from the traditional industrialization revolutions 1.0–3.0 due largely to the integration of the Internet into manufacturing technology. Thus, *Industry 4.0* can be considered to be an open system that does not, as did earlier concepts, rely on a single technology only. Rather, it can be thought of as a toolbox with an almost unlimited variety of different tools that are automatically available. Hence, *Industry 4.0* represents a new paradigm with an intellectual evolution on the manufacturing shop floor, where everything is now programmable through digitization. Thus, the intelligent products manufactured with embedded sensor and communication capabilities are active information carriers which can seamlessly gather and use information over the whole product life cycle. They are addressable and identifiable (see Sect. 6.1.1).

The fourth industrial revolution, *Industry 4.0*, promoted by digitization, has three ancestors through which industrial products always have been given more extensive properties or features, as introduced in Sect. 7.1:

- *First Industrial Revolution*: Enabled by the introduction of the steam engine and the mechanization of manual labor in the eighteenth century.
- *Second Industrial Revolution*: Enabled at the beginning of the twentieth century by mass production made possible by electrification; increased technical features of products such as cars (e.g., mirrors were electrically adjustable and heatable).
- *Third Industrial Revolution*: Enabled through the use of electronics and computer technology for manufacturing and manufacturing automation in recent decades; increased possibilities of technical features of products (e.g., cars could remember various settings for the driver seat and rearview mirrors and restore them with a tap of the finger).
- *Fourth Industrial Revolution*: Enabled by the digitization of the entire value chain and continuous access to information in the form of virtual models. Data

and knowledge will change again anywhere and at any time as will the rules of many industries resulting in increased possibilities of more sophisticated technical features through computerization or through an intellectual revolution on the manufacturing shop floor. Using the example of a car, the following results can be assumed:

- Navigation systems navigate based on the current traffic network data.
- Personal car setup is no longer stored in the car but in the private cloud domain of the user and travels with the user from car to car. It will be interesting to see what happens with customer loyalty, especially in the car rental business.
- Dishwashers and/or washing machines may write a shopping list into the private cloud domain of the user when they run out of detergent and broadcast this to the user's car infotainment or cloud system, leaving a message for the user to drive to the grocery store to buy the needed detergents before driving home.
- Car-2-Car communication talks between cars around (e.g., a car is about to move through the scene of an accident and is warned in time and provided with potential alternative routes).
- Car informs the smart home automation system when the driver is expected to arrive at home and turns on the heating/cooling system as appropriate.
- Error messages are sent directly from the car to the dealer's service workshop to ensure that any spare parts that are needed are available and to confirm the appointment with the dealer workshop for repairs.

In the manufacturing industry of the *Industry 4.0* era, technologies will be developed allowing the manufacturing of small quantities in real time at maximum quality but at a manageably low cost. Manufacturers will be able to profitably serve fluctuating markets and global trends offering a large variety of versions or small series and fulfill individual customer needs on demand. The *SmartFactory*^{KL} project of the *Industry 4.0* era addresses this challenge using adaptive CPS at all levels of manufacturing in order to achieve the introduction, implementation, and operation of integrated manufacturing systems, as shown in Fig. 7.6. Therefore, a special feature is the intelligent analysis of data in smart manufacturing and the usage and service of smart products that leads to smart data from which new or improved products and services can be extracted. Cyber-physical systems and the Internet of Things will be essential features for bringing agility, flexibility, and multiadaptivity to the smart factory of the future.

Therefore, the path to the *Industry 4.0* era will be an innovative and evolutionary one, as it was with previous industrial upheavals in manufacturing technology. In the early 1970s, automation found its way into manufacturing. Customers demanded a greater variety of high-quality products. This became possible through the use of electronics and information technologies in manufacturing. Manual manufacturing steps were taken over by machines, the beginning of the third industrial revolution. Today, we are on the cusp of a fourth industrial revolution, also known as *Digital Manufacturing/Industry 4.0*. Intelligent, networked systems

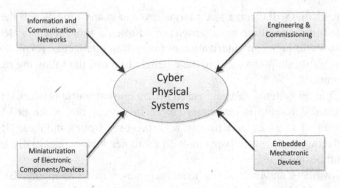

Fig. 7.6 Development toward intelligent technical systems

are expected to usher in this change, the foundation for which was laid years ago by developments in the fields of electronics, software technology, networking, and mechatronics, which resulted in the CPS paradigm.

7.2   Individualized Production

The term mass-customized production was coined in the late 1980s and has become the subject of research concerning operations management (Pine 1993). Therefore, the concept of mass "customerization" in manufacturing and other major trends, such as the growing influence of globalization and the Internet, have large implications for business strategies and for operating manufacturing companies in the twenty-first century.

Mass customerization in manufacturing has been defined by Tseng and Jiao (2001) as "producing goods and services to meet individual customer's needs with near mass production efficiency." In Kaplan and Haenlein (2006), mass customerization in manufacturing is introduced as "a strategy that creates value by some form of company-customer interaction at the fabrication and assembly stage of the operations level to create customized products with production cost and monetary price similar to those of mass-produced products." In McCarthy (2004), mass customerization in manufacturing is defined as "the capability of manufacturing companies to produce a relatively high volume of product options for a relatively large market, which is demanding customized products without tradeoffs in cost, delivery and quality." Implications of mass customerization in manufacturing to the supply chain, concerning information and material flows and the connection between product types and the decoupling point, have been investigated (e.g., by Yang and Burns 2003). Hence, mass customerization manufacturing strategies, such as postponement, can also have effects on customer satisfaction.

In contrast, individualized manufacturing describes a manufacturing process in which all elements of the manufacturing system are designed in such a way that they enable a high level of product variety at mass production costs. Today, manufacturing companies face great challenges as the result of increasing demand for individualized manufacturing. Machinery and systems, such as machine tools, must be able to carry out manufacturing processes with individualized flexibly and to use resources efficiently. At the same time, they need to ensure the quality of the product and its ease of operation.

Up to now, machine tools have mostly required manual adjustments when changing to a manufacturing process for individualized products. Moreover, errors, such as variations in product quality, are not recognized and rectified during the manufacturing process, resulting in machine downtime and production waste. Thus, new technological developments, such as intelligent embedded advanced measurement systems, offer the opportunity to increase the level of machine tool automation in individual manufacturing.

The objective of individualized manufacturing is to develop intelligent machine tools that are capable of autonomously adjusting the machine setup for individual machining processes and checking the quality of the finished workpiece, which requires determining the automation requirements for individual machining processes. This step can be achieved with regard to CPS which allows analysis of a in the future without reducing the reliab workpiece to be processed as well as the work space and automatically adjusts the machine resource for each machining process. The quality of the finished product is checked by automatically comparing it with target data from the computer-aided design (CAD) system. The process can be tested using a demonstration system and then be integrated into the respective machine tools. This will allow machine tools to be more flexible in the future without reducing the reliability of the machining process or the quality of the product. Cyber-physical systems (CPS) offer innovative solutions for the design of manufacturing and the value creation processes and thus the potential for coping with these challenges.

These are the technological approaches of the future that will shape all major manufacturing systems. Thus, in individualized industrial manufacturing, CPS have the potential to cause massive efficiency and productivity gains that are essential for competitiveness in the global market based on the precise integration of all management and control levels within the manufacturing line. Today, the relevant data is automatically passed through from top to bottom. Conversely, data from manufacturing are used to monitor manufacturing processes and to change them if necessary. Depending on the extent of such manufacturing changes, various decision-making levels must be involved. Sophisticated software solutions coordinate the processes and integrate them with a user-friendly interface. The coordination of all manufacturing-relevant processes works across multiple manufacturing sites because the Internet accesses machines, and other manufacturing resources at the manufacturing facilities are interconnected, allowing easy implementation of individualized manufacturing processes. Embedded intelligent sensors not only observe processes and their environment, they also process the data measured and transfer the results obtained back into the loop. This allows monitoring and control of complex manufacturing systems with regard to their specific manufacturing requirements and can also be used to act as an early warning system within individualized manufacturing.

7.3   Networked Manufacturing-Integrated Industry

Technological change, particularly digitization, has dramatically altered the architecture of global manufacturing processes. By facilitating the management and transmission of vast amounts of information, digitization has allowed the codification of highly sophisticated manufacturing processes. Once codified, processes can be split into discrete steps—modules, in effect—and standards to ensure their connectivity can be established.

Modularization, in turn, has permitted activities that once had to be colocated geographically and managed organizationally within the confines of a single company to be spread out across great geographic and organizational expanses. The issue is not that any activity can be done anywhere or that all manufacturing has been completely modularized but rather that new options now exist for structuring activities (Steinfeld 2004).

Thus, networking at the technological and the organizational levels has been recognized as a crucial factor for networked manufacturing in the near future in order to realize efficient value chains. This change in the manufacturing world is characterized by the theme, *Networked Manufacturing-Integrated Industry*, ultimately a self-organizing manufacturing feature. The machines and equipment involved in the manufacturing process, the storage and/or the parts logistics, sensors, and networks communicate with each other. The software used gives the workpieces the necessary intelligence to keep the machines informed of how they are to be processed as well as their actual assembly status within the manufacturing process. This will allow, in the future, unique manufacturing at the cost level of large-scale (mass) production and will be achieved by embedding CPS, the key technology of the *Networked Manufacturing-Integrated Industry*. Processes, products, equipment, and workers will be combined and communicate via the Internet. This future option of *Networked Manufacturing-Integrated Industry* is currently being tested by different manufacturers for application on their manufacturing shop floors.

In order to meet these challenges, one has to take into account the intensive cooperation required between partners from industry, research, and education as well as the role of workers in the networked factory of the future, including the training and continuing education required to provide the necessary technological skills (see Chap. 8). One example is the research project, Open Platform for Autonomous Engineering, Mechatronic Automation Components (OPAK). The objective of this project is to make an assistance function for information and communication technologies available to plant developers and, therefore, allow an engineering work. All of the design steps of a manufacturing plant, beginning with its planning, through commissioning, and all the way to operation and amended commissioning, supported by a suitable novel automation architecture, methods, and tools, should be simplified by the OPAK project (http3 2015). OPAK is supported by the Federal German Ministry of Economics and Energy (BMWi) within the scope of the AUTONOMIK for the *Industry 4.0* program (http4 2015).

The developer will be able to focus on the desired automation process by planning a manufacturing process with the appropriate engineering tools. The automation process will subsequently be turned into a business reality without significant installation, control, and commissioning effort. Instead of working with abstract variables or input-output signals, the engineer will interact directly via a 3D-based engineering interface with perceptible physical plant functions. These entries will be automatically linked to functionally complete, mechatronic automation components which have to be developed. Those components will include everything that is necessary for their operation—from mechanics, electronics, and software—on uniform connections to manual and maintenance information. An electromechanical interface technology for automation components will enable true "plug and produce," also known as the concept of a resilient manufacturing system in the *Industry 4.0* paradigm. This approach will be motivated by a feature-oriented description of the products and capabilities of the manufacturing modules, which enable an optimal plant layout determined through the simulation of the current job situation. By the time an optimization potential is identified, a reconfiguration of the manufacturing system will have been carried out. The plug and produce capability will enable the modules to log into the host computer and transmit their skills. They will be incorporated into the manufacturing process planning, and new capacity, delivery, and performance specifications for manufacturing will be issued. This manufacturing scheduling represents an automated reconfiguration of the manufacturing process depending on the manufacturing orders. In the case of underutilization, available capacity or manufacturing modules can be provided to partner companies for their use with regard to existing agreements. Furthermore, manufacturing lines can be put together even across companies.

Hence, the previously complicated installation, wiring, piping, configuration, and system integration in manufacturing will be reduced to a minimum which represents a paradigm shift from a centralized to decentralized control architecture. This will make future engineering processes more intuitive, faster, and more efficient.

With the advent of the IoT in industrial manufacturing, the large number of components to be networked, with regard to *Networked Manufacturing-Integrated Industry*, requires a corresponding number of Internet addresses. The Internet Protocol, version 6 (IPv6), is the latest IP revision which routes traffic across the Internet. The launch of IPv6 replaces IPv4. IPv6 was developed by the Internet Engineering Task Force (IETF) to overcome the long-anticipated IPv4 address exhaustion problem because IPv6 allows for 128 bit. Hence, IPv6 addresses use eight sets of four hexadecimal addresses (16 bits in each set), separated by a colon (:), like this: xxxx:xxxx:xxxx:xxxx:xxxx:xxxx:xxxx:xxxx (x would be a hexadecimal value). This notation is commonly called string notation.

$$2^{128} = 34\ 02\ 823\ 669\ 093\ 846\ 346\ 337\ 460\ 743\ 176\ 211\ 456\ \text{addresses}$$

This means that the number 34 with 37 numbers behind it marks a significant step into the future. Thus, *Networked Manufacturing-Integrated Industry* is the name of this unprecedented global network. Henceforth, *Networked Manufacturing-Integrated*

Industry supposes that manufacturing machines will be smarter, networked, and collecting the data required to increase productivity and efficiency. Moreover, *Networked Manufacturing-Integrated Industry* is a key title for the convergence of manufacturing industries that communicate using a much stronger interconnection. Interconnection in equipment, material, and specific information exchanges are the characteristics of *Networked Manufacturing-Integrated Industry*.

Smart devices, such as smartphones, tablet PCs, and RFID chips, will become important components in this endeavor. In manufacturing intelligent systems support workers in their manual jobs. The phrase "intelligent components" refers to self-contained, autonomous functioning mechatronic assemblies. In the future, it will be more common for components to organize themselves, thereby taking on jobs at the control level. To realize such intelligent *Networked Manufacturing-Integrated Industry* systems, technologies, such as precision engineering and microsystems technology, have to constantly evolve. In future manufacturing lines, the product will have its data stored in an RFID chip for its lifetime and will be able to pass on its own manufacturing instructions at any time. Each sensor and each actuator within the manufacturing process will have its own IP address and thus will be addressable. Therefore, tailor-made products and highly flexible mass production with the ability to quickly adjust to changing market requirements could be realized. Not only product life cycles getting shorter, but the world outside the factory gates is also changing because the entire logic of the manufacturing process is changing. Intelligent machines and products, storage systems, and resources will consistently interlock with regard to information and communication technology. This will take place along the entire value chain, from logistics to manufacturing and from marketing to service, requiring advanced manufacturing facilities. Moreover, digital networks can work 24/7 around the globe. It is assumed that by 2020, 50 billion components will communicate with each other in *Networked Manufacturing-Integrated Industry* (http5 2015).

Besides all of the positive effects of digitization, however, one has to be aware of the hidden risks and hazards. Already today, increasing prevalence of cybercrime is a serious concern. Last but not least, the evermore complex and interdependent digital technology could lead to new forms of cyberattacks, which have an intrinsic threatening potential. Therefore, adequate answers to these security questions are required without compromising the opportunities of digitization (see Sect. 7.5).

In addition to the development of new technologies, clarification is needed of where humans will be situated in the manufacturing process of the future and how interactions between humans and technology will be organized. This topic includes the issue of training and skills development of the workforce for the future manufacturing world (see Chap. 8).

There are at least four technology elements that provide the foundation for *Networked Manufacturing-Integrated Industry*. These include (but are not limited to):

- *Network*: Cisco research states that only 4 % of the devices on the manufacturing shop floor are actually connected to a network. Many manufacturers have used proprietary networks in the past. A *Networked Manufacturing-Integrated Industry* environment requires a standardized IP-centric network that will enable all devices within a plant to communicate to both operational and enterprise business systems. A standard IP network will also make it easier to connect and collaborate with suppliers and customers to improve supply chain visibility. Manufacturers will need robust networks that can cope with radio frequency challenges in the plant, harsh environmental conditions, and reliability for transmission of alarms and real-time data stream processing. For example, GM implemented a standard-based network architecture, called the Plant Floor Control Network (PFCN), to standardize the design of each plant network and establish a single engineering team that monitors and troubleshoots network operations globally. PFCN helped GM to reduce network downtime by approximately 70 % (Lopez Research 2014).
- *Security*: IT security was the most often cited obstacle to setting up smart factories. Operations managers will need to ensure that safeguards are built into the solution including security procedures, such as hardware encryption, physical building security, and network security for data in transit. The network must also allow secure remote access to systems. Security and networking solutions must also be engineered to withstand harsh environmental conditions, such as heat and moisture, to which typical networks are not subjected. Identity and authentication structures will also need to be updated to support things as well as people.
- *Software systems*: Today's IoT data is different than the data we use to operate systems. It requires collecting a wide range of data from a variety of sensors. These software systems and models must translate information from the physical world into actionable insight that can be used by humans and machines. For example, Toyota is using Rockwell's software for real-time error corrections in the plant. With improved troubleshooting capabilities and error correction, Toyota has minimized rework and scrap rates in its Alabama plant, which has resulted in a tremendous annual cost saving.
- *Smart data instead of big data analytics*: While manufacturers have been generating big data instead of smart data for many years, companies have had limited ability to analyze and effectively use all of the available data. The new smart data concept enabling real-time data stream analysis can provide enormous improvement in real-time problem-solving and cost avoidance. Therefore, it is assumed that smart data and analytics will be the foundation for areas such as forecasting, proactive maintenance, and automation. ConAgra Mills makes 800 different kinds of flour for its customers. It uses "Building Smarter Manufacturing with the Internet of Things" predictive tools and services to forecast pricing, capacity requirements, and customer demand. This allowed the company to maximize revenues through improved margin decisions and increase production capacity utilization by 5 %.

7.4   Open and Closed Manufacturing Lines

As introduced in Sect. 7.1, a manufacturing line is a set of sequential operations established in a manufacturing environment, whereby parts are put through a refining process to manufacture a finished product suitable for consumption or components which are assembled to make a finished product. Manufacturing in large manufacturing environments often involves transporting parts from one manufacturing shop floor to another on carriers. In a case where the number of parts in the manufacturing line is bound by the number of carriers, this manufacturing line is called closed with regard to carriers.

A closed manufacturing line with M machines is shown in Fig. 7.7, where the empty carrier buffer b_0 has a capacity of C_0, and the number of carriers in the closed manufacturing line is S.

Since in a closed manufacturing line the first machine can be starved for carriers and the last be blocked by b_0, the production rate PR_{CML} of the closed manufacturing line is, at best, equal to that of the open manufacturing line PR_{OML}. In a case where C_0 or S or both are chosen inappropriately, the behavior of the closed manufacturing line impedes performance, and, as a result, PR_{CML} can be substantially lower than PR_{OML} (Li and Meerkow 2008). Closed manufacturing lines have been analyzed in the literature under the assumption that the machines obey either the Bernoulli or exponential reliability models (Biller et al. 2008).

A Bernoulli line is a synchronous line with all machines having identical cycle time which can be represented by a vector $(p_1, \ldots, p_M, N_1, \ldots, N_{M-1})$ of machine reliability parameters and buffer capacities. It is a slotted time model with the cycle time τ of the machines. The status of each machine is determined at the beginning, and the state of the buffers is determined at the end of each time slot. The status of a machine is UP with probability p_i and DOWN with probability $(1-p_i)$, and it is independent of past history and the status of the remaining system. An UP machine is blocked if its downstream buffer is full at the end of the previous time slot, and the downstream machine cannot produce. It is starved if its upstream buffer is empty at the end of the previous time slot. At the end of a time slot, an UP machine that is neither blocked nor starved removes one part from its upstream buffer and adds one part in its downstream buffer. The first machine is never starved; the last machine is never blocked (Li and Meerkow 2008).

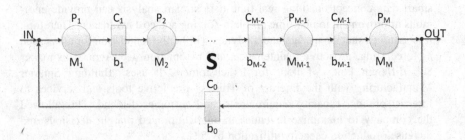

Fig. 7.7 Closed serial manufacturing line

The exponential distribution is a simple distribution with only one parameter and is commonly used to model reliability data.

Let's assume that the processing times at machine i, $i = 1, 2, \ldots, m$ are exponential with parameters μ_i and in front of each machine there is a buffer of size $N_i - 1$. Jobs are circulating on carriers to keep jobs in a fixed orientation and make jobs easier to handle for transportation. As soon as a job is finished at machine m, it is removed from the carrier, and a new job is immediately placed on the carrier after which it returns to machine 1. The number of circulating carriers is $b_j, j = 0, 1, \ldots, I$, as shown in Fig. 7.7. The number of carriers and their capacity affects the throughput of the closed manufacturing line. If the number and capacity of carriers is small, the throughput will be low. If the number and capacity of carriers is large, a high throughput can be expected.

Let us assume that the closed manufacturing line is based on only two machines which operate under the communication blocking protocol (http6 2015) with the constraint that $M_1 \geq M_2$. Thus, this closed manufacturing line can be described by a Markov chain in which k is the number of jobs at machine 1.

The Markov chain itself is a random process that undergoes transitions from one state to another in a state space. It must possess a property that is usually characterized as memorylessness, which means that the probability distribution of the next state depends only on the current state and not on the sequence of events that preceded it. This specific kind of memorylessness is called a Markov property. Markov chains have many applications as statistical models of real-world processes.

Let p_k be the equilibrium probability of state k. In determining these probabilities, one has to distinguish between several cases. If $m \leq M_2$, then there is no blocking at all, as shown in the flow diagram in Fig. 7.8.

Thus, it follows that

$$p_k = \left(\frac{\lambda_2}{\lambda_1}\right)^k \frac{1 - \frac{\lambda_2}{\lambda_1}}{1 - \left(\frac{\lambda_2}{\lambda_1}\right)^{m+1}}$$

where $k = 0, 1, \ldots, n$. If $M_2 < m \leq M_1$, then Machine 1 may be blocked because the buffer of Machine 2 is full. The possible states are $m - M_2, m - M_2 + 1, \ldots, m$ and the probabilities satisfy

$$p_k = \left(\frac{\lambda_2}{\lambda_1}\right)^{k-(m+M_2)} \frac{1 - \frac{\lambda_2}{\lambda_1}}{1 - \left(\frac{\lambda_2}{\lambda_1}\right)^{M_2+1}}$$

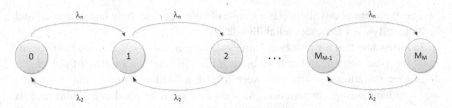

Fig. 7.8 Flow diagram for two machine-closed manufacturing lines with $m \leq M_2$

7.5   Cybersecurity in Digital Manufacturing/Industry 4.0

The protection and sustainable growth of cyberspace is essential for the advancement of society. Cybersecurity can be defined as a body of knowledge with regard to technologies, processes, and practices designed to protect the networks, computers, programs, and data from attack, damage, or unauthorized access. The importance of cybersecurity was illustrated by an article in the *New York Times* in March of 2011 describing how researchers were able to hack a car remotely and take control of the car's critical systems. This was accomplished through the car's embedded communication systems as many of today's cars contain cellular connections and Bluetooth wireless technology (Fahmida 2011). This make it possible for a hacker, working from a remote location, to hijack various features, such as the car door locks and brakes, as well as to track the vehicle's location, eavesdrop on its cabin, and monitor vehicle data.

This hack demonstrates how cyberspace can be used to affect physical processes beyond the cyberspace (Brazell 2014). The same situation can happen in *Digital Manufacturing/Industry 4.0* if a hacker, working from a remote location, hijacks various features of the manufacturing shop floor. This type of hack can have a tremendous influence on the manufacturing process by stopping a whole manufacturing line, canceling assembly steps, or changing the sequence of assembly steps in the manufacturing system. These are just some of the examples which illustrate why cybersecurity is required when coordinating efforts throughout a networked manufacturing system.

The elements of cybersecurity include (http7 2015):

- Application security
- Information security
- Network security
- Disaster recovery/business continuity planning
- End-user education

Nevertheless, one of the most problematic elements of cybersecurity is the fast and constantly evolving nature of security risks. The traditional approach has been to focus the most resources on the most crucial system components and protect against the biggest known threats, leaving some less important system components undefended and systems exposed to some less dangerous risks. Such an approach is insufficient for the current cyber-physical environment. Cybersecurity professionals assume that the traditional approaches to securing CPS information can become unmanageable because the threat environment has become impossibly complex. Hence, CPS has been identified as vulnerable to cyberattacks because of their network-based accessibility which makes them vulnerable to remote. The consolidation of cyber and physical components within CPS enables new categories of vulnerability with regard to interception, replacement, or removal of information from the communication channels resulting in malicious attempts by cyberattackers to capture, disrupt, defect, or fail the CPS operations. The reason for this new vulnerability can be traced to the way in which the cyber and the physical components of CPS are integrated (Karim and Phoha 2014). In this vulnerable space, the cyber component provides computing/processing, control software, and sensory support and facilitates the analysis of data received from various sources and the overall operationalizability of the CPS. As described in Karim and Phoha (2014), the remote network access facilitates highly productive interaction among the various physically distributed or concurrent collaborating units of a CPS, and the efficient system administration is an integral part of the cyber component which allows accessibility. However, this accessibility provides an entrance for launching cyberattacks which can result in:

- Defective operation if the attack affects the control loop
- Denial of service, which is common in the cyber domain
- Destruction and exfiltration
- Information corruption

These attacks not only have tremendous impact on the cyber part but the overall CPS.

In a case of defective operation, a solution for survivability under such a cyberattack has been reported by Cárdenas et al. (2008). Denial of service occurs when a cyberattack creates an artificial mechanism that keeps the targeted system unnecessarily busy, delaying or denying services to legitimate requests (Mirkovic et al. 2005). Destruction and exfiltration, information corruption, and defective operation of a system can be avoided if a system compromise can be detected and eliminated. As reported in Karim and Phoha (2014), numerous static, dynamic, and hybrid solutions are available that analyze patterns and signatures in program codes and the behavior of program executions in order to identify the presence of malicious agents in the system, helping system administrators to disable them (Dinaburg et al. 2008; Sharif et al. 2008; Willems et al. 2007; Moser et al. 2007).

The physical components of a CPS, as dispersed physical infrastructures, are the areas impacted. They are monitored by sensor networks that serve as an early warning system while detecting malfunctions or damage. These sensor networks consist of many tiny components, each of which is subject to physical capture. A cyberattack can remove or destroy the sensors from the region of impact or the region of interest, creating a hole in the security coverage which results in a disruption of the transmission of critical data. The cyberattack can also corrupt or replace sensors and inject erroneous data into the system causing the decision-making sensors that depends on that data to fail. Various schemes, such as the probabilistic dependence graph and anomaly detection, have been proposed for detecting holes in the coverage or identifying compromised sensor nodes by detecting the absence, corruption, or replay of sensor data leading to the detection of manipulation from the outside (Karim and Phoha 2014).

The probabilistic dependence graph approach (He and Zhang 2010) illustrates the connections using a Markov random field (MRF) that is induced by a minimal neighborhood system by inserting an edge between sites that are neighbors (Li 2009). Thus, it utilizes spatially correlated information from phasor measurement units and statistical hypothesis testing. Let's assume that a Gaussian Markov random field (GMRF) is employed to model phasor angles across the integration space (buses which connect the cyber and the physical components) in a way that the phasor angles are evaluated as random variables and their dependencies can be studied (Rue and Held 2005). The pairwise Markov property of a GMRF is exercised in such a way that an MRF is normal with the mean u and the variance J^{-1} where J is the information matrix of the Markov random field; hence, J is zero (Landrum et al. 2014). Thus, the probabilistic dependence graph can be used for fault detection and localization because dependability can be measured to ensure a function as anticipated by the CPS.

Anomaly detection is used to identify cyberattacks for the protection of a CPS. First, the intrusion method must be identified so that the regular operation of the CPS will remain undisturbed. Cyberattacks can be conducted in various ways. To become aware of an intrusion on components or subsystems of a CPS by cyberattackers, an anomaly detection algorithm must be employed. Its benefits include real-time monitoring, analysis of possible effects of intrusion, and approaches for mitigation (Ten et al. 2011). Monitoring the CPS in real time enables the algorithm to rapidly and efficiently determine the status of the computing/processing units, the sensors, and other application-specific components in order to allow a maximum number of connections to be implemented and to authenticate the connections via response time and IP address. One can detect and track abnormalities such as unsuccessful logon attempts in accordance with time and frequency and destructive modifications to files which are vital to the components or subsystem and ascertain behavior. These are characteristics of an intrusion being attempted, and if a cyberattack is suspected, an alarm list of possible attackers is created, and the component or subsystem the intruder is attempting to attack is locked (Landrum et al. 2014).

CPS that are used in *Digital Manufacturing/Industry 4.0* solve many pressing needs but are also vulnerable to conditions that can cause various types of damage or manipulation of mission-critical tasks in the *Digital Manufacturing/Industry 4.0* system. Therefore, mission-critical tasks require a continuously active CPS that will essentially never fail because vulnerability to conditions can cause mission-critical systems to fail. Hence, two main concerns have to be considered for CPS:

- Safety
- Security

Safety is indispensable for mission-critical systems within the overall *Digital Manufacturing/Industry 4.0* system. For optimal safety, one has to focus on the safety of the interaction between the computing/processing devices and the physical components because the interactions of the distributed computing/processing devices interfere with each other. In addition, the influence of physical phenomena resulting from the conditions being monitored could affect the functionality of the cyber devices. These kinds of safety issues need to be addressed and prevented in the design of CPS (Guy et al. 2014).

Security is defined as the *"ability to ensure that both data and the operational capabilities of the system can only be accessed when authorized"* (Banerjee et al. 2012). Hence, unauthorized access to any component of the CPS poses an obvious security threat. With regard to the increasing use of CPS in industry for mission-critical operations in *Digital Manufacturing/Industry 4.0*, it is imperative that security is always at the design forefront.

It is stated, in the cybersecurity report by Harris (2014), that in 2012, 50 % of all targeted cyberattacks were aimed at businesses with fewer than 2500 employees. More significantly, businesses with fewer than 250 employees were the target of 31 % of all cyberattacks. But some small businesses may also have access to their business partner's computing/processing systems as part of an integrated supply chain or to sensitive data and intellectual property. Though it can be argued that the rewards of attacking a small business are less than what can be gained from a large enterprise, this is offset by the fact that most small businesses dedicate fewer resources to protecting their information assets and are, therefore, easy targets for cyberattacks or cybercrime.

Meanwhile, several cybersecurity standards, which are digital security techniques to prevent or mitigate cyberattacks, have been developed because sensitive information is now frequently stored on computing/processing devices that are attached to the Internet. For certain standards, cybersecurity certification by an accredited body can be obtained. There are many advantages to obtaining certification including the ability to get cybersecurity insurance.

ISO/IEC 27001:2013 is part of the growing ISO/IEC 27000 family of standards, information security management system standards, published in October 2013 by the International Organization for Standardization (ISO) and the International Electrotechnical Commission (IEC). ISO/IEC 27001:2013 formally specifies a management system that is intended to get information security under explicit management control. In addition to the ISO/IEC 27001, ISO/IEC 27002 is a high-level guide to cybersecurity. It is most beneficial as a guide for the management of an organization to obtain certification to the ISO 27001 standard. The certification, once obtained, lasts 3 years. Depending on the auditing organization, none or some intermediate audits may be carried out during the 3 years (http8 2015).

ISO/IEC 21827 (SSE-CMM–ISO/IEC 21827) is an international standard based on the Systems Security Engineering Capability Maturity Model (SSE-CMM) that can measure the maturity of ISO control objectives (http9 2015).

The Standard of Good Practice for Information Security, published by the Information Security Forum (ISF), is a business-focused, practical, comprehensive guide to identifying and managing information security risks in organizations and their supply chains. The 2011 Standard of Good Practice is the most significant update of the standard in 4 years. It includes information security hot topics such as consumer devices, critical infrastructure, cybercrime attacks, office equipment, spreadsheets and databases, and cloud computing. The 2011 Standard of Good Practice is aligned with the requirements for an Information Security Management System set out in the ISO/IEC 27000 series standards and provides wider and deeper coverage of ISO/IEC 27002 control topics, as well as cloud computing, information leakage, consumer devices, and security governance (http8 2015).

The North American Electric Reliability Association (NERC) 1300 standard is referred to as Critical Infrastructure Protection (CIP) 002-3 through CIP-009-3. These standards are used to secure bulk electric systems although NERC has created standards within other areas. The bulk electric system standards also provide network security administration while still supporting best practice industry processes (http10 2015).

The International Society of Automation (ISA) Security Compliance Institute (ISCI) operates the first conformity assessment scheme for IEC 62443 Industrial and Automation Control Systems (IACS) cybersecurity standards. This program certifies commercial off-the-shelf (COTS) IACS products and systems, which address securing the IACS supply chain (http11 2015).

The National Institute of Standards and Technology (NIST) special publication 800-12 provides a broad overview of computer security and control areas. It also emphasizes the importance of the security controls and ways to implement them. Initially, this document was aimed at the federal government, although most practices in this document can be applied to the private sector as well (Guttman and Roback 1995).

Henceforth, security in manufacturing CPS requires generalized security methods which can involve compensating for accidentally observable actions to enforce the security properties of information flow in a manufacturing CPS. This means that the relevant manufacturing CPS adheres to an information flow property (IFP) P. However, sometimes there is an event causing a violation of P and thereby causing an effect which can be interpreted as being outside the projected manufacturing schedule. This causes the event to be posted as a potential manufacturing security risk. If this event is identified, the security system immediately inserts a compensating event or a sequence of events into the series of manufacturing processes currently being executed. Any individual process of this correcting event chain may violate P, but the chain is executed quickly enough that the violation is not noticeable. At the conclusion of the compensating event chain, an observer would see that P is adhered to as if the violating event never occurred (Guy et al. 2014).

7.6 Case Studies in Digital Manufacturing/Industry 4.0

This section presents a selection of special case studies that introduce industrial applications and national research agendas which are facing changes in today's manufacturing operations, and, therefore, these institutions/industries have started to think about how to implement *Digital Manufacturing/Industry 4.0*.

7.6.1 Digital Manufacturing/Industry 4.0: The Hannover Centre for Production Engineering (PZH) Approach

With the advent of the Internet of Things in manufacturing, a new industrial age began. Whether it is a revolution or an evolution is yet to be decided. It is certain that with the progressive digitalization in various areas of daily life, a fundamental change is taking place. Therefore, the Internet, in combination with other modern technologies, increasingly impresses the manufacturing industry. Global companies in all sectors and sizes with a variety of topics and approaches are employing digitalization to secure their competitiveness and be able to offer better products in the near future. In this regard, Germany has the potential to become an international market leader in digitalization in conjunction with *Industry 4.0*. For small and medium enterprises, the technology required for this purpose, however, is still in its infancy. Therefore, the recently founded platform *Industry 4.0* has initially formed five working groups, in which representatives from business, academia, associations, trade unions, and ministries are involved (http12 2015):

- WG 1 Reference architectures, standards, and standardization
- WG 2 Research and innovation
- WG 3 Security of networked systems
- WG 4 Legal framework
- WG 5 Work, education, and training

It offers consulting services on issues in the fields of standardization, security, networking, system integration, and the establishment of demonstration centers which are planned by the initiative "Mittelstand 4.0" (http13 2015) of the Federal Ministry of Economic Affairs and Energy. For this purpose, five information and demonstration centers are being created in 2016 throughout Germany (http14 2015). The aim of the new demonstration centers is to support small- and medium-sized enterprises (SMEs) in digital transformation. The new initiative is designed to strengthen the competitiveness of SMEs and craft enterprises and to develop new businesses in the context of digitalization and *Industry 4.0*.

One of those five centers is located in the Hannover Centre for Production Engineering (PZH), which belongs to the Leibniz Universität Hannover. The institutes housed at the PZH and the Institute of Integrated Production Hannover (IPH) bring together their expertise along the entire supply chain production in the center. Additional expertise, such as IT security or law, is incorporated by the

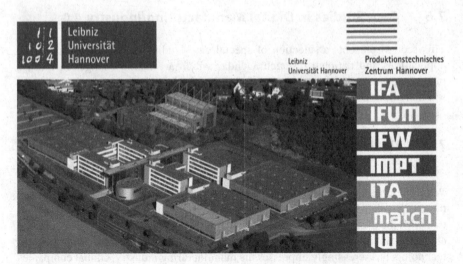

Fig. 7.9 Hannover Centre for Production Engineering (PZH)) Campus with its institutes

partners. The resulting portfolio of the PZH spans the fields of information, qualification, consulting, and implementation. The campus of the Hannover Centre for Production Engineering (PZH) is shown in Fig. 7.9.

The PZH is one of the most important research centers for production technology—nationwide and internationally. Meanwhile seven research institutes of the Leibniz Universität Hannover are located at the PZH with about 250 scientists, mostly from the engineering and natural sciences. Under the umbrella of the PZH, employees of these institutions not only bring in their professional disciplines, they also apply to the engineering service of TEWISS GmbH and the numerous small- and medium-sized, production-related companies, many of which are spin-offs from the PZH institutes. Activities are equitably divided between basic research, diverse research collaborations with industrial partners, and practical teaching. Current examples of research being conducted in the institutes and an overview of activities at PZH are described in the annually published magazine of the PZH, which each also contains the annual report from the previous year (http15 2015).

One of the flagship projects of the PZH is the Collaborative Research Centre CRC 653 "Gentelligent Components in Their Lifecycle—Utilization of Inheritable Component Information in Product Engineering," founded in 2005, where fundamentals of *Industry 4.0* were developed before this designation was introduced in 2011. With regard to the innovative research conducted in the CRC 653, components that intrinsically store information on their own production and find their way through the processing steps without outside control, landing gears that monitor their own condition autonomously and call for an inspection if necessary, will soon become a reality. These components, as well as innovative concepts, methods, and techniques for their manufacturing and utilization in production engineering, will be developed as a result of the interdisciplinary research

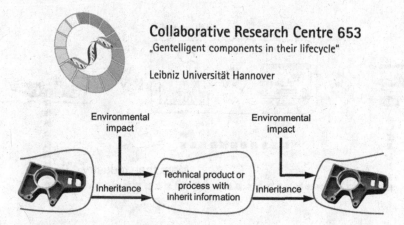

Fig. 7.10 Illustration of the long-term objective in the CRC 653—with permission from the Institute of Manufacturing Engineering and Machine Tools (IFW)

in the CRC 653 at the PZH. The interdisciplinary long-term objective in the CRC 653 is the integration of components with their corresponding information for reproduction as well as stress information from their life cycle. This idea is summarized in Fig. 7.10.

The CRC 653 and most activities concerning *Industry 4.0* are mainly driven by the Institute of Manufacturing Engineering and Machine Tools (IFW), one of the institutes of the PZH. Concrete solutions, such as the "feeling" machine tool the IFW has been developing for more than 10 years, make production planning and control much more efficient. The machine is equipped with additional intelligence to detect manufacturing inaccuracies and own status information which is constantly fed back to production planning and control in order to make these steps more accurate especially during the ramp-up phase of new products being manufactured. The premise of the IFW is that connecting intelligent devices to each other should not be an end in and of itself. So the first step is to analyze the process within an enterprise that should be optimized. The second step is to determine how *Industry 4.0* can help achieve this optimization. This premise is also the basis for the Production Innovations Network (PIN), founded by the IFW in 2015, which should strengthen cooperation between enterprises regarding *Industry 4.0*.

A popular example of a research scientist application from the CRC 653 is the converted formula student race car that stores loads of races around the wheel and makes them very simple to use, among other things, for maintenance planning, as shown in Fig. 7.11.

7.6.1.1 Interactive Maintenance

According to the *Maintenance Efficiency Report 2013*, only 55 % of the turnaround of maintenance activities are value-adding processes. Because of the wait time for tools or components and an insufficient organization of complex processes, the

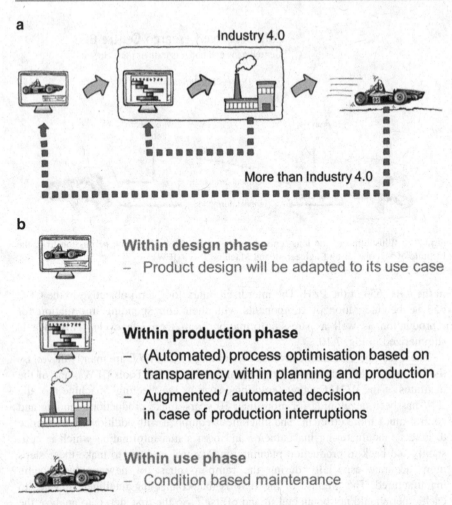

Fig. 7.11 Embedding Industry 4.0 into the student formula racing car approach (**a**) and the respective phases between design, production, and use (**b**)—with permission from the Institute of Manufacturing Engineering and Machine Tools (IFW)

work on the commission is delayed. In order to minimize the amount of additional work, the IFW is developing a mobile platform for interactive maintenance that gathers useful information, such as error causes, and resources, such as tools, for the worker, as shown in Fig. 7.12. Additionally, the personalized display of work steps can be used to train inexperienced workers. After the realization of the interactive maintenance platform, they expect an increase of value-added processes to a minimum of 79 %. This meets the current standards for Best Practice Applications in the industry.

In 2015, the IFW realized that the demonstrator platform, digitalized repair and maintenance assistant (DRIA), identifies machines through QR codes. Workers get

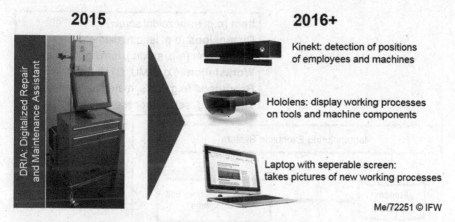

Fig. 7.12 Interactive maintenance platform—with permission from the Institute of Manufacturing Engineering and Machine Tools (IFW)

access to this information through a password. If a problem cannot be solved with the available data, the worker can be supported by colleagues via a webcam-based system. The tools necessary for the solution of a problem are provided in shadow boards.

Currently, the IFW recognizes the following development potential:

1. Kinekt: Precise localization of the exact position of a worker and the machine for a more detailed representation of information for work processes
2. Hololens: Display of tool- or machine-relevant data on the relevant components (e.g., data from the production)
3. Laptop with detachable display: Documentation of operations with an integrated camera
4. Developer kits from Kinekt and Hololens: Programming of new apps (e.g., to make DRIA even more person specific)
5. Database/cloud: Centralized storage of work processes and workers' skills

In the medium term, the IFW plans the programming of specialized manufacturing application software. This software summarizes the current functional combination of different programs (e.g., ILIAS).

7.6.1.2 Process Planning 4.0
In the context of *Industry 4.0*, the IFW is working on the next step, the so-called Process Planning 4.0, as shown in Fig. 7.13. Listed below are the steps outlined to achieve this goal.

Step 1. Problem Definition:
• High level of effort on initial process planning and start-up of the MES (e.g., enter plan data, such as schedule times and work plans).

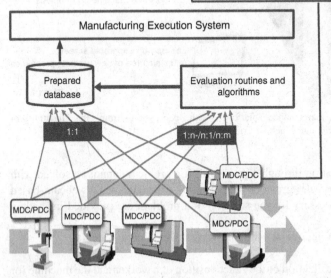

Item (e.g. trapezoidal screw)
Dimensions (e.g. length, diameter)
Operations (e.g. sawing, turning, milling)
Workstation (e.g. DMU, CTV)
Resources (e.g. tools, materials, staff)
Times (e.g. setup, process, recovery)

Fig. 7.13 The Process Planning 4.0 as part of the Industry 4.0 paradigm—with permission from the Institute of Manufacturing Engineering and Machine Tools (IFW)

- Autonomous process planning presupposes knowledge about the production process for individual features.
- Change of plan data over time (e.g., by learning effects, decentralized revision of work plans).
- Static plan data in computer-aided planning systems.
- High level of maintenance effort, imprecise planning, lack of acceptance.
- Data form production data collection (PDC) and machine data collection (MDC) remains unused.

Step 2. Goals:

- Development of a planning approach for dynamic learning process planning to provide the essential foundation for an effective and operational application
- Autonomous, dynamic learning process planning by using production data
- Reduction of the start-up and maintenance effort for the MES
- Increase in the accuracy of planning, increase in acceptance

Step 3. Solution:

- Development of methods for dynamic learning process planning with the help of real-time data and feedback on an MES with no longer sufficient validity

- Detection of production type of individual features on the basis of machine data and NC code
- Repatriation in the ontology or NC programs (e.g., for tool paths)
- Integration of quality characteristics for verification
- Prototypical implementation and operational use at industrial partners

7.6.1.3 Feeling Machine

The subproject N1 represents the central demonstration project of the SFB 653. The findings of the systematic investigation of "Gentelligent Technologies" within functional machine tool components are obtained with regard to the implementation, use, and potential performance. Priorities of the subproject are searching for "Gentelligent Components," the interaction between these components and the inheritance of life cycle information. With the proposed research, the vision is to realize a "feeling or sentient" machine. The objective is to detect any process, machine, or workpiece state with the aid of distributed, diverse, inherent sensory abilities. These features will continuously improve the process model for online process rating and shape error prediction. The approach uses measured machining process data and a simultaneous cutting simulation. The applied data sources can be divided into primary and secondary data sources. Primary sources provide data that is used without further transformation. Process forces, position information of the machine tool, and surface samples provided by a measuring sensor are the primary sources used within this approach. Secondary or augmented data sources utilize primary data to generate additional information that is based on a cutting simulation. The simulation is driven by streaming position values from the machine tool and determines effective cutting conditions, such as depth of cut, width of cut, and material removal rate. All obtained and generated data sets are equipped with a time stamp which allows correlating the system state in discrete steps, as shown in Fig. 7.14.

It can be seen in Fig. 7.14 that the aggregated process data serve as a training data set to build up a multiparameter process model. So far, the model is based on a support vector machine and achieves a mean standard error of 5 % for the parameters considered. After the model is trained, it is used to rate cutting processes online. The model is also used to predict the shape error for simulated processes with simulated forces or no forces at all. The prediction accuracy is continuously increased by refining the model by recent data and process conditions.

7.6.1.4 Conclusion

With regard to the SFB 653 research work objective, its outcome, and the recently founded PIN network, the knowledge transfer is provided on a broader basis, pushing forward the issue of *Industry 4.0*. The SFB 653 spokesman and PIN initiator, Prof. Denkena, is convinced:

"We are doing research since at least ten years at the forefront of the topic—and the interaction of ten different institutions with their respective expertise. If we now can build a strong network of companies, in which not only a bilateral experience, knowledge and

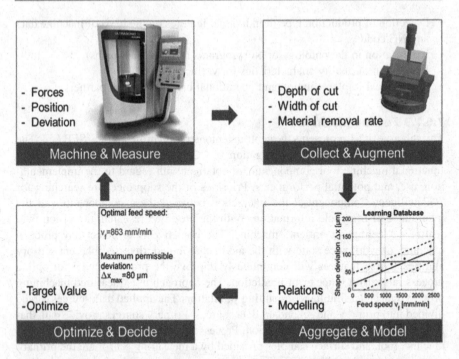

Fig. 7.14 The feeling machine approach—with permission from the Institute of Manufacturing Engineering and Machine Tools (IFW)

research exchange takes place, but the company also stimulate each other, this will be a very strong future boost for everyone involved."

7.6.2 Digital Manufacturing/Industry 4.0: The Steel Industry Approach

Steelmaking is the manufacturing process used to produce steel from iron and ferrous ores and scrap. In steelmaking, impurities, such as nitrogen, silicon, phosphorus, sulfur, and excess carbon, are removed from the raw iron, and alloying elements, such as manganese, nickel, chromium, and vanadium, are added to produce different quality grades of steel. In regard to international competition, the steel industry continues to modernize its manufacturing processes, thus increasing productivity and quality. Hence, investment in modern equipment and employee training has transformed the steel industry.

Over the next decade, the Internet of Things, Data, and Services (IoTDaS) will combine people, processes, data, and things. If all of this comes together, it will open up new possibilities for the future. Cyber-physical system, a system of collaborating computational elements, will be the driver for innovations in the steel industry. Autonomous, self-organizing manufacturing processes, as well as cloud-based data support and services, will constitute the platform for new services and business models. Cyber-physical systems (see Chap. 3) in this sense can be introduced as a strong digital platform—well structured and well integrated and only as complex as absolutely necessary with regard to the designated application of the manufacturing process. Thus, CPS-based manufacturing processes will independently and autonomously control, depending on external requirements, manufacturing production lines. In addition, the chain of economic value-added, tailored, on demand, real-time manufacturing processes can be optimized.

The *Digital Manufacturing/Industry 4.0* paradigm shift will enable solutions in the steel industry based on networking of storage, logistics, and services based, for instance, on appropriate sensor and actor technologies. With the help of *Digital Manufacturing/Industry 4.0*, it will become possible that objects throughout the steel mill supply chain can be identified and located at any time, resulting in reduced loading and wait times and reduced logistics costs. Hence, the information and communication technologies offered by the *Digital Manufacturing/Industry 4.0* paradigm will be able to connect manufacturing equipment and raw material suppliers in real time in the near future. This will enable the automatic exchange of data between steel mill manufacturing processes and raw material suppliers to optimize workflow supply chain. In this regard, the manufacturing process will respond immediately to fluctuations in the delivery chains.

Manufacturing process optimization has been implemented in steel mills for a long time. Extending the automation of manufacturing processes was realized during the last 25 years with the support of modern information and telecommunication technologies. This has resulted in many innovative technical developments for more cost-effective, flexible manufacturing processes, sustainable quality improvements, and other new developments. Furthermore, this has enabled the implementation of increased customer requirements with regard to quality and delivery.

The steel mill activities described thus far can be consolidated in the context of a *Digital Manufacturing/Industry 4.0* smart factory. As described in Sect. 7.1.2, a smart factory represents an intelligent factory. In the center of a smart factory, the interconnectedness of autonomous processes is a key issue, as they situationally control themselves, even configuring knowledge, sensor-based, and spatially distributed manufacturing resources (manufacturing machines, robots, conveying and storage systems, utilities, and more), including their planning and control systems. Thus, a smart factory is characterized by a consistent engineering approach, which incorporates the manufacturing processes, as well as the manufactured product itself, seamlessly interlocking the cyber and the physical worlds through CPS. The output of a smart factory in the near future will result in manufactured products which can control their own manufacturing processes by sharing the utilities of machines through their product codes, showing which requirements are essential and what manufacturing steps are required next. Through this smooth transition, a smart factory, with its smart products, smart data (see Sect. 7.1.2), and smart services, can better utilize manufacturing capacity while saving resources and opening up new possibilities for value-added economic chains and employment.

One of the highest priorities in steel mill manufacturing is to develop the most efficient techniques for producing a given quantity of steel at a high level of quality at the lowest possible cost through horizontal, integrated digital technologies. Steel coil products emerge from the melt shop, the hot rolling, the cold rolling, and the finishing processes in that order. Optimized scheduling is required for all of these manufacturing steps to operate efficiently, but it is too complicated to create schedules for all of these processes at once.

Another key to successfully improving efficiency in steelmaking is the collection, analysis, and interpretation of smart data from the manufacturing processes. The result is input from a horizontal integrated manufacturing process monitoring system. This helps to identify the steel mill's condition as the basis for predictive and regular intervals of maintenance service to avoid unplanned downtimes, unintentional impacts on product quality, and costly equipment damage. Altogether, this will lead to an increase in the overall productivity of steelmaking while also sustaining a high level of product quality and manufacturing operations. As a result of smart data analysis, delivery reliability and lead times can be improved and stock levels reduced. Furthermore, operators can be assured that manufacturing processes and materials are used optimally and that machines are only released for maintenance if no running orders are hampered. Moreover, fewer quality issues will reduce customer complaints and faulty production. In this sense, smart data are the data collected for every production step every time, and they can be aggregated to efficiency coefficients, such as key performance indicators for the manufacturing processes. For this purpose, smart data can be received by leveraging company-defined semantic models to link and manage the diverse data. In this regard, the semantic models allow the data to be linked by the respective business concepts and the metadata to run with the data. Therefore, the advantage is that such detailed analyses of equipment performance and downtimes will enable continuous process improvements along the whole manufacturing chain.

Another issue in regard to resources is saving energy in steelmaking, a major economic and environmental concern because the steelmaking industry is among the most energy-intensive industries, where energy is a major share of the operating costs. Typically, energy costs account for 20–30 % of the total production cost. A large portion of this cost can be avoided by improving energy awareness and associated energy-reduction measures. Against this background, process optimization in the steel industry is highly important because energy costs represent a high percentage of the costs associated with steelmaking. Improvements can be achieved in the context of *Industry 4.0* through optimized, linked process steps (e.g., running the steel slabs in the hot rolling mill at the highest possible temperature level, as shown in Fig. 7.15). For technical reasons it is mandatory for certain alloys and possible for all of the products.

In integrated smelters, processed gases from blast furnaces, steel mills, and coking plants are used to generate electricity and steam. Costs can be reduced through optimal coordination of production and maintenance which can be

Fig. 7.15 Steel slab in the hot rolling mill—with permission from Salzgitter Flachstahl GmbH

achieved by developing a suitable predictive model, based on data analysis, to determine the best time to replace a worn part based on machine or process data. This not only saves time on maintenance and service by reducing the frequency with which worn parts are handled and replaced, it also reduces costs and materials (see Sect. 7.6.3).

Out of individual *Industry 4.0* applications, thoughtful new strategies have been formed which can be applied in the steel industry (e.g., dynamic process optimization, which can be applied to minimize warehousing in steel manufacturing). This depends on whether individual process steps are required due to different requirements for successive products, requiring intermediate warehouses with semifinished material. The constraint in using intermediate warehouses is that they should tie up as little capital as possible. Thus, holistic planning of production processes is possible through concrete digitalization solutions, thus minimizing the capital commitment and accelerating production to make production planning and control much more efficient. One tangible solution can be applied to coils because coils are usually stored in multiple layers, and the objective is to reduce the number of needless coil movements, as shown in Fig. 7.16.

A more general aspect of putting *Industry 4.0* methods into practice in the steel industry is modeling several processes in steel production. Models simplify reality and make complexity manageable. The model type which can be applied is data based. With this type of model, one cannot look into the real steel converter as shown in Fig. 7.17, which does not mean that there is no information about the process available. Based on the huge number of process variables, such as

Fig. 7.16 Manufactured coils in the coil storage—with permission from Salzgitter Flachstahl GmbH

temperature, gases, pressure, and others, the model can be built. The impact of individual production processes on the product properties can be described by empirically derived relationships or physical models. As a result, the individual processes can be optimized more effectively and their limitations can be identified. With regard to process planning and control, the required product quality and

Fig. 7.17 Steel converter—with permission from Salzgitter Flachstahl GmbH

features can be specified with minimal technical and financial effort. This allows product characteristics to be evaluated at an early enough stage to initiate corrective measures.

Another issue is automatic product control, which depends on the results from previous production steps. Each process step affects the product's properties to a certain extent. This also can be described by models or empirical relationships. The subsequent process steps can be adjusted in order to achieve better results than before. This requires a planning tool that can take these constraints into account despite the limitations of the individual processes. For semi-finished products, it can be quickly determined if further processing is useful if another job must be found or if the product must be devalued. This process planning requires a powerful job management and planning tool.

All of the aforementioned characteristics have an intrinsic feature: quality. Quality assurance has to take into account an error analysis of process data. Product defects with simple, recognizable contexts relative to a few process parameters are usually eliminated quickly. Errors that only appear under certain combinations of process parameters (possibly several production steps) are more difficult to identify or analyze. Here, more intelligent machine learning tools are needed that perform data mining according to relevant parameters. A data mining analysis of the digitized process to discover patterns in big data sets involves methods at the edge of artificial intelligence, machine learning, statistics, and database systems. For this purpose, a restriction based on the right parameters is absolutely necessary (i.e., moving from the big data to the smart data paradigm). This is self-explanatory if looking at all of the data collected on the processes in a steel mill, which usually does not provide meaningful results because spurious correlations are not recognized, and nonrelevant parameters with strong outliers to relevant parameters, which are controlled within narrow limits, predominate. Hence, a smart data management concept is required with meaningful parameters with high resolution for fault analysis and condition monitoring (maintenance) in quality-relevant parameters with a respective resolution.

7.6.3 Digital Manufacturing/Industry 4.0: The Bosch Software Innovations Approach

"Industrial Internet: Putting the vision into practice," is the title of a white paper by Bosch Software Innovations GmbH (Bosch 2015) for machine and component manufacturers to drive their service business forward and secure a competitive edge with the Internet of Things (see Chap. 4). In this context, the example of predictive maintenance offers a particularly potent illustration of the principles and benefits of the Industrial Internet. The term *Industry 4.0*, or Industrial Internet, refers to the fourth paradigm shift in manufacturing, in which intelligent manufacturing technology is interconnected (see Sect. 7.1). The increasing interconnection of manufacturing and the Internet offer a wealth of potential economic benefits, particularly for machine and component manufacturers. By connecting their products and expanding their range of services to include novel software solutions, they have an opportunity to leverage new market potential, compete effectively, and, in a best case scenario, gain a measurable edge over their competitors. The description in this section is based on the aforementioned white paper with the permission of Bosch Software Innovations GmbH.

The service business plays a key role in *Industry 4.0*. To combat dwindling service revenues, primarily attributable to increased Xstandardization in the spare parts business, machinery manufacturers need to develop new business models because the services offered by traditional machinery manufacturers are typically the most lucrative side of their business.

New technologies, such as remote access and data analytics, are prompting a focus on the service business. By connecting their machines on the manufacturing floor, companies can access machine data during real-time operation (see Sect. 7.1.2). Intelligent evaluation of this data can offer new insights into issues such as:

- What works in the field?
- What functions might lead to faults in the field?

These insights can provide a basis for developing needs-based services and applications and optimizing product functions for real-life use, which, in turn, will have a positive impact on the product price (see Sect. 7.1.3).

In practice, it is often difficult to know where to start when it comes to implementing *Internet 4.0*. Will new, innovative applications and services genuinely provide significant added value? And when does it become worth investing in an *Industry 4.0* project? Typically, there are a multitude of different ideas within a company on what approach to take and no defined strategy on how to proceed. One of the key challenges is to recognize that the implementation of *Internet 4.0* is not a linear process. In many cases, new business potential will not become apparent until an *Industry 4.0* project is well underway or even after it is finished. Numerous opportunities may arise, and the consequences of each of these are difficult to assess. It is, therefore, sometimes necessary to make a major investment in an *Industry 4.0* project without having a clear initial estimate of profitability because the component or machine manufacturer is entering unexplored territory with their innovation.

One practical and feasible first step into the world of *Internet 4.0* for component and machine manufacturers is to expand and optimize their existing services. Equipping components and machines with sensors and software makes it possible to automatically collect a diverse range of field data. By connecting components and machines, data can be retrieved in near real time and gathered in a central location. In most cases, the knowledge required to interpret this data is already available within the company. This know-how can be modeled as rules and applied to the data automatically. Information previously obtained directly from the respective components and machines on the manufacturing shop floor can now be visualized and monitored on a single platform using customized software.

This creates a tremendous degree of transparency. All of the data is made available in an application-oriented format, making it much easier to identify faults or deviations and determine their exact nature. The result is a significant reduction in response times.

The ability to read the status of machines and manufacturing processes at any point in time and take targeted action when something goes wrong already constitutes a major improvement to a company's service business and a boost in its market position. But manufacturers can go one step further by applying data analytics, allowing them to prepare and analyze the accumulated data in order to transform new insights into concrete services. To reach this stage, it is necessary to equip products with sensors or software to generate the data required in the first place. This stock of data serves as a basis for making decisions on which services will be profitable and should, therefore, be provided by the company concerned.

The continuous process of developing an existing business and new services in the *Internet 4.0* paradigm is illustrated by the *Industry 4.0* innovation cycle, shown in Fig. 7.18. The innovation cycle is comprised of three phases that a company passes through in one continuous process. It is also possible to carry out the phases in parallel.

The purpose of the product feature phase is to equip the machines and components with *Industry 4.0* features, which includes sensors, actuators, and an information processing system and customized application software. In addition, the machines and components require a network interface to provide them with a

Fig. 7.18 Industry 4.0
innovation cycle—with
permission from Bosch
Software Innovations GmbH

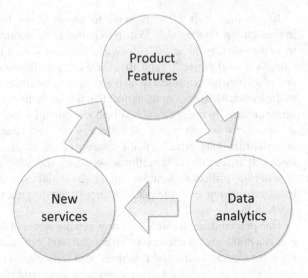

wireless or wired network connection in the area of the application so that they can be accessed.

7.6.3.1 Gaining Knowledge Through Data Analysis

As well as enabling the optimization of existing services, access to machines also opens up the possibility of collecting large quantities of data. It is important to clearly stipulate that data should be collected in order to meet the objectives in each case. Objectives may include reducing maintenance costs by slashing the number of callouts or reducing the cost of deviations in the manufacturing process and more. The accumulated data consists of both historical and current data and forms the basis for the data analytics. It is important not to underestimate the quantity and complexity of the data acquired. A multitude of sensors, components, and machines will typically generate enormous quantities of data, a phenomenon often referred to as big data (see Sect. 7.1.1).

Data analytics is essentially a means of modeling and acquiring knowledge. The goal is to recognize patterns in data and develop predictive models on that basis. A pattern is a representation of an event in the form of data or a series of events in the physical world. In the context of data analytics, a distinction is made between descriptive and predictive analytics. The aim of descriptive analytics is to condense data and identify patterns. These patterns then form the basis for predictive analytics. By drawing on a number of different techniques (e.g., statistical methods, modeling, and machine learning), it is possible to predict what may happen in the future, such as forecasting the probability that a certain event or situation will occur. In order to predict events, the current flow of data is analyzed to detect known patterns. If part of a pattern is identified, then it is possible to predict how likely it is that the rest of the pattern will occur and, thus, a certain event in the physical world.

In an ideal scenario, the newly acquired information can be used to help automate decision-making processes.

Example 7.6.1: Improving Process Quality
Analyzing process data makes it possible to identify deviations in quality within a manufacturing process by identifying previously unknown patterns. This technique enables quality trends to be depicted in a much more subtle and differentiated way to obtain comprehensive insights into the quality of manufacturing processes. It is able to identify trends in quality over time and react to problems before a fault actually occurs.

This detailed form of data analysis has an additional benefit. In cases where no process data analysis is performed, some faults and deviations pass through the entire process. Data analysis helps manufacturing companies to reduce the cost of both faults and scrap. In addition, trends can be analyzed to pinpoint the best ways of optimizing how the machine or component is used.

Example 7.6.2: Analyzing Machine Data to Detect Wear at an Early Stage
The objective of any manufacturing company is to keep parts in operation for as long as possible in order to get the most out of their service life cycle and reduce the use of materials. Two main types of maintenance are currently established in industry: reactive and preventive. In reactive maintenance, machines and components are only repaired when technical problems arise, with the resulting downtimes typically racking up significant costs. In preventive maintenance, costly worn parts are normally replaced at predefined intervals, a form of preventive maintenance in which parts are typically replaced more frequently than necessary, with the resulting waste of resources, manpower, and material.

By developing a suitable predictive model using data analysis, it is possible to determine the best time to replace a worn part based on machine or process data. As well as saving time on maintenance and service by reducing the frequency with which wear parts are handled and replaced, this strategy also reduces the use of materials. The more of these critical parts there are on a manufacturing line, the greater the savings that can be achieved in the maintenance area. At the same time, this approach reduces unplanned downtimes to a minimum by identifying worn part failure at a sufficiently early stage.

7.6.3.2 New Business Models for Maintenance
In the future, machinery and component manufacturers will be able to generate bigger margins in their service business. One example is optimized condition monitoring with a corresponding service agreement. This involves monitoring machines and components via remote access and automatically triggering maintenance and service where required. The data recorded is analyzed by the service provider in order to identify patterns that could indicate that a part is about to wear out or a machine is at risk of imminent failure.

This type of service paves the way for predictive maintenance. Machine condition data provides insights into deterioration and potential failure, while process data allows conclusions to be drawn on a machine's condition and the maintenance or service required. For example, deviations from the stipulated cycle time could indicate that the machine settings are suboptimal.

Connecting products and equipping them with suitable sensors, actuators, and software is an essential prerequisite for this kind of business model. Once access to the machines has been facilitated, traditional services, such as reactive maintenance management, can be offered in an optimized format. This service provides detailed information on faults and deviations and comprehensive documentation of action taken. The manufacturers themselves benefit from these new business models because predictive maintenance enables them to order spare parts just in time, avoiding unnecessary storage costs.

There is no doubt that companies will be able to generate revenue in the future with new business models based on existing *Industry 4.0* technologies. But what form could this kind of integrated business model take in the field of predictive maintenance? The so-called magic triangle developed by St. Gallen University, Switzerland, gives a vivid illustration of how this kind of project could be developed in practice. The model defines, as can be seen from Fig. 7.19, four dimensions of predictive maintenance—who, what, how, and revenue—to take into account both in-house and external factors and create a comprehensive view of all of the issues involved.

The first question, "Who?," is easy to answer. Machinery and component manufacturers and component developers would direct this kind of business model at their existing customer base as well as new customers. Typically, these customers would be industrial manufacturing companies, in other words, users of machines, systems, and components.

The second question, "What?," addresses what is offered to the customer. Predictive maintenance enables the machine manufacturer to determine, at an early stage, when maintenance should be performed at the customer site in response to an imminent machine malfunction. That enables machine manufacturers to offer their customers new services, such as guaranteed machine availabilities, while simultaneously reducing their own resource consumption. As well as assigning fewer employees to preventative maintenance tasks, the machine manufacturer also benefits from only having to replace spare parts when there is a high likelihood of imminent problems. At the same time, users benefit from minimal downtime and a correspondingly higher manufacturing output.

The answer to the third question, "How?," is that the current machine condition is recorded using sensor technology and automatically checked for patterns. This allows possible malfunctions to be detected at an early stage and machine failure to be averted.

All of these factors put together provide the added value or "*revenue*" for the machine manufacturer. They are able to add new services to their existing portfolios in order to create additional ongoing sources of income. In addition, manufacturing companies save money thanks to the optimization measures. This provides another direct benefit to the machine manufacturer because the boost in customer satisfaction safeguards their business and helps the machine manufacturer stand out from

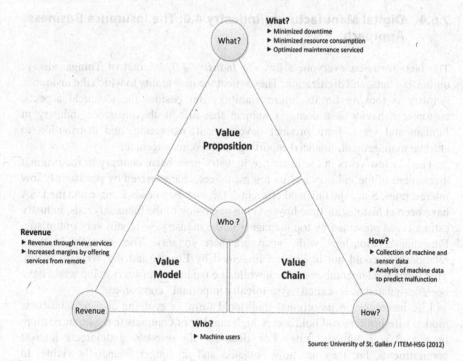

Fig. 7.19 Magic triangle—with permission from Bosch Software Innovations GmbH

the competition. A further advantage is that predictive maintenance requires remote access to machines and systems. This enables maintenance work to be carried out from a distance, which has a positive impact on the manufacturer's margins in the service business.

Therefore, when it comes to *Internet 4.0*, many manufacturing companies are waiting for a key technology to emerge, yet the increasing interconnection of manufacturing and the Internet offers plenty of tremendously promising potential right now. Optimization of existing services is already yielding new business models, especially for machine and component manufacturers. The only step required to apply these models is to make products and systems *Industry 4.0-ready* by incorporating sensors, actuators, and information processing software. Once these foundations are in place, machine and process data can be analyzed and optimization measures implemented on the basis of this analysis. Predictive maintenance is just one example of the numerous possible applications of *Industry 4.0*. In the long term, machine and component manufacturers will benefit from the increasing customer satisfaction and higher turnover with regard to the continuous development of their service models.

7.6.4   Digital Manufacturing/Industry 4.0: The Insurance Business Approach

The buzzwords on everyone's lips are *Industry 4.0*, Internet of Things, always online, big data, and digitization. They reflect the new reality to which the insurance industry is looking for its future viability. But besides the technical aspects, customer behavior is a complex subject that affects the insurance industry in fundamental ways, from product development, marketing, and distribution to enforce management, financial reporting, and risk management.

The last few years in the insurance industry have been, contrary to the original assessment of the industry and its top managers, characterized by persistently low interest rates. Since the financial crisis in 2008, interest rates in Europe and the USA have been at historic all-time lows. At the beginning of the financial crisis, industry estimates, as presented by top managers in the industry, were still very optimistic. Financial relationships with Japan are here to stay. The strength of these relationships could not have been imagined by Europe and the Western world. That was a fundamental error. Meanwhile, the ten largest insurers in the world have been designated as so-called "systemically important" companies.

Life insurance, in its original traditional form, is evolving as a new business model. Life insurers still hold a very high number of contracts that guarantee more than a 3.5 % interest rate. For decades, the models guaranteed interest commitments, but they are now outdated and no longer financially viable. In addition, the life insurers are subject to new legislation that is a consequence of the financial crisis, and they continue to be in a tight spot. The Solvency II Directive and additional interest reserves should suffice as keywords here. A specific emphasis now is on the provisioning of contracts by the Life Insurance Reform Act (LVRG). The consequences arising from LVRG for the existing distributors in Germany are not yet apparent. Policy makers have made statements on the subject that appear to threaten the insurance industry. If the insurance industry is not capable of reducing the commission level on its own, relevant laws will be enacted, creating a whole new situation. Examples of this are well known from events in the Netherlands and Great Britain.

The insurance companies in Germany are trying to counter the situation based on their individual situations. The German Insurance Association (GDV) is applying political pressure in an effort to prevent the worst-case scenario.

Due to failing investment income, the companies are forced to earn a claim/ expense ratio (combined ratio) of less than 100 %. With profit expectations of up to 8 % and more, the combined ratio will be 90–92 %. The profits earned before the financial crisis are generating at least 100 % combined ratios. The majority of insurance companies are in the automotive and residential building mass markets where for years, they had not experienced hardship until a price war began, which the industry has itself repeatedly, and sometimes unnecessarily, fueled. This has resulted in horrendous losses for the insurance companies. For some time, they have implemented a control strategy based on advanced analytics which has yielded positive results.

The reinsurance market is also under pressure today, because previously unknown market participants who are extremely financially sound have entered the business. They are pressing to raise reinsurance premiums which should actually rise due to the increasing number of, and largely unpredictable, natural events (natural disasters NatCat).

The first conclusion that can be made is that the traditional business model of the insurance industry, which has been used for more than 100 years, has eroded. It has primarily been based on an average represented by a male, 40 years old, a non-smoker, with a regular income, and no history of joblessness.

In addition to the previously mentioned general market conditions, there are still challenges, primarily digitization. Digitization is a nebulous concept. What makes up the core of this concept, no one can precisely explain, although all who are asked can clearly define digitization. Therefore, what digitization means to the insurance industry is not sufficiently clear. Digitization is often shortened to mean existing processes and products. Running a Web portal to distribute products seems to pass as digitization.

The buzzwords, "digitization" and "big data," currently dominate discussions in the insurance sector. Large amounts of money are spent to track Web portal strategies and projects related to business intelligence systems.

It should be noted that the insurance sector has not yet defined a standard, despite great effort, especially in terms of funds spent, and has experienced job cuts in recent years, comparable to other industries. The insurance sector, as a whole, still uses concepts that lag years behind the most modern concepts of the industry. By taking advantage of smart/big data digitization concepts, which are primarily used to capture client information, insurance companies could offer more precise practices and better risk forecasting, making their businesses more profitable. This has not yet occurred until today. Looking into today's information handling in social media, a lot of policyholder data can be found everywhere, such as on Facebook pages, which can obtain an incredible amount of information.

Therefore, the second conclusion that can be made is that digitization is not a purely technical issue; digitization is much more of a cultural issue in the insurance industry.

In the insurance sector, the focus has been on so-called dark processing and standardization, particularly in the private sector. In the corporate and industrial customer sectors, there has been difficulty with this approach. Therefore, a production level still exists compared to the one in manufacturing or semi-manufacturing.

With regard to the challenge of digitization and big data, the question for the insurance sector is different from other industries. That would be one way to make the industry, and thus individual companies, ready for the future. This requires rethinking in terms of markets and customers, as well as suppliers, corporate organization, and the behavior and attitude of the staff and management.

Thus, digitization can be understood as political and social development. In modern social and political systems, the issue of communication and information flow is paramount. Digitization fundamentally changes the way the actors communicate in these social and political systems. Actors are all participating people and institutions in these systems. The vehicles of communication are then all technical and physical systems that are specifically developed, built, and operated for this purpose. The most important system in our time is the Internet which enables communication to span the entire globe, referred to as globalization. This means that every actor in this global network has to find his position and role in order to survive in such a world. Therefore, digitization is only another expression for networking, connecting, and linking.

For an insurance company, it is important to find a role and position and redefine itself as well as gain standing, although it makes a difference whether the company operates in regional or global markets. In Germany, the vast majority of insurance companies are regionally active.

With regard to an individual company's situation, an appropriate strategy needs to be developed, to keep the company independent and viable in the marketplace. Therefore, the extraordinary challenge lies in the parallelization and sustainability of all activities in the age of the Internet. It could jeopardize the very existence of a company in the case of missing or insufficient expertise and the transformation of existing information processing systems as mobile technology has the potential to alter the entire claims experience.

Based on formative professionalism due to changes in the behavior of the customers and changes in market conditions, the enterprise architectures will be subject to a profound change. Thus, the traditional, functional statement of views is not important; an execution- or process-oriented perspective is required. Despite all declarations and using the word "process," only in a very few insurance companies, if any, does a process point of view exist, much less a process-oriented way of working. This means that the mode of production, i.e., the production and distribution, does not meet the complex requirements of production lines that are common in today's manufacturing industry. Furthermore, many requirements essential to making the leap into the digital era are missing.

A key point is that the company needs to transition from manufacturing-driven production to industrial-engineering production. This industrial process is based on all relevant aspects that play a role in this industrially oriented process composed of meeting the requirements that are imposed on modern manufacturing companies. The aspects behind these requirements are varied, and each one has its individual meaning. At the moment, data requirements are highlighted. Here, as so often happens in the insurance sector, other aspects are more or less forgotten, or neglected. A holistic approach is not pursued. Using simulation science method to simulate customer behavior under multiple scenarios can help the insuring industry to develop a more holistic understanding of social and political initiated market changes.

But how can the new state be achieved? Here are some considerations. Some of them are already field tested; others still need to be formulated and implemented.

The goal is to implement real-time, controlled processes. The quality of the data and its provision in the respective consistency and coherence at the right time and the right place presents a major obstacle for the process design.

In addition, it is not always clear which issues on the data will be addressed. Only properly selected questions can open up the possibility to get even further with correct answers.

Another technical aspect relates to the processing of the data. First of all, the data must be collected and available. Then they must be quickly processed in accordance with the stated requirements and correctly forwarded, stored, and retrieved. With regard to the current available systems, particularly in the communication between the back-end systems and distributed applications, mismatch is discovered. Dispute is still among experts how the proper balancing between these systems should look like. Still the question arises again, Thin Client vs. Thick Client, an anachronistic question when you look at today's possibilities of modern IT architectures.

Thus, the following six questions must be asked and have been satisfactorily addressed, before beginning with digitization or *Industry 4.0* in the insurance industry:

1. What has to be structured? Methods of information systems design.
2. Who is responsible for the designs? Actors of information systems design.
3. Where is it to be structured? Criteria of the information systems design.
4. How will it be designed? Methods of information systems design.
5. Where will it be designed? Levels of information systems design.
6. What causes have to be structured? Measures of information systems design.

To conclude the considerations for an architectural design that can serve as a basis to orient itself for the challenges of digitization we must show how the actor relationships, especially in terms of contractual arrangements with each other, are embedded. Only when an insurance company has a clear view of how such architecture should look as a balanced construct, from the external and internal perspective, may it move through the difficult path of digitization. This requires new reporting metrics around which insuring companies will be able to develop an effective metrics and communication framework. In PWC (2015), it is reported that a notable characteristic of each of these metrics is that they have common key principles around which insuring companies can develop an effective metrics and communications framework. Only then is it reasonably assured that the significant investment in the future will bear success and will not be lost. The communication framework consists of digital distribution channels through which insurance products and assistance services are offered as part of new integrated solutions provided by the insurance company. In this sense, services will be a major component of the new solution packages to put them in a good position to a growing market. In Fig. 7.20 a general view of the several components of a digitization strategy is shown which includes services.

With respect to possible services, a model for clients and contracts is shown in Fig. 7.21, which refers to a possible claim/harm platform.

As reported in Mayr (2013), assistance service providers could be automatically notified in emergencies; this would then allow them to coordinate both immediate assistance and the claim settlement process. Moreover, products that are based on

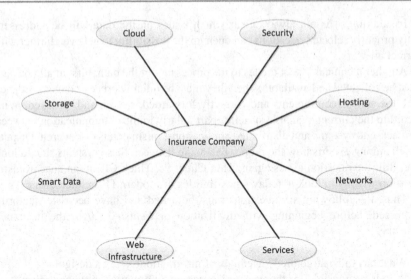

Fig. 7.20 Components required for the digitization of an insurance company

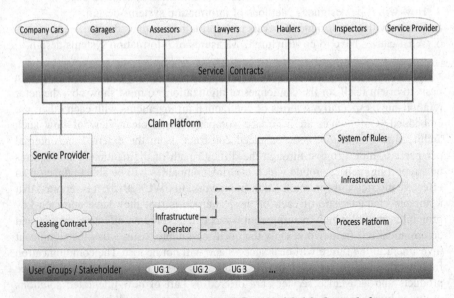

Fig. 7.21 Model for clients and contracts as part of a potential claim/harm platform

networked objects and services offer insurers the opportunity to gain direct and permanent access to their target groups. This enables them to give their customers offers based on their individual needs. Daily use of and interaction with networked products allow companies to obtain valuable information that helps them to understand the current circumstances and requirements of their customers and to offer them further products and supplementary services that match their current situation.

This might also include the products and services of a company's partners. If customers find this experience positive, it reinforces their confidence in the provider. This leads to new business opportunities as reported in Mayr (2013).

7.6.5   Digital Manufacturing/Industry 4.0: The German Industry 4.0 Working Group Approach

The executive summary of the "Recommendations for implementing the strategic initiative INDUSTRIE 4.0, Final report of the Industrie 4.0 Working Group" (Kagermann et al. 2013) illustrates that Germany has one of the most competitive manufacturing industries in the world and is a global leader in the manufacturing equipment sector. Germany's strong machinery and plant manufacturing industry, its globally significant level of IT competences, and its know-how in embedded systems and automation engineering mean that it is extremely well placed to develop its position as a leader in the manufacturing engineering industry. Germany is uniquely positioned to tap into the potential of a new type of industrialization. *Industry 4.0* will involve the technical integration of CPS (see Chap. 3) into manufacturing and logistics and the use of the Internet of Things, Data, and Services (IoTDaS) (see Chap. 4) in industrial processes. This will have implications for value creation through smart manufacturing, business models, downstream services, and work organization.

Smart manufacturing or smart factories (see Sect. 7.1.2) will allow individual customer requirements to be met and will enable even one-off items to be manufactured profitably. They will also result in new ways of creating value and novel business models. In particular, they will provide start-ups and small businesses with the opportunity to develop and provide downstream services. At the same time, it will be necessary to create and serve new leading markets for CPS technologies and products. In order to achieve the goals of this dual CPS strategy, the following features of *Industry 4.0* should be implemented:

- *Horizontal integration through value networks*: models, designs, and implementations of horizontal integration through value networks.
- *End-to-end digital integration of engineering across the entire value chain*: cyber and physical worlds will be integrated across a product's entire value chain also incorporating customer requirements.
- *Vertical integration and networked manufacturing systems*: CPS will be used to create flexible and reconfigurable manufacturing systems.

If *Industry 4.0* is to be successfully implemented, research and development activities will need to be accompanied by the appropriate industrial policy decisions. The Industrie 4.0 Working Group (Kagermann et al. 2013) believes that action is needed in the following eight key areas:

1. *Standardization and reference architecture*: *Industry 4.0* will involve networking and integration of several different companies through value networks. This collaborative partnership will only be possible if a single set of common standards is developed. Reference architecture will be needed to provide a technical description of these standards and facilitate their implementation.
2. *Managing complex systems*: Products and manufacturing systems are becoming more and more complex. Appropriate planning and explanatory models can provide a basis for managing this growing complexity in *Industry 4.0*. Engineers should, therefore, be equipped with the methods and tools required to develop such models.
3. *A comprehensive broadband infrastructure for industry*: Reliable, comprehensive, high-quality communication networks are a key requirement for *Industry 4.0*. Broadband Internet infrastructure, therefore, needs to be expanded on a massive scale within Germany and between Germany and its partner countries in the manufacturing domain.
4. *Safety and security*: Safety and security are critical to the success of smart manufacturing systems. It is important to ensure that manufacturing facilities and the products themselves do not pose a danger either to people or to the environment. At the same time, manufacturing facilities, products, and, in particular, the data and information they contain need to be protected against misuse and unauthorized access. This will require, for example, the deployment of integrated safety and security architectures and unique identifiers, together with the relevant enhancements, training, and continuing professional development (CPD) (see Chap. 8).
5. *Work organization and design*: In smart factories, the roles of employees will change significantly. Increasingly, real-time-oriented control will transform work content, work processes, and the working environment. Implementation of a socio-technical approach to work organization will offer workers the opportunity to enjoy greater responsibility and enhance their personal development (see Chap. 8). For this to be possible, it will be necessary to deploy participative work design and lifelong learning measures and to launch model reference projects.
6. *Training and continuing professional development*: *Industry 4.0* will radically transform workers' jobs and competence profiles. It will, therefore, be necessary to implement appropriate training strategies and to organize work in a way that fosters learning and enables lifelong learning and *Industry 4.0* workplace-based CPD. In order to achieve this, model projects and best practice networks should be promoted and digital learning techniques should be investigated.
7. *Regulatory framework*: While the new manufacturing processes and horizontal business networks found in *Industry 4.0* will need to comply with the law, existing legislation will also need to be adapted to take the new innovations into account. The challenges include the protection of corporate data, liability issues, handling of personal data, and trade restrictions. This will require not only legislation but also other types of action on behalf of businesses to ensure

an extensive range of suitable instruments exists, including guidelines, model contracts, and company agreements or self-regulation initiatives, such as audits.

8. *Resource efficiency*: Quite apart from the high costs, the manufacturing industry's consumption of large amounts of raw materials and energy also poses a number of threats to the environment and security of supplies. *Industry 4.0* will deliver gains in resource productivity and efficiency. It will be necessary to calculate the trade-offs between the additional resources that will need to be invested in smart factories and the potential savings generated.

7.6.5.1 Role of the Internet of Things, Data, and Services

The Internet of Things, Data, and Services (IoTDaS) will make it possible to create networks incorporating the entire manufacturing process, converting factories into a smart environment. Cyber-physical manufacturing systems are comprised of smart machines, warehousing systems, and manufacturing facilities that have been developed digitally and feature end-to-end information and communication technology (ICT)-based integration, from inbound logistics to manufacturing, marketing, outbound logistics, and service. This not only will allow manufacturing to be configured more flexibly but will also tap into the opportunities offered by much more differentiated management and control processes.

In addition to optimizing existing IT-based processes, *Industry 4.0* will, therefore, also unlock the potential of even more differentiated tracking of both detailed processes and overall effects at a global scale, which it was previously impossible to record. It will also involve closer cooperation between business partners, such as suppliers and customers, and between employees, providing new opportunities for mutual benefit.

More in general, it can be stated that the Internet of Things, Data, and Services (IoTDaS) will become the key enabler in the manufacturing industry, because *Industry 4.0* will involve the technical integration of CPS into manufacturing and logistics. This will have implications for value creation, business models, downstream services, and work organization.

7.6.5.2 Potential of Industry 4.0

The Industrie 4.0 Working Group believes that *Industry 4.0* has a huge potential which is outlined in the following topics (Kagermann et al. 2014):

- *Meeting individual customer requirements: Industry 4.0* allows individual, customer-specific criteria to be included in the design, configuration, ordering, planning, manufacture, and operation phases and enables last-minute changes to be incorporated. With *Industry 4.0*, it is possible to manufacture one-off items and have very low production volumes (batch size of 1) while still making a profit.
- *Enabling Germany to further strengthen its position as a manufacturing location, manufacturing equipment supplier, and IT business solutions supplier*: It is encouraging to see that all of the stakeholders in Germany are now working

closely together through the *Industry 4.0* platform in order to move ahead with implementation.

- *Flexibility*: CPS-based ad hoc networking enables dynamic configuration of different aspects of business processes, such as quality, time, risk, robustness, price, and eco-friendliness. This facilitates continuous trimming of materials and supply chains. It also means that engineering processes can be made more agile, manufacturing processes can be changed, temporary shortages (e.g., due to supply issues) can be compensated for, and huge increases in output can be achieved in a short space of time.
- *Optimized decision-making*: In order to succeed in a global market, it is becoming critical to be able to make the right decisions, often on very short notice. *Industry 4.0* provides end-to-end transparency in real time, allowing early verification of design decisions in the sphere of engineering and more flexible responses to disruption and global optimization across all of a company's production sites.
- *Resource productivity and efficiency*: The overarching strategic goals for industrial manufacturing processes still apply to *Industry 4.0*: delivering the highest possible output of products from a given volume of resources (resource productivity) and using the lowest possible amount of resources to deliver a particular output (resource efficiency). CPS allows manufacturing processes to be optimized on a case-by-case basis across the entire value network. Moreover, rather than having to stop production, systems can be continuously optimized during production in terms of their resource and energy consumption or emission reduction.
- *Creating value opportunities through new services*: *Industry 4.0* opens up new ways of creating value and new forms of employment, for example, through downstream services. Smart algorithms can be applied to large quantities of diverse data (big data) recorded by smart devices in order to provide innovative services. There are particularly significant opportunities for small- and medium-sized enterprises (SMEs) and start-ups to develop business-to-business (B2B) services for *Industry 4.0*.
- *Responding to demographic change in the workplace*: In conjunction with work organization and competency development initiatives, interactive collaboration between human beings and technological systems will provide businesses with new ways of turning demographic change to their advantage. In the face of a shortage of skilled labor and the growing diversity of the workforce (in terms of age, gender, and cultural background), *Industry 4.0* will enable diverse and flexible career paths that allow people to keep working and remain productive for longer.
- *Work-life balance*: The more flexible work organization models of companies that use CPS mean that they are well placed to meet the growing need of employees to strike a better balance between their work and their private lives and also between personal development and continuing professional development. Smart assistance systems will provide new opportunities to organize work in a way that delivers a new standard of flexibility to meet companies'

requirements and the personal needs of employees. As the size of the workforce declines, this will give CPS companies a clear advantage when it comes to recruiting the best employees.

• *A high-wage economy that is still competitive*: Industry 4.0's dual strategy will allow Germany to develop its position as a leading supplier and also the leading market for *Industry 4.0* solutions.

However, *Industry 4.0* will not pose an exclusively technological or IT-related challenge to the relevant industries. The changing technology will also have far-reaching organizational implications, providing an opportunity to develop new business and corporate models and facilitating greater employee engagement. Germany successfully implemented the third industrial revolution during the early 1980s by delivering more flexible automated manufacturing through the integration of PLCs into manufacturing technology while at the same time managing the impact on the workforce through an approach based on social partnership. Its strong industrial base, successful software industry, and know-how in the field of semantic technologies mean that Germany is extremely well placed to implement *Industry 4.0*. It should be possible to overcome the current obstacles, such as technology acceptance issues or the limited pool of skilled workers in the labor market. However, it will only be possible to secure the future of German industry if all of the relevant stakeholders work together to unlock the potential offered by the Internet of Things, Data, and Services for the manufacturing industry.

Since 2006, the German government has been promoting the Internet of Things, Data, and Services (IoTDaS) under its high-tech strategy. Several technology programs have also been successfully launched. The Industry-Science Research Alliance is now implementing this initiative at a cross-sectoral level through the *Industry 4.0* project. The establishment of the Industrie 4.0 Platform with a Secretariat provided jointly by the BITKOM, VDMA, and ZVEI professional associations was the next logical step in its implementation. The next task will be to generate R&D roadmaps for the key priority themes. Securing the future of the German manufacturing industry is the goal that the partners in the Industrie 4.0 platform have set for themselves. The platform invites all of the relevant stakeholders to continue exploring the opportunities provided by *Industry 4.0* so that together they can help ensure the successful implementation of its revolutionary vision.

7.6.6   Digital Manufacturing/Industry 4.0: The US Digital Manufacturing and Design Innovation Institute Approach

In February 2014, an Illinois consortium, led by the University of Illinois Labs, a nonprofit research and development group of the University of Illinois, was elected to lead the Digital Manufacturing and Design Innovation Institute (DMDI). DMDI will address the life cycle of digital data interchanged among myriad design, engineering, manufacturing, and maintenance systems and flowing across a networked supply chain. The University of Illinois Labs was awarded $70 million to fund the DMDI, which will leverage $250 million in commitments from leading industry partners, as well as academia, government, and community partners, to form a $320 million institute (Selko 2014).

Digital manufacturing is a competitive game changer, bringing the USA research, engineering, and production communities together in new and exciting ways. Specifically, the combination of advanced materials, high-performance computing resources, modeling and simulation tools, and additive manufacturing practices is allowing large and small enterprises alike to design and build otherwise impossibly complex shapes and systems while significantly reducing manufacturing costs and cycle times. The National Digital Engineering and Manufacturing Consortium, one of the partners, will help firms to leverage high-performance computing (HPC) for modeling, simulation, and analysis (MS&A). This capability helps manufacturers to design, test, and build prototype products or components much more rapidly, enabling them to bring innovations to market more quickly and less expensively (Selko 2014).

DMDI enables US manufacturing to bring innovations to market more quickly to gain an advantage in global competitiveness. The digital thread that integrates and drives modern design, manufacturing, and product support processes can be exploited to reduce cycle time and achieve first pass success and is the only feasible way to deal with constantly increasing complexity in products and manufacturing enterprises.

Visions of the factory of the future (see Sect. 7.1.2) have many different names (advanced manufacturing enterprise, intelligent manufacturing systems, smart manufacturing, Industrial Internet, etc.). They all have a common understanding that the key to success is networked, data-driven processes that combine innovative automation, sensing, and control with a transformed manufacturing workforce at every level, from the shop floor to the factory control level to the global supply chain. Realizing this vision will require precompetitive collaboration on many fronts, and DMDI will be focused on maturing the digital thread for applications in manufacturing and design of electromechanical assemblies and systems. This is of significant interest not only to the defense sector but also to most commercial industrial sectors (including aerospace, transportation, and energy) due to increasing levels of complexity, integration, and cost. Examples include power train, propulsion, and structural components, as well as control subsystems and systems. DMDI will provide the proving ground to link promising information technologies, tools, standards, models, sensors, controls, practices, and skills and then transition these capabilities to the industrial base for full-scale application. The institute will meet the need for cross-disciplinary teams to integrate IT and manufacturing solutions and multi-industry collaboration to promote interoperability in supply chains. The institute will be the intellectual hub that helps US manufacturers be the best in the world at connecting their flexible manufacturing operations, driving them securely with digital data, controlling quality with feedback from sensors and data analysis, maintaining a trusted chain of custody, and delivering products in significantly less time than global competitors.

7.6.6.1 National Economic Impact

DMDI will raise the global competiveness of US small- and medium-sized manufacturers by smart and comprehensive use of the digital thread throughout design, production, and support, thereby erasing any competitive advantage from low-cost, low-skill labor. One strength of the DMDI approach is that the results will be applicable to nearly every manufacturing industry sector and are expected to decrease costs by roughly 10 % across the manufacturing enterprise—not simply for one technology or manufacturing process. Industry has analyzed major economic sectors for potential benefit by implementing a DMD environment in which every machine, facility, and fleet is intelligently connected. The projected savings in commercial aviation alone is $30 billion over 15 years, for example, when each major engine subsystem has the built-in intelligence to predict its performance over its lifetime. Fleet readiness can be more efficiently planned, and part manufacturing can be more effectively managed. The US Department of Defense (DoD) and other economic sectors share similar business cases and opportunities for strong returns on investment.

7.6.6.2 DoD Investment Rationale

The DoD has an enormous stake in ensuring that US manufacturing evolves into a more agile, connected, collaborative, and efficient industry (see Sect. 7.1.1). The department requires complex, highly integrated systems to gain technological advantage, but it lacks the open market or volume to push costs or cycle times lower. Proving and moving intelligent electromechanical design and manufacturing capabilities from the laboratory to prototype factory environments would deliver commercial production efficiencies at lower than DoD production rates and reduce the prefinal design missteps that greatly increase time and cost, approaching the production decision milestone in the defense acquisition system process. Thus, DMDI will help drive a paradigm shift in the development, production, and sustainment of complex weapon systems by reducing acquisition lead time and costs through the application of digitally networked and synchronized processes and tools, resulting in an open and highly collaborative environment. The institute will also establish and integrate processes that sustain and enhance retention of supply chain knowledge and improve capacity and capability of both the organic and commercial industrial bases to affordably manufacture low-volume, varying demand, complex systems for the DoD in support of national security.

7.6.6.3 Description of Activities

For the institute to be a resource for industry that reduces the risk of adoption and provides a pathway for commercialization of new technologies, it must address technology advancements specific to intelligent Digital Manufacturing and those that cut across all Digital Manufacturing initiatives. The technology areas that are crosscutting to all advancement initiatives include the ability to demonstrate the technologies in a representative environment, the development of materials and opportunities for upgrading the skills of the workforce to support the new technology in the marketplace, and ensuring the cyber-physical security of the network and information. Technology advancements specific to the domain of intelligent elec-tromechanical design and manufacturing include the following:

- *Digital Manufacturing Enterprise*: Encompasses agile and robust manufacturing strategies and integrated capabilities that dramatically reduce the cost and time of producing complex systems and parts (see Sect. 7.1.1). This includes the development and implementation of modeling and simulation tools to allow faster time to market and efficient production of complex systems. It also includes a focus on tools and practices to minimize the multiple designs, prototypes, and test iterations typically required for product or process qualifi-cation, all connected via the digital thread to enable the designer, analyst, manufacturer, and maintenance collaboration.
- *Intelligent Machines*: Involves the development and integration of smart sensors, controls, and measurement, analysis, decision, and communication software tools for self-aware manufacturing that will provide continuous improvement and sustainability. Intelligent machines realize the first part of this philosophy by incorporating equipment with plug-and-play functionality and allowing equip-ment to use manufacturing knowledge from planning and processing components, including big data analytics (see Sect. 7.1.1).
- *Advanced Analysis*: Capitalizes on advances in high-performance computing to develop physics-based models of material performance with design for manufacturing in mind. This includes developing and integrating smart design tools to help reduce overdesign in order to reduce manufacturing cost.
- *Cyber-Physical Systems Security*: Focus on methods and technologies to provide a secure and trusted infrastructure for the management of information assets in a highly collaborative manufacturing environment. In addition to the known vulnerabilities of networked business systems and transactions used in manufacturing, the factory of the future needs to address the new vulnerabilities of CPS (see Sect. 7.5) in intelligent machines, sensors, and control systems.

There are examples of islands of success in the development and application of the digital thread to the industrial base in the USA. Even so, significant hurdles exist to the integration of intelligent electromechanical design and manufacturing across the US DoD and broader commercial industrial enterprises. Hurdles include establishing true interoperability, the effective and balanced management of intel-lectual property interests, maintaining network technology and security, and advancing machine intelligence, workforce skills, and new organizational cultures that embrace and leverage the digital thread to maximize US industrial competitiveness.

7.7   Exercises

What is meant by the term *manufacturing*?
Describe the architectural structure of a manufacturing system.
What is meant by the term *work in progress*?

Describe the equation for work in progress.
What is meant by the term *assembly line*?
Describe the structure of an assembly line.
What is meant by the term *agile manufacturing*?
Describe the four key elements of agile manufacturing.
What is meant by the term *smart factory*?
Describe the vision of a smart factory.
What is meant by the term *Industry 4.0*?
Describe the vision of *Industry 4.0*.
What is meant by the term *smart data*?
Give an example of smart data.
What is meant by the term *automation pyramid*?
Describe the tasks of the different levels of the automation pyramid.
What is meant by *individualized production*?
Give an example of individualized production.
What is meant by the term *networked manufacturing*?
Give an example of networked manufacturing.
What is meant by the term *closed manufacturing line*?
Give an example of a closed manufacturing line.
What is meant by the term *cybersecurity*?
Give an example of cybersecurity in manufacturing.

References

(Askin and Selfridge 1993) Askin, R. G., Standridge, A., C.: Modeling and Analysis of Manufacturing Systems. John Wiley and Sons, 1993

(Banerjee et al. 2012) Banerjee, A., Venkatasubramanian, K. K., Mukherjee, T., Gupty, S. K. S.: Ensuring Safety: Security and Sustainability of Mission-Critical Cyber-Physical Systems. Proceedings IEEE Vol.100(1), pp. 283–294, 2012

(Biller et al. 2008) Biller, S., Marin S. P., Meerkow, S. M., Zhang, L.: Closed production lines with arbitrary models of machine reliability. In: IEEE International Conference Proceedings on Automation Science Engineering, pp. 466–471, 2008

(Bosch 2015) Industrial Internet – Putting the vision into practice. White Paper. Bosch Software Innovations GmbH. 2015; www.bosch-si.de

(Brazell 2014) Brazell, J. B.: The Need for a Transdisciplinars Approach to Security of Cyber Physical Infrastructure. In: Applied Cyber-Physical Systems, Chapter 2, pp. 3–14, Eds.: Suh, S. C., Tanik, U. J., Carbone, J. N., Eroglu, A. Springer Publ. 2014

Cárdenas, A.A., Amin, S. and Sastry, S.: Secure control: towards survivable cyber-physical systems. Proc. 28th International Conference on Distributed Computing Systems Workshops, pp. 495–500, 2008

(Christensen 2015) Christensen, J.: Digital Business: in The Digital Age. Books on Demand GmbH, Copenhagen, 2015

(Dinaburg et al. 2008) Dinaburg, A., Royal, P., Sharif, M., Lee, W.: Ether: Mailware Analysis via Hardware Virtualization Extensions. Proceedings of the 15th ACM Conference on Computer and Communication Security, pp. 51–62, 2008

(Fahmida 2011) Fahmida, R. Y.: GM's OnStar, Ford Sync. MP3, Bluetooth possible Attack Vectors for Cars. In: eweek.com. IT Security and Network Security News, March 16th, 2011; http://www.eweek.com/a/Security/GMs-OnStar-Ford-Sync-MP3-Bluetooth-Possible-Attack-ectoras-for-Cars-420601/.

(Görlich et al. 2007) Görlich, D., Stephan, P., Quadflieg, J.: Demonstratíng remote operation of industrial devices using mobile phones. In: Proceedings of the 4th International Conference on Mobile Technology, Applications and Systems, pp. 482–485, 2007

(Guttmann and Roback 1995) Guttmann, B., Roback, E.: An Introduction to Computer Security: The NIST Handbook. National Institute of Standards and Technology: Special Publication 800-12. 1995

(Guy, S., Boyle, E., Hu, F.: Cyber-Physical System Security. Chapter 9, pp. 125–133, In: Cyber-Physical Systems, Ed.: Hu, Fei, CRC Press, 2014

(Harris 2014) Harris, K. D.: Cybersecurity in the Golden State. Privacy Enforcement and Protection Unit, California Department of Justice, 20134

(He and Zhang 2010) He, M., Zhang, J.: Fault Detection and Localization in Smart Grid: A Probabilistic Dependence Graph Approach. Proceed. 1st IEEE Internt. Conf. on Communications, pp. 43–48, 2012

(Iacocca 1991) 21st Century Manufacturing Enterprise Strategy. An Industry-Led View, Volumes 1 & 2. Iacocca Institute, 1991

(Kagermann et al. 2013) Kagermann, H., Wahlster, W., Helbig, J. (Eds.): Recommendations for implementing the strategic initiative Industry 4.0 – Final Report of the Working Group Industry 4.0. Office of the Industry-Science Research Alliance Secretariat of the Platform Industrie 4.0, 2013

(Kaplan and Haenlein 2006) Kaplan, M. M., Haenlein M.: Toward a parsimonious definition of traditional and electronic mass customization. Journal of Product Innovation Management, Vol. 23, Issue 2, pp. 168–182, 2006

(Karim and Phoha 2014) Karim, M. E., Phoha, V. V.: Cyber-physical Systems Security. Chapter 7, pp. 75–83, Applied Cyber-Physical Systems, Eds.: Suh, S. S., Tanik, U. J., Carbone, J., N., Eroglu, A., Springer Publ. 2014

(Kidd 1994) Kidd, P. T.: Agile Manufacturing: Forging New Frontiers. Addison-Wesley, 1994.

(Landum et al. 2014) Landrum, R., Pace, S., Hu, F. : Cyber-Physical System Security – Smart Grid Example. Chapter 10, pp. 135–152. In: Cyber-Physical Systems, Ed.: Hu, F. CRS Press, 2014

(Li and Meerkow 2008) Li, J., Meerkow, S. M.: Production Systems Engineering, Springer Publ., 2008

(Li 2009) Li, S. Z.: Markov Random Field Modeling in Image Analysis. Springer Publ., 2009

(Little 1961) Little, J. D. C.: A Proof for the Queuing Formula L=AW. Operations Research Vol. 9, Issue 3, pp. 383–387, 1961

(Lopez Research 2014) Lopez Research: Building Smarter Manufacturing With The Internet of Things (IoT), 2014

(Mayr 2013) Mayr, M.: Eight reason why insurers should be engaging with the Internet of Things. Bosch Connected World Blog, 2013, blog.bosch-si.com/catego-ries/internetofthings/2013/08/eight-reasons-why-insurers-should-be-engaging-with-the-internet-of-things/

(McCarthy 2004) McCarthy, I. P.: Special Issue Editorial: The What, Why and How of Mass Customization. Production Planning and Control, Vol. 25, Issue 4, pp. 347–351, 2004

(Mirkovic et al.2005) Mirkovic, J., Dietrich, S., Dittrich, D., Reiher, P.: Internet Denial of Service: Attack and Defense Mechanisms. Prentice Hall, 2005

(Möller 2014) Möller, D. P. F.: Introduction to Transportation Analysis, Modeling and Simulation – Computational Foundations and Multimodal Applications, Springer Publ., 2014

(Moser et al. 2007) Moser, A., Krügel, C., Kirda, E.: Exploring Multiple Execution Paths for Malware Analysis. IEEE Security and Privacy, pp. 231–245, 2007

(Pine 1993) Pine, B. J. II: Making Mass Customization Happen: Strategies for the New Competitive Realities. Planning Review, Vol. 21, Issue 5, pp. 23–24, 1993

(PWC 2015) PWC 2015: The insurance industry in 2015. PricewaterhouseCoopers LLG, 2015

(Rue and Held 2005) Rue, H., Held, L.: Gaussian Markov Random Fields – Theorys and Applications. Chapman & Hall/CRC Press, 2005

(Sanchez and Nagi 2001) Sanchez, L. M., Nagi, R. : A review of agile manufacturing systems, In: International Journal Production Research, Vol. 39, No. 16, pp. 3561–3600, 2001

(Schroer et al. 2007) Schroer, B. J., Harris, G. A.,, Möller, D. P. F.: Simulation to Evaluate Several Critical Factors Effecting Manufacturing, In Proceed. SCSC 07: Moving Towards the Unified

Simulation Approach, Ed.: G. A. Wainer, H. Vakilzadian, pp. 587–592, SCS Publ. San Diego, 2007; DOI: 10.1145/1357910.1358002

(Selko 2014) Selko A.: Chicago is New Home of Digital Manufacturing and Design Innovation Institute. In: Industry Week's Expansion Management, Feb. 26th 2014

(Sharif et al. 2008) Sharif, M. I., Lee, W., Cui, W., Lanzi, A.: Impeding Malware Analysis using Conditional Code Obfuscation. In: NDSS, 2008

(Steinfeld 2004) Steinfeld, E. S.: China's Shallow Integration: Networked Production and the New Challenges for Late Integration. In: Worl Development, Vol 32, No. 11, pp. 1971–1987, 2004

(Stephan et al. 2013) Stephan, P., Eich, M., Neidig, J., Rosjat, M., Hengst, R.: Applying Digital Product Memories in Industrial Production. In: SemProM, pp. 283–304, Ed.: Wahlster, W., Springer Publ. 2013

(Suri 1998) Suri, R.: Quick Response Manufacturing – A Companywide Approach to Reducing Lead Times. Productivity Press, 1998

(Ten et al. 2011) Ten, C., Hong, J., Liu, C.: Anomaly Detection for Cybersecurity oft he Substations. IEEE Transact. Smart Grid Vol. 2(4), pp. 865–873, 2011

(Terwiesch and Ganz 2009) Terwiesch, P., Ganz, C.: Trends in Automation. In: Handbook of Automation, pp. 127–143. Ed.: Nof, S. S., Springer Publ., 2009

(Tseng and Jiao 2001) Tseng, M. M., Jiao, J.: Mass Customization, Chapter 25, pp. 684–709. In: Handbook of Industrial Engineering, Technology and Operation Management. (3rd ed.). Ed.: Salvendy, G. John Wiley Publ. 2001

(Veugelers and Sapir 2013) Veugelers, R., Sapir, A.: Policies for manufacturing EU growth, Chapter 7, pp. 161–168, In: Manufacturing Europe's Future, Ed. Veugelers, R., Breugel Blueprint Series, 2013

(Westerman et al. 2014) Westerman, G., Bonnet, D., McAfee, A.: Leading Digital, Harvard Business Review Press, 2014

(Willems et al. 2007) Willems, C., Holz, T., Freiling, F. C.: Toward Automated Dynamic Malware Analyses using CW Sandbox. IEEE Security and Privacy, Vol. 5, pp. 32–39, 2007

(Yang and Burns 2003) Yang, B., Burns, N.: Implications of postponement for the supply chain. International Journal of Production Research, Vol. 49, Issue 9, pp. 2075–2090, 2003

(Zühlke 2008) Zühlke, D.: SmartFactory – A Vision becomes Reality. In: Proceedings of the 17th World Congress IFAC'08, pp. 14101–14108, 2008

Links

(http1   2015)   http://www.its-owl.de/fileadmin/PDF/News/2014-01-14-InIndustrie_   4.0-Smart_Manufacturing_for_the_Future_German_Trade_Invest.pdf

(http2 2015) http://www.leanproduction.com/agile-manufacturing.html

(http3 2015) http://www.fortiss.org/en/research/projects/opak/

(http4 2015] http://www.autonomik40.de/

(http5 2015) http://www.vdma.org/video-item-display/-/videodetail/2996989

(http6 2015] http://xmpp.org/extensions/xep-0191.html

(http7 2015) http://whatis.techtarget.com/definition/cybersecurity

(http8 2015) https://en.wikipedia.org/wiki/ISO/IEC_27001:2013

(http9 2015) http://www.iso.org/iso/catalogue_detail.htm?csnumber=34731

(http10   2015)   https://files.sans.org/summit/scada08/Stan_Johnson_NERC_Cyber_Security_Standards.pdf

(http11 2015) https://www.isa.org/intech/201406standards/

(http12   2015)   http://www.plattformi40.de/I40/Navigation/DE/Plattform/Aktivitaeten/aktivitaeten.html;jsessionid=2C7F7C0B9B5CA36127F25F1F469A9FFE

(http13 2015) http://www.mittelstand-digital.de/DE/Foerderinitiativen/mittelstand-4-0.html

(http14 2015) http://www.mittelstand-digital.de/DE/mittelstand-digital,did=726302.html

(http15 2015) (ww.pzh-hannover.de/pzh-publikationen.html)

Social Impact on Work Lives of the Future

8

This chapter is a brief introduction to the social impact on the work lives of the future. Section 8.1 introduces the changes in manufacturing skills from those historically required to the modern global, digital working environment. Section 8.2 refers to the economic, social, and organizational challenges of the future of work with regard to the requirements of digitized and automated manufacturing processes. Section 8.3 focuses on the changing demands in the world of work and attempts to answer questions resulting from the effects of Digital Manufacturing/Industry 4.0. Section 8.4 introduces the principal aspects of greater product individualization and the shifting factors of global influence with regard to Digital Manufacturing. Section 8.5 contains comprehensive questions regarding the social impacts on work lives in the future, and section "References" includes references and suggestions for further reading.

8.1 Introduction

Today's global, digital work environment is increasingly characterized by computers imitating human activities and skills, leaving human workers to wonder if they will eventually be relieved of their duties or replaced. This is especially true when considering skills such as sensory perception, pattern recognition, and declarative knowledge. Although these ultimately possess subtleties that are intangible and unpredictable, they can be reduced to sequences of statements in the context of many lines of code in a software program. Because a software program essentially consists of a series of specific written instructions—do this, then that, then this—with which the hardware, the computer, obtains skills that depend on declarative knowledge, and the case-dependent skills of the respective applications can be divided into well-defined steps. The human ability to assess situations and to make decisions quickly comes mostly from the fuzzy area of tacit knowledge. Here is where the most creative and artistic abilities of humans reside.

© Springer International Publishing Switzerland 2016
D.P.F. Möller, *Guide to Computing Fundamentals in Cyber-Physical Systems*,
Computer Communications and Networks, DOI 10.1007/978-3-319-25178-3_8

With regard to the progress of automation in the context of *Digital Manufacturing/Industry 4.0* processes as part of the fourth industrial revolution, referred to as an "evolution" by some authors, the concurrent digitization of the workplace leads to further challenges, including new forms of work and work models. Also, the impact on workers must be considered, given the previously mentioned acquisition of human activities and skills by fully automated and partially or fully autonomous machines which can run 24/7 cost-effectively.

In this regard, the social challenges of the digital and global work environments and their impact on employees' work lives need to be determined and addressed as society continues to work toward a self-determined human and the promise of employment for humans in the global, digital era. The reason for this is that new technologies are unleashing humans from the constraints that once bound them. As reported in Westerman et al. (2014), employees can increasingly work where they want and during the hours they choose. They can communicate as they wish, with a few friends or hundreds of friends, sharing sensitive information easily with others inside and outside their organization. To many employees, this sounds like freedom. To many managers, it sounds like chaos. Therefore, communicating openly about the opportunities and risks of the future of work lives in the global, digital working environment becomes an important issue as professionals are an important component of all economic growth, innovation, and prosperity. Henceforth, the job qualifications applicants and employees must meet will involve in ever shorter development cycles adapted to the requirements of digitization because the skills race won't slow down anytime soon. Having the right digital skills is a competitive advantage and a key enabler of the digital transformation. Hence, companies that build skills faster will get ahead faster, as reported in Westerman et al. (2014).

To win the digital skills race in *Digital Manufacturing/Industry 4.0*, the transformation of manufacturing processes must take place under adequately adapted automated manufacturing systems. This involves the use of digital models and methods of manufacturing planning and control linking them to real manufacturing subsystems, manufacturing components (hardware), and tools (software). The central feature is the networking of the virtual world (cyber) with the manufacturing components world (physical) through cyber-physical systems (CPS). Cyber-physical systems can be introduced as a strong digital platform—well structured and well integrated and only as complex as is absolutely necessary with regard to the designated application of the manufacturing process. Thus, CPS-based manufacturing processes are largely able to control, independently and autonomously, depending on external requirements; optimize; and configure what outruns on an outstanding automation level.

When thinking in terms of introducing semi- or fully automated CPS-based manufacturing processes, a wide range of industrial process innovations and labor conditions are involved. This is of great sociological and labor policy importance. In the event of the broad implementation of such semi- and fully autonomous systems, the existing landscape of work sustainably in *Digital Manufacturing/Industry 4.0* will change. Not only internal change processes will be affected but also intercompany processes, as the structures can change entire supply or value chains. This applies to all direct and indirect value-added activities in *Digital Manufacturing/Industry 4.0* integrated solutions. It refers to the operational and executive levels of the manufacturing staff; the strategic levels of planning, management, and control; lower and middle management in *Digital Manufacturing* integrated solutions; and technical experts, specifically systems and software engineers (Hirsch-Kreinsen 2014). Finally, this means that besides company financial and human resources, strong digital governance is needed because the new demands of digital capabilities and new risks from digital-technology-driven activities have made digital governance essential. Governance can help in implementing new solutions faster than ever before, while also managing the challenges of security, regulatory compliance, and legacy system integration. It helps companies gain a more integrated view of operations, collaborate more effectively, and make policies work better (Westerman et al. 2014).

During the first and second industrial revolution, however, governance was focused on employees and machinery in order to blend both into a tightly controlled perfect manufacturing unit. As part of the technological improvement, which was introduced as the third industrial revolution, the increase in complexity of the new machines reduced the activities of employees to monitoring and operating the machines. At the same time, the embedded computing systems of the new machines adopted a dual role by being used on one hand for monitoring and process control and on the other hand by taking over diverse activities previously carried out by employees. This was made possible through application-specific software programs whereby the automated manufacturing processes became increasingly standardized to avoid deviation during automated manufacturing. Hence, the computer has been integrated into *Digital Manufacturing/Industry 4.0* processes as a flexible, powerful tool. The term computer also includes the application-specific software program without which computers could not execute the applications required in *Digital Manufacturing/Industry 4.0* integrated solutions.

From the above, it should not be concluded that computers have now gained implicit knowledge, meaning that they can think like humans. This is not the case, even though artificial intelligence software algorithms have become more sophisticated. The term artificial intelligence properly expresses artificial intelligence and not human intelligence. In the case of demanding tasks in the integrated *Digital Manufacturing* industry, the computer can reach the same goals as humans but not by using human means. When a sensor reports to a computer that an assembled workpiece is incomplete, it does not rely on a source of intuition or skill; rather it follows the respective software program. Moreover, the underlying strategies in manufacturing are fundamentally different while the same results can be achieved. However, computers execute instructions specified by the software program more quickly.

8.2   Economic, Social, and Organizational Challenges

An analysis of the effect of the new digital technology with the resulting personnel and organizational changes requires a look at the entire manufacturing system and its interdependencies. Analyzing the impact of new digital technologies in manufacturing automation requires extensive technological and sociological research on the interaction of digital technologies as well as employees' activities.

Looking at the possible and imagined conversion trends of digital technologies in today's manufacturing processes will determine the types of digitized automation systems required. It is clear that the diffusion of these leading digital technologies, in comparison with the previous forms of manufacturing organization, in particular the currently known patterns of work organization and the deployment of personnel, will permanently change work lives. This situation can be investigated using the socio-technical system method, an approach to complex organized work design that recognizes the interaction between humans and technology in workplaces (http1 2015). In Geels (2004), socio-technical systems are described as encompassing manufacturing, diffusion, and the use of technology and are, from a more functional sense, the linkages between elements necessary to fulfill societal functions. As technology is a crucial element for fulfilling those functions in modern society, it makes sense to distinguish the manufacturing, distribution, and use of technologies as subfunctions. To fulfill these subfunctions, the necessary elements are characterized as resources. Socio-technical systems thus consist of artifacts, knowledge, capital, labor, cultural meaning, and more.

Following the socio-technical system method, one can precisely identify the relevant dimensions of change-leading digital work in semi- or fully automated manufacturing systems (Hirsch-Kreinsen 2014):

- Immediate man-machine interaction and the directly related skill requirements
- Task structures and activities of employees directly employed in the system
- Work organization as the division of labor, structuring tasks and activities in horizontal and hierarchical ways, with associated designs of cooperation and communication between the system and the employees

These dimensions are some of the initial findings from various disciplines of research. The most important finding is that the focus is on the operational level of action of current systems, while the parent-level management system planning and system management is on the fringe.

An important role can be seen in the onset of the *Digital Manufacturing/Industry 4.0* paradigm shift in human technology and environmental interactions, with new forms of collaborative work in virtual factories and mobile work environments, which do not necessarily need to take place at the manufacturing site. Intelligent assistance systems with multimodal, easy-to-use user interfaces will support employees in their *Digital Manufacturing/Industry 4.0* work. But it is not only new technical, economic, and legal aspects that determine future competitiveness. New social work infrastructures can ensure much stronger employee involvement in structuring innovative processes. An important role will be the shift in human-computer and human-environment interactions. Intelligent assistance systems with easy-to-use, multimodal user interfaces support employees in their work. Critical to a successful change is the addition of comprehensive qualifications and training and the organization and design of work models. These models should combine a high degree of self-responsible autonomy with decentralized management and control forms.

The key to successful change that meets with a positive response by employees, in addition to comprehensive qualification training and further training measures, is the organization and design models of work. They should be models that combine a high degree of self-responsible autonomy with decentralized management and control forms (Helbig 2013). In this regard, process responsibility relocates again to the place of actual value in the manufacturing line.

Therefore, the ultimate goal is to split the functions and responsibilities so that an optimal exploit not only focuses on the speed of the computer but also the employee's adequate, active, and intuitive involvement within the process. In this model, the employee is part of the working model and no longer excluded. This results in a technology and human comprehensive hybrid perspective for the socio-technical configuration in *Digital Manufacturing/Industry 4.0*, which is the distribution of activities and autonomy degrees to the subject and makes this viable (Hirsch-Kreinsen 2014: Rammert 2003).

As described in the final report of the Industry 4.0 working group, entitled "Implementing Recommendations for the Future Project Industry 4.0" (in German), *there are not only new technical, economic, and legal aspects that determine the competitiveness in the future but also new social infrastructure work in Industry 4.0, which can ensure a much stronger structural involvement of employees in innovation processes.*

The key to successful change that is positively accepted by employees, in addition to comprehensive qualifications and training, is the organization and design of work models. The employees should call for extended decision-making and participation in determining the scope and possibilities for load regulation. This socio-technical approach to projected future Industry 4.0 new spaces will be opened to much needed innovation, based on an extended awareness of the importance of human labor in innovation processes (Kagermann et al. 2013).

8.3 Changing Demand in the World of Work

It is common today to accept that technological innovations represent a sufficient, reliable basis for progress. Therefore, new digital technologies become a desirable destination. Considering the *Digital Manufacturing/Industry 4.0* paradigm, it is evident that the partitioning of labor in complex, automated manufacturing systems is tailored with regard to the capabilities of computers and application-oriented software programs as well as the Internet. The CPS platform increases productivity, reduces labor costs, and avoids human error. In this regard, *Digital Manufacturing/*

Industry 4.0 transfers as many control activities as possible to the software. Therefore, the more powerful the software is, the more competencies it gets. For the human operator, the only responsibility left is for those tasks which could not be automated by the developer, such as the control for anomalies or as helpers in a power failure. The approach described so far is referred to as technology-centric automation, which can be achieved by intelligent sensors. These sensors not only observe processes and their environment but also process the data measured and feed the results obtained, in the case of the networked manufacturing integrated solutions approach, back into the automated manufacturing system. This allows monitoring and/or controlling of complex automated manufacturing systems with regard to the specific area of concentration which can act as an early warning system. An example is intelligent vibration sensing in wind turbines in regenerative energy systems, which switch off the system if the vibrations of the tower reach dangerous high amplitudes.

The use of wireless sensor networks has the potential to significantly improve the efficiency of *Digital Manufacturing/Industry 4.0*. Wireless sensor networks are spatially distributed autonomous sensors to monitor physical or environmental conditions and to cooperatively pass their data through the network to a main location (see Chap. 4). Sensor networks composed of a large number of sensor nodes are allocated either inside the manufacturing process under test or very close to it, capable of performing some processing, gathering sensory information, and communicating with other connected nodes in the network. They belong to the fastest-growing technologies with the potential of affecting human work life enormously. The sensor node itself is a component of a larger network of sensors. Each sensor node in the network is responsible for collecting data about the environment around it and sending that data to processors in the network. Sensor nodes perform many activities in manufacturing processes.

Wireless sensors are widely deployed in a huge number of novel application domains and have been made feasible by the convergence of technologies:

- Data processing (including hardware, software, middleware, algorithms) .
- Microelectromechanical systems (MEMS) technology (see Chap. 5)
- Wireless communication

Wireless sensor networks can:

- Collaboratively monitor an area of interest
- Execute a task
- Interact with others though a gateway

Wireless sensor networks have enabled the development of:

- Low cost
- Low power
- Multifunctional sensor nodes

These are devices which are small in size, communicate independently within a short distance, and have enabled the idea of sensor networks based on the collaborative effort of a large number of nodes (Akyildiz et al. 2002).

Wireless sensor networks require solutions for.

- *Routing*: how to route information from one node to another one and so forth, how to route to the gateway, and how to arrange energy efficiently at each node, as well as performance.
- *Clustering*: nodes will easily be able to communicate with each other. Considering energy as an optimization parameter, clustering is imperative.
- *Self-organizing*: these collaborative, dynamic, and distributed devices will have the capability of distributing a task among them for efficient computation.
- *Energy scavenging*: how to receive energy from the environment.
- *Task definition*: how to define the task the sensor has to fulfill as a device within the sensor node.

Standards, such as the IEEE 802.11 Wireless Networks Standard, have gained in popularity, providing users with mobility and flexibility in accessing information (http2 2015). This standard has been differentiated as follows:

- *IEEE 802.11™-2012*: IEEE Standard for IT; telecommunications and information exchange between systems; local and metropolitan area networks specific requirements; Part 11: Wireless LAN Medium Access Control (MAC) and Physical Layer (PHY) specifications
- *IEEE 802.11aa™-2012*: IEEE Standard for IT; telecommunications and information exchange between systems, local and metropolitan area networks specific requirements Part 11: Wireless LAN Medium Access Control (MAC) and Physical Layer (PHY) specifications. Amendment 2: MAC enhancements for robust audio video streaming
- *IEEE 802.11ae™-2012*: IEEE Standard for IT; telecommunications and information exchange between systems; local and metropolitan area networks specific requirements Part 11: Wireless LAN Medium Access Control (MAC) and Physical Layer (PHY) specifications. Amendment 1: prioritization of management frames

Currently, collecting manufacturing data for manufacturing planning and management is achieved mostly through wired sensors. The equipment and maintenance cost and time-consuming installations of existing sensing systems prevent large-scale deployment of real-time data traffic monitoring and control. Small wireless sensors with integrated sensing, computing, and wireless communication capabilities offer tremendous advantages in terms of low cost and easy installation in the Digital Manufacturing—integrated industry, including sensor technologies, energy-efficient networking protocols, and applications of sensor networks for digitized automated manufacturing processes. Therefore, human operators in the *Digital Manufacturing/Industry 4.0* are being squeezed out of the cycle of action, feedback, and decision-making more and more. Because these operators are omitted from determining the operation of Digital Manufacturing—integrated industry systems—they are only doing auxiliary functions. Their skills and abilities and implicit/explicit knowledge are being ignored.

In this regard, the research union in Germany stated that Germany can only be a successful location for manufacturing if it succeeds in mastering the driving force of the Internet in the fourth industrial revolution and develops and operates autonomous, self-controlling, knowledge- and sensor-based manufacturing processes (http3 2015).

Many problems associated with automated systems are the result of ignoring the development of human-machine interactions based on the principles of a human-centric human-machine interaction. As a consequence, only some of the anticipated benefits of automation have, in fact, materialized, primarily those related to the improved precision and economy of operations, i.e., those aspects of system operation that do not involve much interaction between employees and machines. Another reason for the observed, unanticipated difficulties with automation was the initial focus on quantitative aspects of the impact of digital technology. Expected benefits included reduced workload, reduced operational costs, increased precision, and fewer errors. Anticipated problems included the need for more training, less pilot proficiency, too much reliance on automation, and the presentation of too much information, as reported in Sarter et al. (1997).

Systems and software engineers exacerbate the problem further if they hide the workflows of their constructs from the operator and make every system an impenetrable black box. The idea behind this view is to avoid errors by allowing employees to be kept in the dark and to not confuse them with complexity, thereby relieving them of personal responsibility. The probability that such an error can now occur, however, is higher because an unaware operator is a dangerous operator (Carr 2014).

Therefore, experts for cognitive science require systems and software engineers to abandon the technology-centered approach and advise them to deal instead with human-centered automation. In human-centered automation, the recommended design construction does not begin with an assessment of the capabilities of the machine but with a careful evaluation of the strengths and limitations of the employees who will operate the machine and interact with it. This reverses technical development back to the human-centered principle. The aim is to split the functions and responsibilities such that they not only use the speed and precision of the computer optimally but also keep the employees active, aware, and alert. Hence, the human is included and not excluded (Billings 2000; Carr 2014) (see Sect. 8.2).

In Carr (2014), an interesting case is reported with regard to potential failures of pilots when they are less directly active while in flight as they are using the autopilot rather than flying manually. On January 4, 2013, the US Federal Aviation Administration (FAA) published a one-sided message on Safety Alerts for Operators (SAFO). This document suggests that airlines commit to prompting their pilots to fly manually whenever possible. The FAA determined, based on hand indications from studies after crashes, reports on accidents, and cockpit studies, that pilots have become strongly dependent on the autopilot and other computerized systems on board the aircraft. Excessive use of the autopilot, according to the FAA, could lead to a deterioration of the ability of the pilot to lead the aircraft quickly away from an undesirable state. The warning concluded with the recommendation that airlines circulate an operational directive to pilots to fly less on autopilot and more manually (Carr 2014; SAFO 13002 2013).

8.4   Greater Product Individualization and Shifting Factors of Global Influence

The fourth industrial revolution (*Digital Manufacturing/Industry 4.0*)—not a revolution because of the innovations in technologies—was initiated by the Internet of Things (see Chap. 4), which offers a competitive advantage through increased efficiency. However, *Digital Manufacturing/Industry 4.0* also creates economic, social, and organizational challenges, as described in the preceding sections. In this regard, changing demands in the world of work, greater individualization of products, and shifting factors of global influence are all playing an important role.

Manufacturing of the future faces major challenges, which are described in detail in the report in Spath et al. (2013), whereby a competitive advantage comes from the ability to be faster and more reliable with tailor-made products on the market, as this is required for global competition. For the manufacturing industry, this means:

- Manufacturing of many variants of customized products in small series or even reliable manufacturing of individual units
- Successfully dealing with the complexity of products needed to control processes and supply materials
- Ensuring short delivery times, despite market volatility and unreliable sales forecasts

The wide use of computers, the Internet, and mobile phones in the private and public sectors will lead to new ways of working. These new technologies have not penetrated the areas of manufacturing to date. A striking example involves the use of smartphones and tablet PCs in the manufacturing industry. Is the productive use of mobile devices still limited to a few applications in logistics, maintenance, and repair? The perception that these new technologies can be used in direct manufacturing process areas will be strengthened (Spath et al. 2013). One reason for this is that today's delivery reliability in the manufacturing industry can, at best, be achieved with a high level of manual effort for short-term control of the flexible capacity. Therefore, from the *Digital Manufacturing/Industry 4.0* paradigm, a shift is expected from central control toward a more flexible, decentralized coordination of autonomous processes in the manufacturing industry. In this context, CPS, with its networking capabilities, is the platform for the paradigm shift to monitoring processes and ensuring transparency and providing intelligent agents to configure autonomous manufacturing processes. This will also have a strong influence on the flexibility of employees in manufacturing, since they will be deployed adaptively in real time largely dispensing with idle time.

The networking of objects, controls, and sensors with users to an Internet of Things, Data, and Services, representing the Internet of Everything, will bring a new level of information availability to manufacturing processes in real time. The data models are up to date, and the total manufacturing activities are more transparent. It will be important to have a true picture of the manufacturing processes. We are doing so today; when we have an idea of the manufacturing process which may be fuzzy, we are deterministically planning for the processes to go within a tenth-of-a-minute accuracy. Facing this difficulty, today we are in the order processing stage of manufacturing planning (Spath et al. 2013).

The information-technology-based approaches in manufacturing processes try to increase the transparency of manufacturing events where the information is collected and stored in decentralized systems and is available via Internet protocols. The vision of the *Digital Manufacturing/Industry 4.0* paradigm shift is to enable machines to broadcast their state via Internet technologies to different user groups—the maintenance staff, machine owner, producers of the machine, and more. Today, these are rather individual standalone solutions, which must be laboriously maintained. Henceforth, the decentralized planning approach in the manufacturing industry is of importance, as stated by the logistics Professor Michael ten Hompel: "The measure of volatility has grown strongly in the sense of the Internet and e-commerce. Our predictions for the logistics of tomorrow are often worse than those for the weather. We need to create systems that are convertible based on preconceived boundaries. The central question is: How can a system look like one that you cannot plan? An answer is provided by the Internet itself, following the motto: "It is better to make a sensible decision locally within a limited time as a supposedly optimal one too late. One possible future is, for example, the application and implementation of swarm intelligence in the management of our systems" (Spath et al. 2013).

Another not to be underestimated advantage of decentralized planning in manufacturing is the ability to address the executive level, i.e., the masters and employees on the manufacturing shop floor, to be included in the planning. They bring their experience and knowledge to the planning, on one hand; and on the other hand, they have to be considered as active players in the manufacturing plan. This is important because a short response time is aligned with exacting customer demands which can only be achieved if adequate human capacity is included in the planning and control. With regard to greater product individualization, automation is possible for smaller and smaller series at mass production prices, which means that employees' work remains an important part of the manufacturing process. This has been described in detail in eight main statements for work in the manufacturing industry in the study reported in Spath et al. (2013):

1. Flexibility is still the key factor for manufacturing work in Germany—in the future but still short term than it is today.
2. Flexibility has to be targeted in the future and systematically organized—discount flexibility "is no longer enough."
3. Industry 4.0 is more than cyber-physical systems networking. The future includes intelligent data acquisition, storage, and distribution by objects and employees.
4. Decentralized control mechanisms, like full autonomy of decentralized, self-controlling objects, are there but not in the foreseeable future.
5. Safety aspects (safety and security) must be taken into account already in the design of intelligent manufacturing.
6. Traditional manufacturing tasks are blending manufacturing and employee's knowledge in regard continues to grow together.
7. Tasks of manufacturing and knowledge workers continue growing together.
8. Employees must be qualified for shorter term for less predictable working activities on the job.

As described in the final report of the working group Industry 4.0 with the title "Recommendations for implementing the strategic initiative industry 4.0 – Final Report of the Working Group Industry 4.0," we have to be aware that Industry 4.0 will address and solve some of the challenges facing the world today, such as resource and energy efficiency, urban production, and demographic change. Industry 4.0 enables continuous resource productivity and efficiency gains to be delivered across the entire value network. It allows work to be organized in a way that takes demographic change and social factors into account. Smart assistance systems release employees from having to perform routine tasks, enabling them to focus on creative, value-added activities. In view of the impending shortage of skilled employees, this will allow older employees to extend their working lives and remain productive longer. Flexible work organizations will enable employees to combine their work, private lives, and continuing professional development more effectively, promoting a better work-life balance (Kagermann et al. 2013). Therefore, the realization of the digital transformation in manufacturing is not only seen in the automation of manufacturing processes but rather in the skills of the employees. Their expertises have to be preserved sustainably at a high level of qualification. But from a more general perspective, there exist a difference between younger and older employees. Younger employees are more familiar with the digital transformation. Sometimes they are called digital natives. They are more concerned about working in an innovative environment and having access to completely new career profiles and opportunities. Older employees sometimes need help to get over their fears of digitization and connectivity because today employees automatically receive the right information when they need it working well in *Digital Manufacturing/Industry 4.0*—the way in which machines, products, and humans will interact in the near future. Visiting trade fairs, research labs, universities, or pilot plants at industrial companies, one can see that all are working on the cutting edge of a future paradigm shift in automation called *Digital*

Manufacturing/Industry 4.0 where manufacturing machine systems operators will develop into manufacturing machine systems managers. Thus, in the near future, there will be different skilled work available in manufacturing than there is now. The result is that digitization and networking in *Digital Manufacturing/Industry 4.0* will transform today's professions. If capabilities of digitization ever faster grow in what pace humans have to adapt in their future work lives? There are risks and opportunities but the big question is how to organize technological progress.

8.5   Exercises

What is meant by the term *automation* in the context of *Digital Manufacturing/ Industry 4.0*?

Describe the structure of automation in a *Digital Manufacturing/Industry 4.0* system.

What is meant by the term *Industry 4.0*?

Describe the efforts of Industry 4.0.

What is meant by the term *digital and global work environments*?

Describe the effects of the digital and global work environment.

What is meant by the term *digital governance*?

Give an example of digital governance.

What is meant by the term *software program*?

Describe the structure of a software program.

What is meant by the term *social challenge in the work life*?

Describe the effects of the social challenges in the work life.

What is meant by the term *organizational challenge in Digital Manufacturing*?

Give an example of an organizational challenge in Digital Manufacturing.

What is meant by the term *socio-technical systems concept*?

Describe the structure of a socio-technical systems concept.

What is meant by the term *changing demand in the world of work*?

Describe the effects of the changing demand in the world of work.

What is meant by the term *wireless sensors*?

Give an example of wireless sensors.

What is meant by the term *wireless sensor network*?

Describe the structure of a wireless sensor network.

What is meant by the term *sensor node*?

Give an example of a sensor node.

What is meant by the term *greater production individualization*?

Describe the effects of greater production individualization.

What is meant by the term *manufacturing control work*?

Give an example of manufacturing control work.

What is meant by the term *decentralized planning in manufacturing*?

Describe the effects of decentralized planning in manufacturing.

References

(Akyildiz et al. 2002) Akyildiz, I. F., Su, W., Sankarasubrmaniam, Y., Cayirie, E.: Wireless Sensor Networks: A Survey. Computer Networks Vol. 38, pp. 393–422, 2002

(Billings 2000) Billings, C. J.: Aviation Automation: The Search for a Human Centered Approach. In: IEEE Transactions on Systems, Man and Cybernetics, Vol. 30, Issue 3, pp. 286–297, 2000

(Carr 2014) Carr, N.: The Glass Cage, Automation and Us. W. W. Norton Publ., 2014

(Geels 2004) Geels, F. W.: From sectoral systems of innovation to socio-technical systems insights about dynamics and change from sociology and institutional theory. In: Research Policy, Vol. 33, pp. 897–920, 2004

(Helbig 2013) Helbig, J.: The Vision: industry 4.0 as part of a networked, intelligent world. In: Implementing recommendations for the future project industry 4.0 Final Report of the Working Group Industry 4.0, Chapter 2, pp. 22–30, Office of Research Alliance at the Donors' Association for German Science, 2013

(Hirsch-Kreinsen 2014) Hirsch-Kreinsen, H.: Wandel von Produktionsarbeit – "Industrie 4.0" (in German). Faculty of Economic and Social Sciences, University of Dortmund, 2014.

(Kagermann et al. 2013) Kagermann, H., Wahlster, W., Helbig, J. (Eds.): Recommendations for implementing the strategic initiative Industry 4.0 – Final Report of the Working Group Industry 4.0. Office of the Industry-Science Research Alliance Secretariat of the Platform Industrie 4.0, 2013

(Rammert 2003) Rammert, W.: Technik in Aktion: Verteiltes Handeln in soziotechnischen Konstellationen (in German). In: Autonome Maschinen, pp. 289–315, Eds.: Christaller, T., Wehner, J., Westdeutscher Verlag, Wiesbaden, 2003

(Sarter et al. 1997) Sarter, N.B., Woods, D. D., Billings, C. E.: Automation Surprises. In: Handbook of Human Factors and Ergonomics, Ed. Salvendy, G., Wiley Publ. 1997

(SAFO 13002 2013) Federal Aviation Administration, SADO 13002, January 4th 2013, faa,gov/other_visit/aviation_industry/airline_operators/airline_safety/safo/all_safo/media/2013/SAFO13002.pdf

(Spath et al. 2013) Spath, D., Ganschar, O., Gerlach, S., Hämmerle, M., Krause, T., Schlund, S.: Produktionsarbeit der Zukunft – Industrie 4.0 (in German). Fraunhofer-Institut für Arbeitswirtschaft und Organisation IAO. Fraunhofer Verlag, 2013

(Westerman et al. 2014) Westerman, G., Bonnet, D., McAfee, A.: Leading Digital – Turning Technology into Business Transformation. Harvard Business Review Press, 2014

Links

(http1 2015] https://en.wikipedia.org/wiki/Sociotechnical_system
(http2 2015) http://standards.ieee.org/about/get/802/802.11.html
(http3 2015) http://www.forschungsunion.de/veroeffentlichungen/

Glossary

A

AADL Architecture Analysis and Design Language is a modeling language that supports early and repeated analyses of systems architecture with regard to performance-critical properties through an extendable notation. Moreover, it is a tool framework with precisely defined semantics.

ACATECH National Academy of Science and Engineering in Germany represents the German scientific and technology community. As a working academy, ACATECH supports policy makers and society by providing qualified technical evaluations and forward-looking recommendations.

AC/DC Alternating current (AC) is an electric current in which the flow of an electric charge periodically reverses direction, whereas direct current (DC) is the flow of an electric charge in only one direction.

ACM Association for Computing Machinery, the world's largest educational and scientific computing society. It provides the computing field's premier digital library and serves its members and the computing profession with leading-edge publications, conferences, and career resources.

ADC Analog-to-digital converter, a device that converts a continuous physical quantity into digital numbers that represents the quantity's amplitude.

ADL Architecture development language, used in several disciplines, such as systems engineering, software engineering, enterprise modeling, and engineering.

ALU Arithmetic logic unit, a digital electronic circuit that performs arithmetic and bitwise logical operations.

ANSI American National Standards Institute, a premier source for timely, actionable information on national, regional, and international standards.

ARP Address Resolution Protocol, used by the Internet Protocol IPV4 to map IP network addresses to the hardware addresses used by a data link protocol.

ARPANET Advanced Research Projects Agency Network, the network that became the basis for the Internet.

ASIC Application-specific integrated circuit, a microchip designed for a special application.

ASP Application-specific processor, a component used in system-on-a-chip design.

AUP Agile Unified Process, a simplified version of the Rational Unified Process.

B

BITKOM Federal Association for Information Technology, Telecommunications and New Media, the trade association of German information and telecommunications industry.

BMBF German Federal Ministry of Education and Research.

BMWi Federal German Ministry of Economics and Energy.

BPEL Business Process Execution Language is a standard executable language for specifying action with business processes with web services.

BPL Broadband over powerlines, a method that allows relatively high-speed digital data transmission over the public electric power distribution wiring.

B2B Business-to-business refers to business that is conducted between companies, rather than between a company and individual clients.

B2C Business-to-customer refers to business that is directly between a company and customers who are end users of products and/or services.

B2MML Business to Manufacturing Markup Language is an XML of the ANSI/ISA-95 family of standards which consists of a set of XML schemas.

C

CAD Computer-aided design, the use of a computer system to assist in a design.

CAE Computer-aided engineering, the usage of computer software to aid in engineering analysis tasks.

CAN Controller Area Network, a vehicle bus standard designed to allow microcontrollers and devices to communicate with each other.

CEPCA Consumer Electronics Powerline Communication Alliance, a nonprofit industrial organization whose mission is to ensure coexistence between high-speed powerline communication systems for use in the smart home.

CIP Critical infrastructure protection, a concept that relates to the preparedness and response to serious incidents that involve the critical infrastructure in an area of concentration.

CISC Complex instruction set computing refers to computers designed with a full set of computer instructions where a single instruction can execute several low-level operations.

CLB Configurable logic block, a fundamental block in field-programmable gate array technology.

ConOps Concept of operations, document describing the characteristics of a proposed system from the individual viewpoint of the system user.

CO_2 Carbon dioxide, the primary so-called greenhouse gas emitted through human activities.

COTS Commercial off-the-shelf, an adjective that describes hardware or software products that are ready-made and available for sale.

CPD Continuing professional development, method by which people maintain their knowledge and skills to their professional lives.

CPES Cyber-physical energy system, the cyber and physical combination of components in energy grids.

CPLD Complex programmable logic device, a combination of a fully programmable AND/OR array and a bank of macrocells.

CPMS Cyber-physical manufacturing system, a system of collaborating computational manufacturing elements controlling physical entities.

CPPS Cyber-physical production systems collect data through integrated sensors and measuring systems in real time, store them, and analyze them for the purpose of modeling. They can be understood as digital communication devices to each other and are connected with the Internet of Things and Services.

CPS Cyber-physical system, an integration of computation, networking, and physical processes.

CPSEF Cyber-physical system engineering framework, used for apprehending the development and operation of highly efficient CPS.

CPU Central processing unit, the central unit in a computer containing the logic circuitry that performs the instructions of a computer program.

CS Computer Society, an organization of computing professionals.

CSCi Computer software configuration items mean an aggregation of software that satisfies end user functionality and are designated for separate configuration management by the buyer.

CTC Critical to customers is the input to the quality function development activity for the customer requirements' side of the analysis.

CTQ Critical to quality represents the internal critical parameters that relate to the wants and needs of the customer.

CUDA NVIDIA CUDA parallel computing platform.

D

DAC Digital-to-analog converter, a device that converts digital numbers into continuous physical quantities.

DARPA Defense Advanced Research Projects Agency, a division of the US Department of Defense (DoD) which funds developing technologies in support of national security.

DC/AC Direct current (DC), the flow of an electric charge which is only in one direction, whereas alternating current (AC) is an electric current in which the flow of an electric charge periodically reverses direction.

DCS Distributed control system, an automated control system distributed through a machine to provide instructions to a different part of the machine.

DHS Digital home standard is a standard of the universal powerline association (UPA) for home networking via powerline (PLC).

DICT Distributed information and communication technology.

DIP Dual in-line, the most common type of package for small- and medium-scale integrated circuits.

DLL Dynamic-link library, a module that contains functions and data that can be used for another module.

DMAIC Define, measure, analyze, improve, control, a data-driven quality strategy used to improve processes and an integral part of a Six Sigma initiative.

DMDI Digital Manufacturing and Design Innovation Institute, a US research institute for applying cutting-edge technologies to reduce time and cost in manufacturing, strengthen the capabilities of the supply chain, and reduce acquisition costs.

DoD US Department of Defense.

DPM Digital Product Memory stores data collecting along the product life cycle.

DRIA Digitalized repair and maintenance assistant.

DSDM Dynamic Systems Development Method, an agile project delivery framework primarily used as a software development method.

DSP Digital Signal Processor, specialized processor architecture optimized for the operational needs of digital signal processing.

E

ECG Electrocardiogram is used to monitor the heart.

ECPC Expansion-clustering-projection-contraction.

ECS Embedded computing system, a special-purpose system in which the computer is completely encapsulated by the device it controls.

EEPROM Electrically erasable programmable read only memory, a user-modifiable read-only memory (ROM) that can be erased and reprogrammed.

eFFBD Enhanced functional flow block diagram represents the items or the data interaction aspect of the behavior of a system.

EIA Electronic Industries Alliance, a US alliance of trade organizations that lobbies in the interest of companies engaged in the manufacture of electronic-related products.

EMC Electromagnetic compatibility means that a device is compatible with its electromagnetic environment.

EMF Electromagnetic force, based on electromagnetic interaction of electrically charged or magnetically polarized particles or bodies.

EMR Electronic medical records contain the standard medical and clinical data gathered in one provider's office.

EPC Electronic Product Code, a universal identifier that gives a unique identity to a specific physical object.

EPROM Erasable programmable read-only memory, a programmable read-only memory that can be erased and reused.

ERP Enterprise Resource Planning, business process management software that allows an organization to use a system of integrated applications to manage the business.

EssUp Essential Unified Process is a method to software process improvement.

EU European Union.

F

FAA Federal Aviation Administration, US agency responsible for the advancement, safety, and regulation of civil aviation, as well as overseeing the development of the air traffic control.

FAN Fabric area network, a wireless communications infrastructure to enable ubiquitous networking and sensing on intelligent clothing.

FDD Feature-driven development is an integrative and incremental software development approach.

FPGA Field-programmable gate array, a semiconductor device that is based around a matrix of configurable logic blocks (CLB) connected via programmable interconnects.

FPLA Field-programmable logic array is a kind of programmable logic devices used to implement combinational logic circuits.

G

GAL Generic array logic device consists of a reprogrammable programmable array logic (PAL) matrix and a programmable output cell.

GCN Global Communication Networks, one of the oldest established international networks of owner-managed advertising agencies.

GDP Gross national product measures the value of goods and services that a country's citizens produced regardless of their location.

GDV German Insurance Association.

GMRF Gaussian Markov random fields, simply multivariate Gaussian random variables which are extensively used in statistical models.

GPP General-purpose processor, designed for general-purpose computers, such as PCs, laptops, workstations, and more.

GPS Global positioning system, a satellite-based navigation system consisting of a network of orbiting satellites, 11,000 nautical miles in space.

GPU Graphical processing unit, a programmable logic chip that renders images, animations, and video for the computer screen.

GSMA Groupe Speciale Mobile Association represents the interests of mobile operators worldwide.

GUI Graphical user interface, a computer program that enables a user to communicate with a computer through the use of symbols, visual metaphors, and pointing devices.

H

HACMS High-Assurance Cyber-Physical Military System, a US DARPA program with the scope to create technology for the construction of high-assurance cyber-physical systems where high assurance is defined as functionally correct and satisfying appropriate safety and security properties.

HAL Hardware array logic is programmable array logic.

HCI Human-computer interaction, focusing on the interface between users and computers by observing the ways in which users interacts with computers and the respective design technologies for interacting.

HDL   Hardware Description Language, used to model, describe, and test electronic circuits.

HDTV   High-definition television provides much better resolution than current televisions based on the NTSC standard.

HMI   Human-machine interface, focusing on the interface between users and the machine based on the electronics required to signal and control the state of the industrial automation machine.

HPC   High-performance computing, the use of parallel processing for running advanced application programs efficiently, reliably, and quickly.

HRTS   Hard real-time system, hardware or software that must operate within the confines of a stringent deadline.

HWCI   Hardware configuration items are components that need to be managed in order to deliver an IT service.

I

IACS   Industrial Automation and Control Systems.

IATA   International Air Transport Association.

ICAM   Integrated computer-aided manufacturing, a US Air Force program that develops tools and processes to support manufacturing integration.

ICT   Information and communication technology refers to technologies that provide access to information through telecommunications.

IDEF0   ICAM DEFinition for function modeling.

IEC   International Electrotechnical Commission authors international standards for all electrical, electronic, and related technologies.

IEEE   Institute of Electrical and Electronics Engineers, the world's largest professional association dedicated to advancing technological innovation and excellence for the benefit of humanity.

IETF   Internet Engineering Task Force, mission is to make the Internet work better from an engineering point of view.

IFP   Information flow property software engineering-based approach for verification.

IFW   German Institute of Manufacturing Engineering and Machine Tools.

INCOSE   International Council on Systems Engineering.

I/O   Input/output means communication/interaction of an information system with its outside world.

IOB   Input/output blocks components in programmable logic circuits.

IoE   Internet of Everything expands the concept of the IoT in that it connects not just physical devices but quite literally everything by getting them all on the network.

IoT   Internet of Things, the network of physical objects, so-called things, embedded with electronics, software, sensors, and network connectivity, which enables these objects to collect and exchange data.

IoTDaS Internet of Things Data and Services expands the IoE focusing on data and sophisticated data analytics and services and data-based sophisticated services.

IP Internet Protocol, protocol by which data is sent from one computer to another on the Internet and has at least one IP address that uniquely identifies it from all other computers on the Internet.

IPv4 Internet Protocol version 4, the fourth revision of the IP and a widely used protocol in data communication over different kinds of networks.

IPv6 Internet Protocol version 6, launched to replace IPv4; it is often referred to as the next-generation Internet because of its expanded capabilities and it's growth through recent large-scale deployments.

IR Infrared is invisible radiant energy with longer wavelengths than those of visible light.

ISA International Society of Automation.

ISCI ISA Security Compliance Institute, a nonprofit automation controls industry consortium managing the ISASecure® conformance certification program.

ISF Information Security Forum, an independent nonprofit organization dedicated to investing, clarifying, and resolving key issues in information security and risk management by developing best practice methodologies, processes, and solutions.

ISM Industrial, scientific, and medical.

ISO International Organization for Standardization, an independent nongovernmental membership organization and the world's largest developer of voluntary international standards.

IT Information technology is a generic term for information and data processing, as well as for the necessary hardware and software.

J

JIRA Japanese Industrial Robot Association is a trade association made up of companies in Japan that develop and manufacture robot technology.

L

LAB Logic array block contains dedicated logic for driving signals to its logic modules.

LCA Logic cell array, a group of high-performance, high-density, digital integrated circuits.

LCD Liquid crystal display is a display based on liquid crystals that affect the polarization direction of light, when a certain level is applied to electrical voltage.

LCM Life cycle management enables superior product management from concept to retirement.

LCP Life cycle process, a concept for seamless integration of all information generated during the life cycle of a product.

LEAP Locations, entities, and arrival processes, a simulation approach which embeds the model components as locations, entities, arrivals, or processes.

LED Light-emitting diode is a semiconductor device that emits visible light when an electric current passes through it.

LIN Local interconnect network is a sub-bus system based on a serial communication protocol. The bus is a single master/multiple slave bus that uses a single wire for data transmission.

L^{2}CAP Logical Link Control and Application Protocol, supports higher-level protocol multiplexing, packet segmentation and reassembly, and the conveying of quality of service information.

LMP Link Management Protocol, used to maintain control channel connectivity, verify the physical connectivity of data links, correlate the link property information, suppress downstream alarms, and localize link failures for protection/restoration purposes in multiple kinds of networks.

LVRG Life Insurance Reform Act.

M

MAC Media access layer, one of two sublayers that make up the data link layer of the OSI model.

MADP Mobile application development platform supports the need for current and future projects.

MARG Magnetic, angular rate, gravity is a system to integrate and calibrate signals from magnetometers, gyroscopes, and accelerometers to implement a magnetic, angular rate and gravity sensor system.

MC Microcontroller device is a compact microcomputer design to govern the operation of embedded systems.

MDC Machine data collection means automatic recording of real-time production data in the production area.

MEMS Microelectromechanical system, a technology that combines computers with tiny mechanical devices, such as sensors, valves, and actuators, embedded in semiconductor chips.

MES Manufacturing execution systems, computerized systems used in manufacturing.

MIT Massachusetts Institute of Technology.

MRF Markov random field, a graphical model of a joint probability distribution.

MS&A Modeling, simulation, and analysis, a comprehensive, state-of-the-art, and technically correct treatment of all important aspects of a simulation study.

M2M Machine-to-machine, used to describe any technology that enables networked devices to exchange information and perform actions without the manual assistance of humans.

MVS Multivariable system, a system with more than one control loop.

N

NASA National Aeronautics and Space Administration, the US government agency responsible for the civilian space program as well as aeronautics and aerospace research.

NEMS Nanoelectromechanical system, a class of devices integrating electrical and mechanical functionality on the nanoscale.

NERC North American Electric Reliability Association.

NFC Near field communication is a set of protocols that enable electronic devices to establish radio communication with each other to a distance typically 10 cm or less.

NICB National Insurance Crime Bureau, the US first unrecovered stolen vehicle database, a free service to the public.

NIST National Institute of Science and Technology labs conducting world-class research, often in close cooperation with industry, that advances the US technology infrastructure and helps companies continually improve products and services.

NICT Networking, information, and communication technologies.

NRE Nonrecurring engineering cost is the one-time up-front cost for new product development.

NSF National Science Foundation, an independent federal agency created by the US Congress and the only federal agency whose mission includes support for all fields of fundamental science and engineering, except for medical sciences.

O

OBEX Object Exchange, a transfer protocol that defines data objects and a communication protocol of two devices that can be used to exchange those objects.

OLED Organic light-emitting diode, a flat, light-emitting technology made by placing a series of organic thin films between conductors.

OPAK Open Platform for Autonomous Engineering Mechatronic Automation Components is an assistance function in the form of information and communication technologies available for plant developers and, therefore, allows engineering-like work.

OpenCL Open Computing Language is a framework for writing programs that execute across heterogeneous platforms.

OpenUP Open Unified Process is an open-source process framework developed within the Eclipse Foundation.

OS Operating system is a computer program that manages the system resources of a computer.

P

PAD Physical architecture diagram is used for an enterprise management environment.

PAL Programmable array logic, a prefabricated chip with macrocell structures which are connected through a switch matrix and used to implement combinational logic circuits.

pASIC Programmable application-specific integrated circuit, a nonstandard integrated circuit constructed for one specific purpose, application only.

PCAST President's Council of Advisors on Science and Technology.

PDA Personal digital assistant is a handheld device that combines computing, telephone/fax, Internet, and networking features.

PDC Production data collection is an integral part of the production management, production planning, and inventory management process.

PFCN Plant floor control network supports communication between programmable logic controllers and plant floor computers to monitor and control the manufacturing process.

PFD Process flow document, used primarily in process engineering where there is a requirement of depicting the relationship between major components only.

PHY Physical layer is the first (lowest) layer in the ISO/OSI model.

PI Proportional-integral is an algorithm that computes and transmits a controller output signal every sample time to the final control element.

PIC Programmable interface controller is electronic circuits that can be programmed to carry out a vast range of tasks.

PID Proportional-integral-derivative is an algorithm that computes and transmits a controller output signal every sample time to the final control element commonly used in industrial control systems.

PIN Production Innovations Network links university research and business.

PLC Power Line Communication, a communication technology that enables sending data over existing power cables.

PLC Programmable logic controller is a computer used for automation of typical industrial electromechanical processes.

PLD Programmable logic device, an electronic component used to build reconfigurable digital circuits.

PR Production rate is the number of goods that can be produced during a given period of time.

PROFIBUS Process field bus is a standard for field bus communication in automation technology.

PWM Pulse Width Modulation is a method for generating an analog signal using a digital source.

PZH Hannover Centre for Production Technology.

Q

QFD Quality function deployment, a method to transform qualitative user demands onto quantitative parameter to deploy the function forming quality.

QR Quick response code, a type of 2D barcode that is used to provide easy access to information through a smartphone.

R

RAM Random access memory, a computer data storage device allowing data items to be accessed as read or write.

RIA Robotics Institute of America conducts basic and applied research in robotics technologies relevant to industrial and societal tasks.

RISC Reduced instruction set computer, a microprocessor that is designed to perform fewer types of computer instructions so that it can operate at a higher speed.

RF Radio frequency, any frequency within the electromagnetic spectrum associated with radio wave propagation.

RFID Radio frequency identification refers to small electronic devices that consist of a small chip and an antenna.

ROM Read only memory, computer memory on which data has been prerecorded.

RoRo Roll-on/roll-off are vessels designed to carry wheeled cargo.

RVA Requirements validation approach.

S

S + H Sample and hold circuit captures an analog signal and holds it during some operation.

SAFO Safety Alerts for Operators, an information tool that alerts, educates, and makes recommendations to the aviation community.

SBS System breakdown structure is used for breaking down a system into easily manageable components.

SCADA Supervisory control and data acquisition, a category of software application program process control, gathering of data in real time from remote locations in order to control equipment and conditions.

SDL Specification and description language, a language targeted at the unambiguous specification and description of the behavior of reactive and distributed systems.

SECOE Systems Engineering Center of Excellence. SHE^{2}MS: Smart home environment energy management system.

SIMILAR State, investigate, model, integrate, launch, assess, reevaluate.

SISO Single-input, single-output system is a single-variable control system with one input and one output.

SITA Société Internationale de Télécommunications Aéronautiques, a multinational information technology company providing IT and telecommunication services to the air transport industry.

SLD System-level design is an electronic design methodology that focuses primarily on the higher abstraction level concerns.

SME Small and Medium Enterprises are businesses whose personal numbers fall below certain limits.

SMS Short message service (text message).

SMT Satisfiability Modulo Theories, about checking the satisfiability of logical formulas over one or more theories.

SOA Service-oriented architecture, an approach used to create an architecture based upon the use of services.

SOAP Simple object access protocol, a messaging protocol that allows programs that run on disparate operating systems to communicate using HTTP (hypertext transfer protocol) and XML (extensible markup language).

SoC System on a Chip, the packaging of all necessary electronic circuits and parts for a system on a single integrated circuit.

SoGP Standard of Good Practice, a business-focused, practical, and comprehensive guide to identifying and managing information security risks in organizations and their supply chains.

SPI Serial Peripheral Interface is an interface bus commonly used to send data between microcontrollers and small peripherals.

SPLD Simple programmable logic device, a programmable logic device with complexity below that of a complex programmable logic device.

SPP Single-purpose processor is a digital circuit designed to execute exactly one program.

SQL Standard Query Language, a standard interactive and programming language for getting information from and updating database.

SRAM Static random access memory, a random access memory (RAM) that retains data bits in its memory as long as power is being supplied.

SRD Stakeholder requirements document identifies the user's requirements.

SRS Software Requirements Specification represents the view of those at the business or enterprise operation level.

SRTS Soft real-time systems.

SSE-CMM Systems security engineering capability maturity model.

SysML System Modeling Language, a general-purpose graphical modeling language for specifying, analyzing, designing, and verifying an engineered system.

T

TCP Transmission Communication Protocol is a standard that defines how to establish and maintain a network conversation through which application programs can exchange data.

TDMA Time Division Multiple Access is a technology used in digital cellular telephone communication to increase the amount of data that can be carried.

TML Task management layer is a design technique in which the application is divided into three parts: the presentation layer, the business logic layer, and the data layer.

TUI Tangible user interface is built upon those skills that situates the physically embodied digital information in a physical space.

U

UAV Unmanned aerial vehicle, an unscrewed vehicle which can either be remote controlled, remote guided, or autonomous and which is capable of sensing its environment and navigating on its own.

UC Unit cost is the cost incurred by a company to produce, store, and sell one unit of a particular product.

UGV Unmanned ground vehicle is a vehicle that operates while in contact with the ground and without an onboard human presence.

UML Unified Modeling Language, a standard visual modeling language intended to be used for modeling business and similar processes, including analysis, design, and implementation of software-based systems.

UPA Universal Powerline Association, the first global and universal PLC association to cover markets and all PLC applications.

UPnP Universal plug and play, a standard that uses Internet and Web protocols to enable devices, such as PCs, peripherals, intelligent appliances, and wireless devices, to be plugged into a network and automatically know about each other.

UPC Universal product code is a unique 12-digit number assigned to retail merchandise that identifies both the product and the vendor that sells the product.

URL Uniform resource locator, the unique address for a file that is accessible on the Internet.

USB Universal serial bus, an interface that connects the computer to the peripheral devices.

UUV Universal serial bus is a four-pin standardized I/O bus, which serves as a serial interface for computer ports.

UWB Ultra-wideband is a wireless technology for transmitting large amounts of digital data over a wide spectrum of frequency bands with very low power for a short distance.

V

VDMA Association of German Machinery and Plant is the industry-wide network of the capital goods industry in Europe. It groups 37 specialized associations who work closely with the market.

VHDL Very High Speed Integrated Circuits Hardware Description Language describes the behavior and structure of electronic systems but is particularly suited as a language to describe the structure and behavior of digital electronic hardware designs.

VOC Voice of the customer is a term used in business and IT.

VoIP Voice over IP is the technical basis for Internet telephony.

V2I Vehicle-to-infrastructure, the wireless exchange of critical safety and operational data between vehicle and roadway infrastructure.

V2V Vehicle-to-vehicle communication that warns the driver but does not take control of the vehicle.

V&V Verification and validation, independent procedures that are used together for checking that a system, service, or product meets requirements and specifications and that it fulfills its intended purpose.

W

WBAN Wireless body area network, a specialized network concept in the context of the OEEE802.15 for a single person of communication devices.

WEF World Economic Forum is a Swiss nonprofit foundation.

Wi-Fi Wireless Fidelity is a trademark of the WiFi Alliance.

WIP Work in progress refers to all materials and partly finished products that are at various stage of the production.

WLAN Wireless local area network refers to a local radio network, whereby mostly a standard of the IEEE 802.11 family is meant.

WSDL Web Service Description Language, an XML-based interface definition language that is used for describing the functionality offered by a web service.

WSN Wireless sensor network, a group of specialized transducers with a communications infrastructure for monitoring and recording conditions at diverse locations.

X

XML Extensible markup language, a simple, very flexible, text format, originally designed to meet the challenges of large-scale electronic publishing but is also playing an increasingly important role in the exchange of a wide variety of data on the Web.

XP Extreme programming implements a simple, yet effective, environment enabling teams to become highly productive.

Index

© Springer International Publishing Switzerland 2016
D.P.F. Möller, *Guide to Computing Fundamentals in Cyber-Physical Systems*,
Computer Communications and Networks, DOI 10.1007/978-3-319-25178-3

Printed in the United States
By Bookmasters